Evolution:
an introduction

http://www.oup.co.uk/best.textbooks/biology/evolution

Evolution:
an introduction

Stephen C. Stearns
University of Basel

and

Rolf F. Hoekstra
Wageningen University

OXFORD
UNIVERSITY PRESS

Great Clarendon Street, Oxford OX2 6DP

Oxford University Press is a department of the University of Oxford.
It furthers the University's objective of excellence in research, scholarship,
and education by publishing worldwide in

Oxford New York

Athens Auckland Bangkok Bogotá Buenos Aires Calcutta
Cape Town Chennai Dar es Salaam Delhi Florence Hong Kong Istanbul
Karachi Kuala Lumpur Madrid Melbourne Mexico City Mumbai
Nairobi Paris São Paulo Taipei Tokyo Toronto Warsaw

with associated companies in Berlin Ibadan

Oxford is a registered trade mark of Oxford University Press
in the UK and in certain other countries

Published in the United States
by Oxford University Press Inc., New York

First published 2000

A catalogue record for this book is available from the British Library.

Library of Congress Cataloging in Publication Data
(Data applied for)

ISBN 0 19 854968 7

Typeset by J&L Composition Ltd, Filey, North Yorkshire
Printed on acid-free paper by
Snœk-Ducaju & Zoon Printers,
Gent, Belgium

Preface

This book introduces what is essential and exciting in evolutionary biology to university students in a book that can be read straight through, one chapter per session, in one semester. Concepts and examples are featured prominently. We state what is important. We have not tried to cover everything.

Students would benefit from a prior or parallel course in introductory genetics. We assume they know something about genes, alleles, meiosis, DNA, RNA, and proteins. Mathematics has been held to a minimum. We use little more than algebra. Students need not have had calculus; but, because the consequences of variation are a major theme, some understanding of probability and statistics is needed; however, we do not assume much.

The questions at the end of each chapter are arranged roughly in order of increasing difficulty. For some of the more advanced questions there is not necessarily a single best answer.

Concepts introduced in bold are also defined in the Glossary.

We thank Dieter Ebert, Tad Kawecki, Michael Döbeli, the students in Evolution-Ecology-Behavior at the University of Basel in winter semester, 1998-99, with whom the book was tested, and the referees who read the whole book in draft, including Anders Angerbjörn and David Reznick. Their remarks substantially improved the presentation. Ingrid Singh did the scientific illustrations; Dafila Scott did the sketches. We thank them both for their professionalism and cheerful cooperation. Nick Barton, Bengt Olle Bengtsson, Paul Brakefield, Pam O'Neil, Annie Schmitt, John Maynard Smith, Jeremy Jackson, Al Stearns, and Peter van Tienderen suggested topics or called examples to our attention.

It was our good fortune to have Cathy Kennedy as our editor at Oxford University Press. Cathy cares about and supports her authors, reads what they write and gives them prompt, detailed, constructive feedback. She had a vision for this book that kept us on track. We owe her a lot.

Contents

http://www.oup.co.uk/best.textbooks/biology/evolution

Prologue

Sexual cannibalism

After a long and dangerous search for a mate, a male Australian red-back spider finally encounters a female. She is suspended upside down in the middle of her web, waiting for prey, and is much larger than he is. Preparing to copulate, he spins a small sperm web, ejaculates a globule of semen onto it, then dips his highly modified forelegs, his palps, into the semen. The semen is stored in reservoirs in each palp, from which a blood-engorged penis-like organ carrying the semen is extended during copulation to penetrate through the female genital opening and deposit the semen in the female's sperm-storage chamber, the spermatheca. After signaling that he is a male of her species and not prey, he approaches the female and copulates with her in the venter-to-venter position.

The Australian red-back spider belongs to a genus (*Latrodectus*) whose bite can occasionally kill a human. It includes the black widow spider of North America and the Mediterranean area.

Male Australian red-back spiders commit suicide by getting their mates to eat them.

Intent on mating, the male appears to ignore the fatal bite of the female. As soon as sperm is being transferred, and with his copulatory organ still inserted, he does a head-stand and somersaults, positioning his abdomen directly under the female's mouth, from which digestive fluids appear almost immediately. While insemination proceeds, the female chews the male's abdomen and may stab it with her poison-laden fangs. When copulation with the first palp is over, the male moves a few centimeters from the female, re-enacts some courtship maneuvers, and moves back to the female. He inserts his second palp despite his mutilated abdomen. Again, he does his headstand and somersault, and again the female chews some of his abdomen. When he makes his second withdrawal, the female wraps her silk around him, storing him for later consumption.

If the female is not hungry, the male may not be eaten; about one-third escape after copulating with one palp; but even those that escape die of their injuries within 2 days. The somersault that brings the male's abdomen into contact with the female's mouth parts is unique within the genus *Latrodectus*. If a male red-back spider is mated to a female of another species in the same genus, he makes the somersault, but she does not try to eat him.

Copulatory suicide and sexual cannibalism evolved because they increased the reproductive success of both the male and the female.

The copulatory behavior of the male Australian red-back spider is stereotypic; the pattern of the male's movements clearly increases the probability that the female will eat him. Male somersaults and female sexual cannibalism have evolved and are genetically programmed (Forster 1992). How could evolution have produced behavior that reduces the chance of survival? Males are selected to be eaten during copulation because their chances of surviving to mate with another female are low; by being eaten they contribute to the nourishment of their offspring and thereby increase their lifetime reproductive success more than they would, on average, if they survived the first mating attempt but died before finding a second mate. Survival is only important if it contributes to reproduction.

Sexual cannibalism is found in snails, crustaceans, insects, and arachnids, occurring most frequently in scorpions and spiders. Despite reports to the contrary, it is not common in mantids, and it is not necessary for the female praying mantis to bite off the male's head to complete copulation (Elgar 1992).

Rapid evolution

Guppies (*Poecilia reticulata*) are small, freshwater fish from Trinidad and north-eastern South America whose colorful males make them popular in the aquarium trade. In nature, they often encounter two species of predatory fish. One is a large cichlid, *Crenicichla alta*, that can eat guppies of all sizes, juvenile or adult. The other is a small killifish, *Rivulus hartii*,

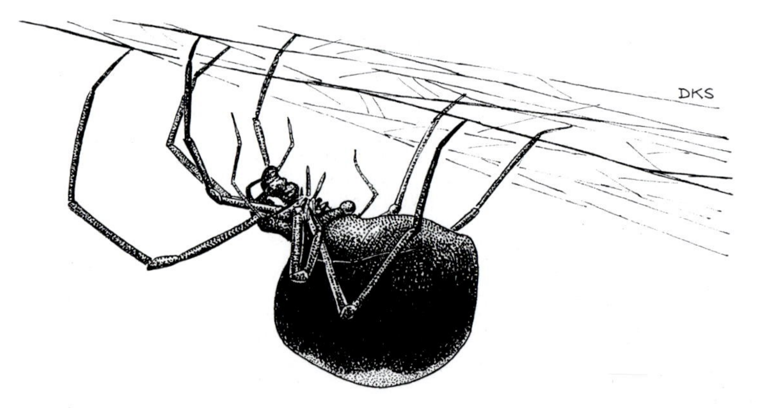

Fig. P1 A female Australian red-back spider, *Latrodectus hasselti*, eating a male after copulation. The male is many times smaller than the female. (By Dafila K. Scott.)

that preys almost exclusively on juvenile guppies. At the north end of Trinidad is a mountain range dissected by numerous streams flowing into the Aripo River. The large cichlid occurs primarily in the Aripo River itself and in the lower reaches of the streams; the small killifish occurs in many of the smaller streams, including their upper reaches. By raising guppies from populations that occur either with the large cichlid or the small killifish, it was discovered that in the populations that had coexisted with the large cichlid, the guppies matured earlier, at a smaller size, and gave birth to more numerous, smaller offspring. Those guppies also showed greater innate tendencies to avoid predators, and the males were less colorful and had less elaborate courtship behavior than did males from populations that had coexisted with the less dangerous, smaller killifish.

Guppies in Trinidad have evolved very rapidly. Evolution does not take millions of years. It can happen before our eyes.

Thus the populations differed genetically in important and interesting traits—they had evolved adaptations to local conditions. How rapidly might such differences evolve? To answer that question, David Reznick, John Endler, and their colleagues took some guppies from the Aripo River, where they had coexisted for a long time with the large, dangerous cichlid, and introduced them to a stream that contained the small, less dangerous killifish, but no other guppies (it was above a waterfall). They then returned year after year to follow the genetically based changes in the guppy population, which they checked by taking fish back to the laboratory and raising them under the same controlled conditions each time. After just 11 years, or 18 guppy generations, the introduced guppies, whose ancestors had lived with the cichlid, had evolved traits like those in guppies that had coexisted with the killifish for a long time (Reznick *et al.* 1990). Selection was strong because the traits studied were directly involved with reproduction and survival: age and size at maturity and size and number of offspring. Evolution can be

Fig. P2 A male and female guppy, *Poecilia reticulata,* with the killifish *Rivulus* (above) and the cichlid *Crenicichla* (below) lurking in the background. (By Dafila K. Scott.)

rapid when selection pressures are strong and populations are large and variable.

Molecular resolution of old systematic problems

The relationships of pentastomids are a systematic puzzle . . .

It has long been difficult for biologists to determine the relationships of highly specialized groups of parasites, for in adapting to their hosts they have often lost most of the features that they had shared with their relatives. For example, the tongue-worms, or pentastomids, are long, flat, glassy, worm-like animals that occur mostly in the tropics or subtropics, where they live in the nasal sinuses or lungs of their primary hosts, which include dogs and crocodiles. They usually have three larval stages. The first stage develops within the unhatched eggs in the stomach of an intermediate host, usually a herbivore, such as a rabbit or a fish. The second stage hatches from the egg in the stomach of the intermediate host and looks like a tardigrade, a group of minute arthropods called water bears because of their locomotion—they move deliberately on four pairs of unjointed, stumpy legs. From the stomach, the tiny, tardigrade-like second larval stages of tongue-worms swim and bore with their mouths into the lungs or liver of their intermediate host and enter a third stage, encapsulating to form cysts. There they wait for their host to be eaten by a carnivore, their definitive host, such as a dog or crocodile. In the mouth or stomach of the carnivore, they leave their cyst, embed themselves again in nasal sinuses or lung tissues, grow and mature, and the cycle begins again (Margulis and Schwartz 1982).

There are not many clues in the simplified morphology of pentastomids as to their systematic relations. Like arthropods, they have a segmented, chitinous exoskeleton, consisting of about 90 rings, but there is no internal segmentation. Otherwise, they look like flat, transparent worms with hooks near the mouth for holding onto the host, a simple, straight intestine, and a body filled with reproductive organs, like most parasites. Sexes are separate. Except for the hooks and the shape of their mouths, there are few useful morphological features in the adults. Various authors have treated them as relatives of tardigrades, mites, onychophorans, annelids, or myriapods. In most zoology texts they are described as belonging to their own phylum, related to the arthropods, and parasitology texts describe them variously as an independent phylum, a class of the Mandibulata, or an order of the Arachnida. The morphology of their sperm suggests a relationship to branchiuran crustaceans.

. . . that has been solved with information on DNA sequences.

The mystery of pentastomid relationships was cleared up recently by analysis of DNA sequences from pentastomids, branchiuran fish lice, other crustaceans, and representative annelids, chelicerates, myriapods, and insects (Abele *et al.* 1989). They are clearly relatives of branchiuran fish lice, and therefore belong well within the Crustacea, rather than having a phylum of their own. Thus molecular evidence can resolve relationships when morphological evidence is lacking, and an endoparasite with a complex life cycle involving two hosts can evolve from an ectoparasite with a single host.

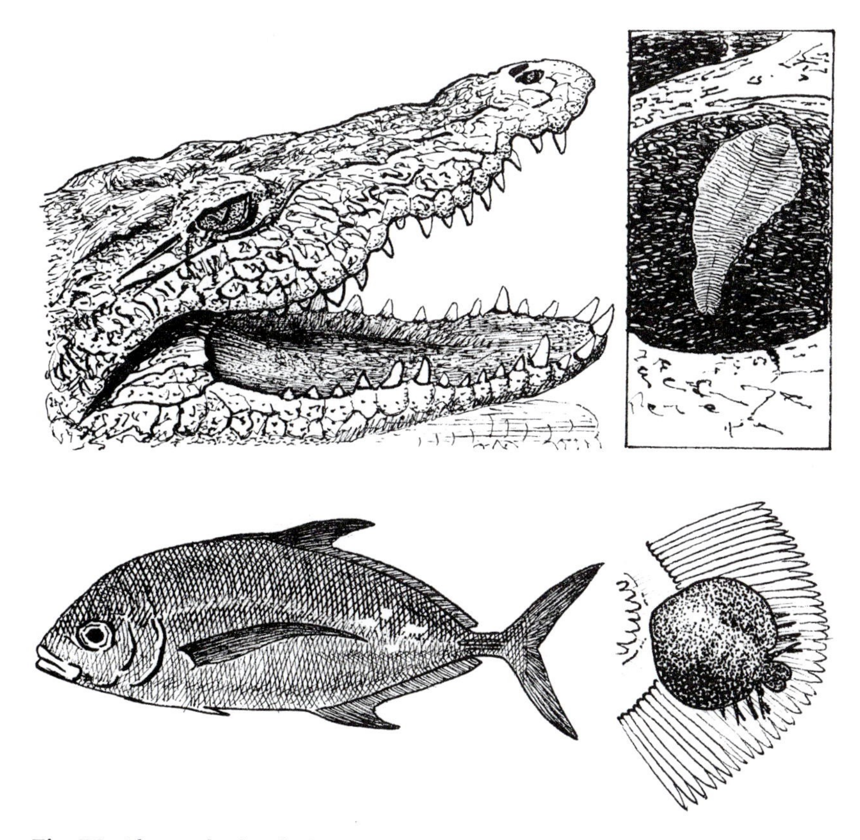

Fig. P3 Above: the head of a crocodile with a close-up of its right nostril, showing the parasitic pentastomid *Linguatula serrata*. Below: the brachiuran fish louse, *Argulus*, on the gill arch of the jack, *Carangoides orthogrammus*. The fish louse is the closest relative of the pentastomids. (By Dafila K. Scott.)

Irreversibility: prior adaptation as subsequent constraint

Plethodontid salamanders have lost their lungs, and some of them have lost their larvae. Such evolutionary changes are hard to reverse.

The plethodontids are a family of salamanders found primarily in Central America. Their ancestor had aquatic larvae, metamorphosis, and semiterrestrial adults, a life cycle still found in some species; but other plethodontids now mature as larvae, and some have embryos that develop directly into adult forms. Plethodontids have lost their lungs and breathe through their skins. The bones and muscles previously associated with lung breathing have moved forward in the thorax, where they help to construct an efficient protrusible tongue used in capturing food. In species that also switched from larval to direct development, the structures previously used in larval gills have found new application in adult structures. Thus plethodontid salamanders are constrained in the sense that it would be hard to select for reversal to lung breathing and larval development because the structures that ancestral salamanders used for those two processes have been employed elsewhere and could not be recovered without killing the developing animal (Wake and Larson 1987).

Similarly, the bones of the mammalian inner ear were originally breathing aids as gill arches in fish, became feeding aids in the jaws of amphibians and reptiles, then became hearing aids in mammals (Romer 1962). To select mammals to relocate the incus, malleus, and stapes from the inner ear into the jaw again would cause drastic hearing loss and probably death as embryos.

Thus key innovations, such as lunglessness and direct development, imply irreversibility because they free up morphological elements to be used in other structures with different functions, functions so important that a return to the original state would involve costs too high to pay.

The problem

Natural selection and history are the great themes of evolutionary biology.

The first two examples, of sexual cannibalism in the Australian red-back spider and of rapid evolution in the Trinidadian guppy, illustrate the power of natural selection. The second two examples, the puzzle of pentastomid relationships and irreversibility in the evolution of plethodontid salamanders, illustrate the importance of history. Natural selection and history are the two great themes of evolutionary biology, the two ways to explain the evolution of biological patterns. Organisms do not know about such distinctions, which we invent to help us analyze them, and every organism is a mosaic, some of whose parts reflect the role of recent selection, others of which recall its phylogenetic history. One of the chief aims of evolutionary biology is to attribute to selection and history their proper roles in the determination of any biological process or trait that interests us. Much progress has recently been made in developing methods to do just that; this book describes some of that progress. We return in Chapter 17 to some problems that have not yet been solved.

Fig. P4 The plethodontid salamander, *Bolitoglossa subpalmata,* capturing a spider with a tongue-flip. (By Dafila K. Scott.)

Chapter 1
The nature of evolution

Introduction

In 1859 Charles Darwin finished his masterpiece, *The origin of species by means of natural selection,* as follows:

> It is interesting to contemplate a tangled bank, clothed with many plants of many kinds, with birds singing on the bushes, with various insects flitting about, and with worms crawling through the damp earth, and to reflect that these elaborately constructed forms, so different from each other, and dependent upon each other in so complex a manner, have all been produced by laws acting around us. . . . Thus, from the war of nature, from famine and death, the most exalted object which we are capable of conceiving, namely, the production of the higher animals, directly follows. There is grandeur in this view of life, . . . that, whilst this planet has gone cycling on according to the fixed law of gravity, from so simple a beginning endless forms most beautiful and most wonderful have been, and are being, evolved.

The structure, diversity, and adaptation of life can be explained by processes we can observe now.

Darwin's daring message was that we can understand life by studying contemporary processes that we can observe and test. He thought evolution through natural selection was as magnificent as Newtonian physics. Events have proven him right. Evolution is now a well-established science with impressive explanatory power.

This book describes our current understanding of evolution. It starts with the mechanisms that cause evolutionary change—natural selection, inheritance, and gene expression—then moves through the evolution of sex, life histories, and sex ratios to sexual selection. After discussing multilevel selection and genetic conflict, it uses speciation to make the transition to phylogenetics and the history of life. The key events in evolution are then assessed, followed with insights from molecular analysis and the comparative method. We finish with thoughts on the status and prospects of evolutionary biology.

This chapter gives a brief overview of the whole subject.

A brief description of evolutionary biology

Evolutionary biology asks what changes in traits, populations, and species tell us about adaptation, history, and relationships.

Evolutionary biology is a rich collection of well-developed approaches to the interpretation of biological diversity and organismal design. Part of it studies how **natural selection** produces **adaptations**. Another part studies what **genealogies** and **phylogenies** tell us about relationship and history. Some methods used to reconstruct history use the observation that much of the variation in DNA sequences is **neutral** with respect to selection—that part of evolutionary biology is not about natural selection. Evolutionary biology is also the study of **conflicts**, conflicts between hosts and parasites, between parents and offspring, among brothers and sisters, between genes with different transmission patterns. Participants in conflicts must make the best of a bad situation that they often cannot escape. Another part studies genetic and phenotypic dynamics, regardless of whether they lead to adaptation or not. Sometimes they do not, and sometimes they cannot. Other parts of the field study the relationships of organisms and patterns in the fossil record that reveal the history of life. No ideological monolith, evolutionary biology is rich in alternatives that can be played off against each other to provide a self-critical, well-tested, and reliable interpretation of the natural world.

How evolutionary biologists think

Evolutionary biologists want to understand how variation in reproductive success arises, what causes the correlation of traits with reproductive success and thus natural selection, how the genetic variation that enables a response to selection originates and is maintained, and how that response is constrained by geography, time, inheritance, conflicts, development, and history. They want to unravel the history of life and understand the relationships among all living things. They investigate all organisms, from viruses to humans, from fungi to trees.

Evolutionary biologists ask many types of questions and use several approaches to answer them. Here are some of the more important ways of thinking about evolution.

Population geneticists think about changes in the frequencies of genes within populations.

Population and quantitative geneticists think about **microevolution**, which occurs within populations over relatively short periods of time; about the effects of changing the frequency of the different forms that one gene can take (its **alleles**) or of holding these frequencies at a stable, intermediate level. Their classical problem is to understand what maintains genetic variation. Among the candidate explanations are a balance between natural selection and **mutation**, **gene flow**, and the **drift** of neutral genes (Chapters 3–5). **Population geneticists** tend not to worry about the design of phenotypes.

Evolutionary ecologists think about the design of phenotypes for reproductive success.

Evolutionary ecologists think about the design of **phenotypes** for **reproductive success**. This involves traits such as age and size at maturity, number and size of offspring, life span and aging (life history evolution, Chapter 8), strategies for investing in sons or daughters (sex-allocation theory, Chapter 8), and the consequences of competition for mates and of choosing mates (sexual selection, Chapter 9). They tend to avoid genetic details.

Molecular evolutionists think about history recorded in DNA sequences. Some examine parts of the genome that are not transcribed into RNA or translated into proteins, parts that have little influence on the phenotype. They view adaptive change against a background of history that they can best infer from the parts of the genome that have not adapted, for adaptation can obscure history (Chapters 12 and 15).

Molecular evolutionists think about history recorded in DNA and protein sequences.

Systematists—many of whom are also molecular evolutionists—think in terms of evolutionary trees, give great weight to history, and focus on variation among species. For them, the major problem is to infer relationships among species reliably, so that they can reconstruct the history of life on the planet (Chapter 12), not to understand why gene frequencies change or how phenotypes are designed for reproductive success.

Systematists think about evolutionary trees and concentrate on variation among species.

Paleontologists are also historians of life. They think in deep time and concentrate on large-scale trends and major events, such as adaptive radiations, mass extinctions, and irregularities in the rate of evolution. They often cannot see processes occurring within periods of less than 100 000 years, but they see the big picture—**macroevolution**—with particular clarity. Paleontology, for example, tells us that rates of evolution in many lineages are irregular, with long periods of little or no change—**stasis**—interrupted by short periods of rapid change—**punctuation**. Paleontology also connects evolutionary history to continental drift and to ancient climates, often with fascinating insights (Chapter 13).

Paleontologists think in deep time about major patterns and trends.

Each approach is a different way of thinking about evolution, with its own advocates, its own school, and its own focus. Good evolutionary biologists are not constrained by these categories. They have the background and flexibility to use whatever methods are needed to answer a question. They know that these approaches all legitimately simplify a complex process, and that none of them retains all the important features of that process. Therefore, when adopting one approach, they check the consistency of assumptions, interpretations, and predictions against those made by other approaches.

Answers to the questions, 'How will gene frequencies change?', 'What is the state of the phenotype at evolutionary equilibrium?', 'How can we reconstruct the history of some species through their relationships?' require different methods.

Evolutionary change: adaptive and neutral

Microevolution describes how populations change in the relative abundance of genes or of phenotypes. The logical conditions necessary for microevolutionary change are described below (Chapters 2–10 put flesh on these bones).

Two concepts and a link between them explain microevolution. The two concepts are heritable variation in traits and variation in reproductive success among individuals within a population. The link is the correlation between the two types of variation. These three elements explain both adaptive and neutral evolution. When the correlation

Three conditions are necessary for adaptive evolution: individuals must vary in reproductive success; some variation in the trait must be heritable; the trait must be correlated with reproductive success.

between reproductive success and a trait is positive or negative, natural selection is operating on that trait, for natural selection consists precisely of variation in reproductive success correlated with variation in a trait. When that correlation is zero, natural selection disappears, even though variation in reproductive success may remain, and what is left is neutral evolution.

You may have heard that evolution is concerned with the survival of the fittest. That is a misleading half-truth. Survival is important, but only in so far as it contributes to reproductive success, to the number of offspring produced per lifetime that survive to reproduce. And to express the central process as the survival of the fittest is to think in a circle and beg the question, for if fitness is defined by survival, the statement is empty. The action of selection is located in the many mechanisms that connect variation in reproductive success to variation in traits and genes. There is no logical circularity in tracing such connections.

The production of surviving offspring is achieved through the number of offspring born, their survival, the survival of the parents to reproduce again, the number of offspring they have in their second and subsequent breeding attempts, the survival of those offspring, and so forth. Variation in reproductive success is made up out of variation in all these components.

There is always some variation in reproductive success in natural populations.

On the one hand, if there were no variation in reproductive success, neither the distribution of genes nor the distribution of phenotypes would change. (This statement excludes mutations, which affect the variation in reproductive success of molecules.) Note that there is always some variation in reproductive success in natural populations. For example, in Newton's (1988) study of sparrowhawks in southern Scotland, 72% of the females that fledged died before they could breed, 4.5% tried to breed but produced no young, and the remaining 23.5% produced between 1 and 23 young apiece (Fig. 1.1).

On the other hand, if there is no heritable variation in a trait, there will also be no evolutionary change—even if there is variation in reproductive success—for the differences in performance exhibited by the parents will not be reflected in the offspring. Only if there is some variation both

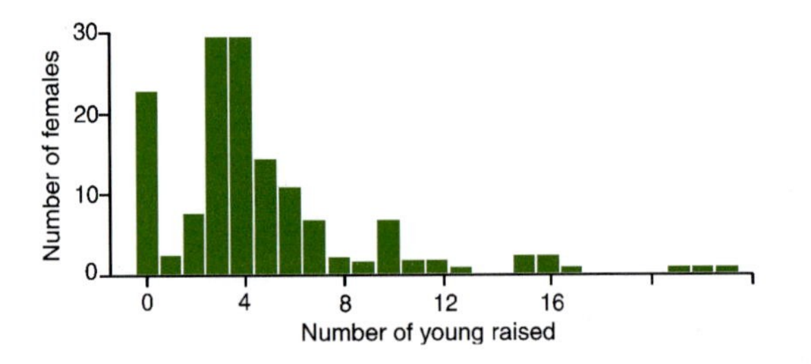

Fig. 1.1 Lifetime reproductive success measured as offspring that fledged for 142 female sparrowhawks from 1971 to 1984. Twenty-three females produced no fledglings, 1 produced 21, 1 produced 22, and 1 produced 23. The potential for natural selection was great. (From Newton 1988.)

in the trait and in reproductive success can there be a correlation between the two producing natural selection. Both conditions are necessary for adaptive change. Both are also necessary for neutral change, in which case the correlation between the two must be near zero.

Heritable variation of traits in natural populations is common.

Many traits in natural populations display heritable variation. To determine whether variation is heritable we can see whether offspring resemble their parents. O'Neil (1997) did this in her study of an introduced plant, the purple loosestrife (*Lythrum salicaria*). She measured the number of seeds per capsule and found considerable variation in this trait among both parents and offspring. To see whether the variation was heritable, she plotted the values for the offspring against the average for the two parents (the midparent value) (Fig. 1.2). Parents that had above-average numbers of seeds per capsule had offspring that were above-average, and parents with below-average numbers of seeds per capsule had offspring that were below-average. The slope of the straight line that best described the relationship (the *heritability*, see Chapter 4), 0.44, was significantly greater than zero. Thus she found heritable variation on which selection could act.

When there is both heritable variation and variation in reproductive success, it is the correlation between them that determines the type of evolutionary change that occurs. If there is little or no correlation between heritable and reproductive variation, then the things that are inherited and that do vary, whether genes or traits, will fluctuate randomly in the population within the limits of the available variation. This is **neutral evolution**. If the correlation between reproductive success and heritable variation is strong, then evolutionary change in the gene or trait will not be random but will move in the direction of increasing adaptation. This is **adaptive evolution**. The engine of adaptive evolution, **natural selection**, consists of two of the three parts of the evolutionary

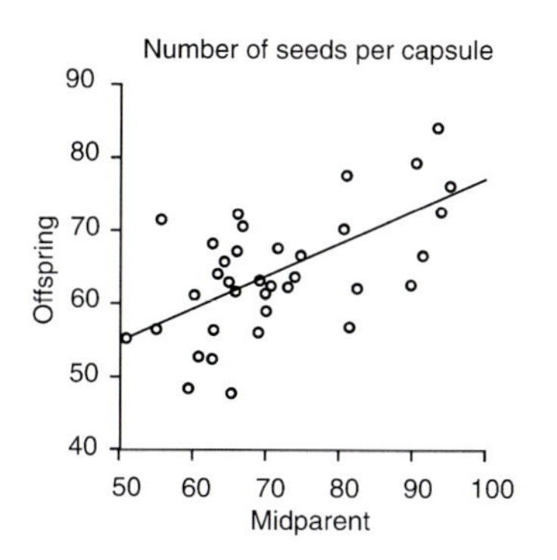

Fig. 1.2 The resemblance of offspring to parents for number of seeds per capsule in the purple loosestrife, an introduced plant that grows in wetlands. There is much heritable variation for this ecologically important trait in a population growing along the Kennebec River in Maine. Midparent = average of the two parents. The potential response to selection is great. (Data courtesy of Pamela O'Neil.)

mechanism: variation in reproductive success and the correlation between reproductive success and the trait under consideration. It does not include the heritable variation that enables a response to selection.

In natural selection, nothing consciously chooses what will be selected. Genes and traits increase or decrease in frequency because of their correlation with the reproductive success of the organisms that carry them. There is no long-term goal, for nothing is involved that could conceive of a goal. There is only short-term relative reproductive success, producing both short- and long-term change.

The correlation between traits and reproductive success determines whether evolution is adaptive or neutral. If the correlation is close to zero, evolution is neutral. Otherwise, it is adaptive.

Note that both adaptive and neutral evolution require variation in reproductive success. The same variation in reproductive success that causes adaptive evolutionary change in one trait, where *some* genetic variation is correlated with reproductive success, will cause neutral drift in another trait, where *none* of its genetic variation is correlated with reproductive success. To see how this works, consider a neutral gene or trait in a large population in which there is a lot of variation in reproductive success, where some females have many surviving offspring whereas others have few or none. The neutral gene or trait increases or decreases erratically, depending on whether it occurs in a large or a small family. Because it is not correlated with family size, occurring in many different family sizes at random from generation to generation, it does not change steadily in any particular direction. Also central to neutral evolution is the fair segregation of alleles into gametes at meiosis, which resembles the flipping of a coin. Fair segregation combines with random variation in family size to cause neutral drift in populations of any size, large or small.

This way of distinguishing between adaptive and neutral evolution focuses attention on the correlation between a trait and reproductive success. The strength of that correlation determines whether a trait will undergo adaptation or perform a random walk through time. All three components of the evolutionary mechanism—heritable variation, variation in reproductive success, and the correlation between the two—are important. The generation and maintenance of heritable variability and the patterns that result from neutral evolution are just as much a part of evolutionary biology as is the study of natural selection and its resulting adaptations.

Distinguishing between adaptive and neutral evolution by using the correlation between a trait and reproductive success clarifies and unifies the basic evolutionary processes. The loose application of 'natural selection' to the entire process, including the response enabled by heritable variation, is confusing. We therefore use 'natural selection' to refer just to the correlation of reproductive success with a trait. When we want to refer to the complete process, including the genetic response to selection, we use the phrase 'adaptive evolution'.

Information replicators and material interactors

Most of what gets inherited is not matter but information, a set of instructions coded in genes that specify how to build an organism. The design of organisms for reproductive success can only be changed by changing the stored genetic instructions. Genetic instructions are changed by natural selection when organisms with different genetic instructions vary in their reproductive success. Thus evolution occurs both in information (in genotypes) and in matter (in phenotypes). Genes function as information **replicators** while organisms function as material **interactors**, interacting with their environments and with each other to survive, reproduce, and get their genes into the next generation (Dawkins 1982; Williams 1992).

Evolution occurs both in information and in matter. Genes function as information replicators. Organisms function as material interactors.

Adaptation

A response to selection occurs whenever heritable variation in a trait is correlated with reproductive success. The result is improvement in reproductive performance. If this improvement continues for enough generations, a *process* called adaptation, it results in a *condition* in the trait that we also call an adaptation. Adaptation can be hard to demonstrate, but as a starting point we can define it as a condition that suggests to us that it evolved because it improved survival and reproductive performance. Here are a few examples of conditions that suggest adaptation.

Heritable variation in reproductive success improves reproductive performance, producing a state called an adaptation.

The striking precision of adaptations is demonstrated by the accurate, coordinated timing of reproduction in marine organisms that rely on the predictability of the moon and the tides. For example, the palolo worm lives most of the year hidden on the sea floor in shallow water in the western Pacific. As the reproductive season approaches, it differentiates a special reproductive organ on the posterior part of its body that looks like an individual worm. Reproduction is triggered by a specific phase of the moon, detected by millions of individuals scattered across a large area. On just a few nights of the year, the reproductive organ splits off the worm and swims to the surface, where it encounters millions of others, forming a massive swarm that spawns and dies. If the timing were not precise, reproduction would often fail, for lack of synchrony would lead to lack of partners. Meanwhile, the adults survive and grow another reproductive organ for the next year.

The reproduction of palolo worms in the South Pacific,

Similarly, the grunion, a fish found along the coast of California, spawns with the highest tide of the month. It rushes in with the waves just as the tide is turning and throws itself out of the water, depositing eggs and sperm in pockets in the wet sand where the eggs will not be disturbed again by waves until a month later, at the time of the next spring tide. When that tide arrives and the waves disturb the eggs, the young hatch explosively and swim out with the receding water. The timing of reproduction, development, and hatching are coordinated precisely with the rhythm of the tides.

of grunion in California,

and the ability of bats to find prey in the dark, using echolocation, are precise adaptations.

The precision of adaptation is also illustrated by the ability of bats to find prey in the dark, using echolocation. The brilliant work of Spallanzani in the late eighteenth century on blinded bats that flew without difficulty was completely ignored by science, which waited for Griffin to rediscover bat echolocation in 1938. Bats produce high-frequency, short wavelength cries that are reflected by small objects. There is a physical constraint on this method of 'seeing'. Sound waves lose energy rapidly with distance. Lower-frequency waves penetrate further but can only detect large objects, and many bats use echolocation to detect small, rapidly moving, flying insects (Fig. 1.3). Griffin (1958) discovered that when a bat approaches to within 1 or 2 m of an obstacle, it increases the number and raises the frequency of the ultrasonic pulses that it emits. Griffin also tested the ability of bats to fly through grids of fine wires. They had no problem with wires that were 0.4 mm in diameter or larger and had some success with wires down to 0.2 mm in diameter, which reflect very little sound energy.

Later experiments (Simmons 1973) demonstrated that a flying bat could discriminate target distances as small as 1 cm by detecting differences of as little as 60 microseconds in the pulse-to-echo interval. It could also discriminate a stationary from a vibrating target where the vibration was as small as 0.2 microsecond, implying a difference in pulse-to-echo interval of just 1 microsecond. For comparison, the duration of a single action potential in the bat's auditory nerve is about 1 millisecond, a thousand times as long as the interval being discriminated. These astounding abilities to discriminate the distance and nature of an object in complete darkness result from selection for ability to locate flying insect prey whose wing beats convey information on both distance and direction of movement.

Similarly impressive abilities to detect faint signals that carry critical information are found in the noses of migrating salmon, which can

Fig. 1.3 A bat catching a moth in complete darkness. Bats have exquisite adaptations that help them to catch moths, and moths have equally exquisite adaptations that help them to avoid bats. (By Dafila K. Scott.)

detect as little as a single molecule characteristic of their native stream, and in the dark-adapted eyes of nocturnal mammals, which can detect as little as a single photon of light.

Thus natural selection evidently has great power to shape precise adaptations. It operates whenever there is variation in reproductive success and the variation in reproductive success is correlated with heritable variation in the trait. When these conditions are fulfilled, the process of adaptation begins. Whether or not it will ever result in the state we call adaptation, illustrated above, depends on whether other factors constrain the response to selection—whether it can occur at all, and whether only certain phenotypes can be produced and not others. Even if the response to selection can occur and if a certain phenotype can be produced, other factors, such as population size, gene flow, and the frequency with which selection occurs, affect the precision with which a trait can be shaped for reproductive success.

Because there is always some variation in reproductive success, and some trait is usually correlated with reproductive success, natural selection has almost always acted and is usually acting in all populations, including our own. Because natural selection acts on all variable traits that contribute to survival and reproduction, if such a trait is not in the state best for survival and reproduction, then something must be limiting its evolution. Three such limiting factors are particularly important: gene flow, sufficient time, and **tradeoffs**. Gene flow and sufficient time are discussed next. Tradeoffs are introduced later under Principles of phenotypic design (p. 22).

If a trait contributing to reproductive success is not well adapted, something must be constraining its evolution.

Limits to adaptation: gene flow

Organisms move genes selected in one area to places where they may not be appropriate and can produce local maladaptations.

Genes 'flow' from one place to another when organisms born in one place reproduce in another, introducing their genes into the local gene pool. When natural selection favors different things in different places, organisms can transport genes that have been successful in one place to other places where they may not be so successful. Gene flow, like mutation, introduces new genetic variants into local populations, and it can produce local maladaptations. For example, in the south of France a small bird, the blue tit, breeds both in downy oak, which is deciduous, and in holm oak, which retains its leaves in winter (Fig. 1.4).

Birds breeding in holm oaks start to breed at the same time as birds breeding in downy oaks. They should start to breed later because the peak of insect abundance on holm oaks comes later in the season; but because more individuals are recruited into the population from downy oaks than from holm oaks, the response to selection reflects the greater frequency with which genes have been selected in the downy oak environment. The birds are maladapted on holm oaks, where clutches laid later in the season would result in more surviving offspring. Because adult birds move freely between downy and holm oaks, and because there are not enough nesting territories on downy oaks, forcing some birds to nest on holm oaks, gene flow is strong, leading blue tits in

Fig. 1.4 Blue tits nesting in downy and holm oaks in southern France have greater success on downy oaks and lay clutch sizes adapted to downy oaks in both habitats, even though they would do better with smaller clutches on holm oaks. (By Dafila K. Scott.)

southern France to breed inappropriately early on holm oaks (Blondel *et al.* 1992).

Local adaptations do evolve when selection is strong.

Despite gene flow, local adaptations can evolve when selection is strong. A classic example is heavy metal tolerance in plants. Plants on mine tailings grow on toxic soil and rapidly evolve adaptations to deal with it. Antonovics and Bradshaw (1970) sampled plants from a transect across a zinc-mine tailing and into an uncontaminated pasture. Patterns of continuous changes in trait values or gene frequencies along a geographical transect are called **clines**, and the steepness of a cline measures geographic differentiation. Antonovics and Bradshaw found an extremely steep cline in zinc tolerance at the boundary between the mine tailing and the uncontaminated pasture. The index of tolerance changed from 75% to 5% in less than 10 m. The plants that were zinc tolerant flowered later and were smaller, meaning that tolerance had costs. They also suffered less from inbreeding than those that were not zinc tolerant, which suggests an evolutionary response to local mating with relatives. Even more rapid evolution and stronger selection was suggested by another study of plants growing near galvanized-steel power poles, for the power poles appeared much more recently than the mine. Plants growing within 10 cm of a power pole had significantly higher zinc tolerance than those just 20 cm from it (Fig. 1.5).

For gene flow to prevent local differentiation, selection must be weak and the distance that genes move in each generation must be large.

Thus strong selection can produce local adaptation despite gene flow, and species often consist of a patchwork of genetically different populations, each displaying different adaptations. For gene flow to prevent local adaptation, natural selection must be weak and the mean distance that genes move in each generation must be large (Endler 1977).

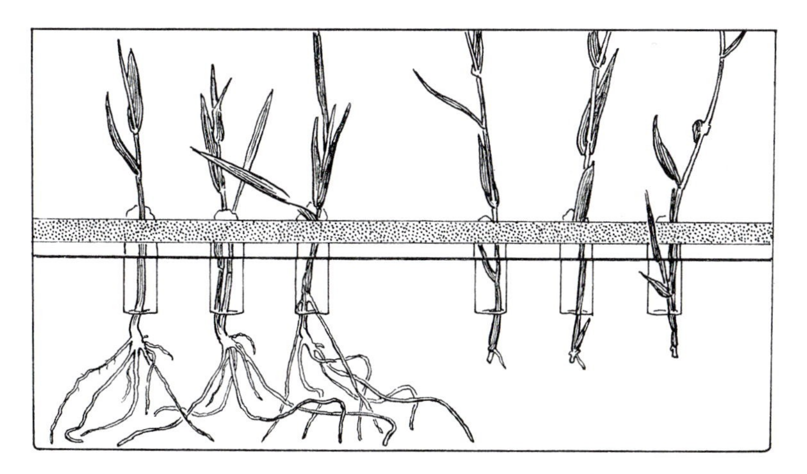

Fig. 1.5 Grass that has adapted to zinc from power poles (on the left) grows much more impressive roots in zinc solution than does grass grown from seed, collected from just a few meters away, that is not adapted to zinc (on the right). (By Dafila K. Scott from a photograph by A. D. Bradshaw.)

Limits to adaptation: sufficient time

Even without gene flow, it takes time for a population to adapt to an environmental change. Consider the absorption of the sugar in milk, lactose, by adult humans. Like other mammals, human children come equipped with the enzymes needed to digest milk, and most children lose that ability at the age at which they used to be weaned—at about 4 years. However, some humans retain the ability to digest fresh milk into adulthood, including the populations of northern Europe and some in western India and sub-Saharan Africa (Simoons 1978). This genetic difference may explain why dairy products are more prominent in French than in Chinese cooking. The ancestral condition was the inability to digest fresh milk after the age of four, and the recently evolved condition is that ability.

How long would it take that ability to evolve? The domestication of sheep and goats occurred about 10 000–12 000 years ago, the subsequent origin of dairying can be traced to 6000–9000 years ago, and the ability to digest fresh milk after the age of four has a simple genetic basis: it behaves as a single dominant allele on one of the normal chromosomes (an **autosome**), rather than on a sex chromosome. **Dominant alleles** increase in frequency under selection more rapidly than do **recessive alleles**, and knowing that the gene is on an autosome simplifies the prediction of how its frequency will change under selection. Imagine a stone-age human population in which goats and sheep had been domesticated and milk production had begun. In that population, a new mutation arose that allowed people to continue to utilize fresh milk after the age of four. It had an advantage over the ancestral state, for people who drank milk but could not absorb lactose suffered from flatulence, intestinal cramps, diarrhea, nausea, and vomiting, which reduced their reproductive performance. Moreover, those who could absorb lactose benefitted from an additional high-quality food source, rich in calcium and phosphorus, especially at times when other food was scarce. This was especially important to nursing mothers and growing children. Suppose that the ability to absorb lactose conferred a selective advantage of 5%, so that for every 100 surviving and reproducing children of non-absorber parents, the same number of absorber parents produced 105.

Even for a gene under strong selection, time is a constraint.

At the beginning, the gene would have been rare, and simply because it was rare, it could only increase slowly, for very few people carried it, enjoyed its advantages, and produced a few more surviving children than did those who did not carry it. As the gene increased in frequency, and more and more people carried it, it began to spread more rapidly through the dairying culture. However, when it became common, its rate of spread decreased, for then most people carried it, and there were very few who suffered from the disadvantage of not having it. How long did it take to increase from a single new mutation to a frequency of 90%? The answer, which comes from genetic models (Chapter 4), is 350–400 generations or 7000–8000 years. This estimate would increase dramatically if we assumed a weaker selective advantage. If the estimated age of adult milk-drinking is accurate, then milk-drinking would appear to have increased fitness substantially.

That argument assumes a causal connection between the ability to absorb lactose and individual fitness, an ability that arose with the origin of agriculture. As we will see in Chapter 15, agriculturists evidently had an advantage over hunter–gatherers for several reasons, not just dairying and the ability to absorb lactose. They spread not only their culture but their genes, including the gene for lactose absorption, from the Near East over all of western Europe. Thus the rapid rise in frequency of the lactose absorption gene in Europe did not depend just on its digestive advantages; it may have hitch-hiked with a culture that expanded and dominated.

Even for a gene under strong selection, evolutionary change takes time.

Principles of genetic information transmission

In asex, the entire genome is transmitted as a unit. In sex, offspring are a much more diverse mixture of genes from two parents.

Evolutionary change requires genetic change. Here is a short summary of basic principles of genetic information transmission.

Sex versus asex

Information is transmitted in quite different ways in asexual and sexual organisms. (We describe here only the commonest of many types of asexual and sexual reproduction.) In asexual organisms, the entire genome is transmitted as a unit, and the only genetic differences between parents and offspring, and among offspring, are the result of mutations. In sexual organisms, offspring are a 50:50 mixture of genes derived from each parent—they are genetically diverse. A sexually reproducing population generates much more genetic variation in each generation through recombination than does an asexually reproducing population through mutation. Because the response to selection depends on the amount of genetic variation available, sexual populations may evolve more rapidly than asexual populations. They do so by combining from different parents beneficial mutations that will increase and disadvantageous mutations that will decrease under natural selection (Crow and Kimura 1965). In asexual populations favorable mutations must accumulate one after the other, in the same lineage, and disadvantageous mutations cannot be eliminated so efficiently (Fig. 1.6).

The sex–asex contrast is one of several major differences in systems of genetic information transmission discussed in Chapters 4, 7, and 10.

Meiosis preserves both DNA sequences and gene frequencies

The information stored in the genes is copied precisely both at the level of the DNA sequence, where replication is very accurate, and at the level of the population, where gene frequencies do not change from generation to generation in large populations if there is no selection, mutation, or gene flow. Information transmission in **meiosis** is astonishingly

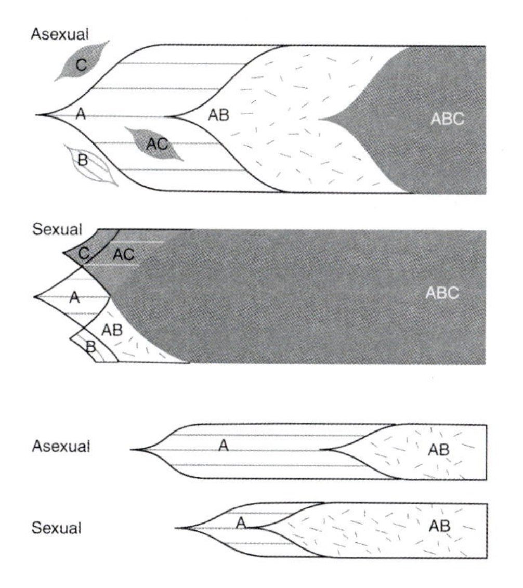

Fig. 1.6 Sexual reproduction can combine beneficial mutations that arise independently in different organisms; with asexual reproduction the mutations must be accumulated sequentially in the descendants of a single mutant individual. This advantage of sexual reproduction is much greater in large than in small populations, for the probability of two beneficial mutations arising at nearly the same time is much greater in a large population. A, B, C = mutations (From Crow and Kimura 1965.)

accurate and fair. Each of the two copies of a gene in a diploid organism has precisely the same chance of being placed in a successful gamete. The fairness of meiosis and accurate replication of genes make information transmission in sexual populations fundamentally conservative. The meiotic mechanisms that ensure fairness probably evolved for other reasons (Chapter 10).

Mendelian inheritance preserves genetic variation, forming a stable foundation for evolutionary change.

Why **gene frequencies** do not change from generation to generation—if there is no selection, mutation, nonrandom mating, gene flow, or genetic drift—is explained in Chapters 4 and 10. Here the important point is not *why* they do not change, but *that* they do not change. With Mendelian inheritance, the genetic variation necessary for a response to selection is preserved, not destroyed (Fisher 1930; Ewens 1993). If gene frequencies often changed for arbitrary reasons, an incremental response to natural selection would be impossible, and adaptive evolution could not occur. Thus the conservatism of Mendelian inheritance at the level of the population is the foundation of evolutionary change in sexual organisms.

Natural selection can rapidly produce highly improbable states

Mendelian populations are conservative when they are not under selection, but under strong selection they can rapidly produce combinations of genes that are, at first glance, extremely unlikely. We demonstrate this with two examples. The first, an analogy using the letters of the alphabet, exaggerates the power of selection because it pre-specifies the target. This is not how natural selection works, but the example makes an important principle clear. Because the second, more realistic example is more complicated, we approach it through the first example.

The 31 letters in THEREISGRANDEURINTHISVIEWOFLIFE can be compared to a sequence of 31 genes each with 26 different possible versions. If evolution assembled such sequences completely at random, it would have to sort through 26^{31} different possible combinations of letters to hit on this one. However, natural selection causes favored gene combinations to increase in frequency, and accurate replication in both molecules and populations preserves those increases—it 'remembers' what worked before. Here strong selection artificially retains the correct letter whenever it occurs. If we start with any random sequence of 31 letters and retain all the letters that happen to be correct, then repeat the process by generating new letters at random for the ones that are not yet correct, we get to the right sequence in about 100 trials, about 30 orders of magnitude faster than a random search (Dawkins 1986; Ewens 1993). For comparison, the age of the universe, about 10–20 billion years, is only 20 orders of magnitude longer than a millisecond.

That example is usefully clear but misleading because natural selection does not aim at any particular final state. It does not aim at anything. It just produces something that works better from among the variants currently available. The next example moves from letter-play to meaningful experimentation.

Because inheritance preserves improvements and selection sorts them, improvements accumulate rapidly.

Ribonucleic acid (RNA), a macromolecule that stores genetic information, can evolve rapidly in test tubes. RNA seems to have originated before DNA, which only later became the common genetic molecule. RNA remains the genetic molecule in many viruses, and the fact that its mutation rate is much higher than that of DNA has important consequences. Human immunodeficiency virus (HIV), for example, is an RNA virus, and its high mutation rate is one reason that it is hard to find a vaccine against it. The high mutation rate of RNA is also one reason that the following experiment worked so quickly.

From a type of virus that infects bacteria (a **phage**), one can extract an enzyme that replicates RNA (an RNA replicase). Given an RNA molecule as substrate, an energy source, and a supply of the four necessary building blocks—the nucleotides—from which RNA is made, this enzyme rapidly produces a large population of RNA copies in solution in a test tube. By transferring a drop of the solution into a new test tube every 30 minutes, one selects the copies present at highest frequency, which are most likely to be transferred in the drop. The molecules that are copied most rapidly are at highest frequency and have a selection advantage. Replication is good but not exact: in about 1 in 10 000 cases the wrong nucleotide is substituted. Thus mutations occur, and some mutations are replicated faster than others—they have greater reproductive success.

Two types of molecules have an advantage: small ones, which can be replicated rapidly, and those that fit especially well to the replicating enzyme. After more than 100 transfers, a large, complex molecule dominates the population; which one depends on details. One that occurs frequently is 218 nucleotides long. Hitting on such a molecule at random has a probability of 4^{218} or 10^{131}. Since there are about 10^{16} molecules in a test tube just before transfer, the procedure screens about 10^{16} molecules every half hour. If it were random, it would take about 10^{110} years to find the best one. Instead, the procedure produces something very close to the best one in about 2 days. The response to selection is efficient because each step leads to a molecule that is better than the previous one, and because the improvements are inherited, they accumulate (Maynard Smith 1998).

Remember this example if you encounter the argument that natural selection cannot work because it starts with random variation. That argument is false. Natural selection can produce highly adapted states rapidly. They only *appear* to be improbable. The efficiency of natural selection makes them probable.

Population genetics answers important questions

Population geneticists study the consequences of genetic information transmission for populations. Some of their important discoveries are listed above: evolution is usually faster in sexual populations, Mendelian populations are conservative, and natural selection can produce improbable states rapidly. Population genetics also answers other important questions, including:

1. When is inbreeding a problem (why avoid incest)?
2. Is evolution faster in small or in large populations?
3. What are the chances that the children of a given couple will have an inherited disease?
4. What can the geographical distribution of gene frequencies in a population tell us about its history?

These issues are developed in Chapters 3, 4, 5, and 15.

Principles of phenotypic design for reproductive success

Fitness is relative reproductive success

The key to adaptation is how phenotypes achieve their lifetime reproductive success.

Population genetics, which reduces the evolutionary process to the analysis of the factors that change the number of copies of a gene in a population from one generation to the next, has great simplicity and power. However, if we only think about changes in gene frequencies, we cannot explain phenotypic evolution. To understand why organisms are designed in some particular way for reproduction and survival, we must analyze both the replication of the genes that the organism carries and the organism as an interactor that manages to reproduce while overcoming problems posed by the environment.

An allele that increases the lifetime reproductive performance of the organisms in which it is found, relative to other alleles of the same gene, will increase in frequency in the population. We say that it has a higher fitness than the other alleles because it increases in frequency at their expense and because it improves the relative reproductive performance of the organisms that carry it. The allele that wins is only fitter, not necessarily fittest, for another mutant could come along and displace it.

After this process has continued for a long time in a large population in a stable environment, the probability of finding fitter mutants decreases, for most of the single mutations have already occurred many times. The versions that have survived are in some sense 'close to fittest', and most single mutations have become detrimental. This argument does not apply to double or triple mutations, which are much rarer. Traits for which several simultaneous mutations are necessary have not been 'saturated' by mutations and can still be improved. The same effect can be accomplished by reorganizations of the genome, such as chromosomal inversions, that 'freeze' beneficial combinations of genes that previously recombined independently.

Components of natural selection: individual, sexual, and kin selection

The analysis of reproductive success begins with the factors determining the number of surviving and reproducing offspring produced by a single individual over its lifetime. This is the most basic and general component of reproductive success, individual fitness. Selection driven by variation in offspring number per lifetime is called individual selection. It occurs in both asexual and sexual organisms, and it is often all that is needed to account for many adaptations. When parental state affects offspring performance, then the measurement of fitness involves the next generation as well, and reproductive success is the number of surviving and reproducing grandchildren.

The number of offspring per individual lifetime is individual fitness.

In sexually reproducing organisms, reproductive success depends on success in interacting with a partner of the opposite sex to produce offspring. This component of natural selection is called **sexual selection**. Sexual selection can improve mating success so much that lifetime reproductive success increases although survival decreases. For example, the male peacock's large and beautifully colored tail improves his reproductive success by making him attractive to females, but it reduces his chances for survival by making it harder for him to fly—and tigers do eat peacocks.

Sexual selection improves mating success and can reduce survival.

Sexual selection involves the two sexes in a complex interaction, with surprising results. When females have preferences for certain male traits and by mating with such males transfer the preferred traits to their sons and their preferences to their daughters, a complex coevolutionary interaction between the sexes is generated (Chapter 9).

Organisms that interact regularly with relatives experience a third kind of selection. Because an allele's reproductive success consists of the number of copies that exist in the next generation, relative to the copies of other alleles, it does not matter through whose reproductive activities those copies were replicated—directly, by the individual that carries it, or indirectly, by relatives that also carry that allele. Thus if an individual can influence the reproductive success of its kin, it should do so if the benefits exceed the costs—if the increase in its indirect fitness through the increased reproductive success of relatives exceeds the reduction in its direct fitness through the reduction of its own reproduction (Hamilton 1964). This is **kin selection**, a powerful tool for understanding the evolution of apparently altruistic behavior (Chapter 14). Its success, and the success of the evolutionary theory of aging (Chapter 8) and sexual selection (Chapter 9), which also emphasize the success of genes at the cost of the organisms that carry them, have convinced many evolutionary biologists that much of evolution is gene centered (Williams 1966; Dawkins 1976).

An individual should sacrifice its own reproductive success to improve that of kin when the genetic benefits exceed the genetic costs.

Traits do not evolve for the good of the species

People unfamiliar with evolution sometimes still say that things evolve for the 'good' of species, so that species can avoid extinction. This explanation is fundamentally wrong. Traits evolve because they improve the reproductive success of individuals and their kin, and if the species to which those individuals belong happens to survive longer because of those changes, this is a by-product of the essential process and not the reason for it.

Because selfish mutants can usually invade, selection on individuals is usually stronger than selection on groups or species.

This insight was achieved in a fascinating episode that can be summarized in three words: selfish mutants invade. If a trait arose that benefitted the species at the cost of the individual, a mutant that was selfish, profiting at the expense of the altruistic individuals, would invade and take over the population. Selection on individuals is much stronger than selection on species. Individuals have much shorter generation times than species, and there is much greater variation in reproductive success among individuals within a species than there is among species within a lineage. In the time that it takes for new species to form and go extinct, a process spanning thousands of individual generations, hundreds of millions of the individuals that form those species have lived and died, giving selection much greater opportunity to sort among individuals than to sort among species. That is why selfish mutants can invade.

Species selection cannot shape adaptations (Maynard Smith 1964; Williams 1966), whose precision is inexplicable by a process that happens infrequently and is poorly correlated with individual reproductive performance. The reproductive timing of palolo worms and grunion, or the ability of bats to fly in the dark, can only be explained plausibly by individual selection. However, species selection can affect large-scale patterns, such as the phylogenetic distribution of sexual species and asexual populations: sex appears to reduce the probability of extinction. The extinction of major groups, such as trilobites, ammonites, and dinosaurs, also affected the subsequent course of evolution of the groups that survived, by providing individual selection with a scope of action that would otherwise not have been available.

Trade-offs

A trade-off exists when a change in one trait that increases reproductive success causes changes in other traits that decrease reproductive success.

A trade-off exists when a change in one trait that increases reproductive success is linked to changes in other traits that decrease reproductive success. The reasons for such linkages are not always well understood, but we know they are frequent. If there were no trade-offs, then natural selection would drive all traits correlated with reproductive success to limits imposed by constraints. Because we find many traits that are clearly correlated with reproductive success varying well within such limits, trade-offs must exist. A common trade-off is that between reproduction and survival. For example, fruit flies selected to lay many eggs early in life have shorter life spans (Rose 1991). They cannot both reproduce a lot early in life and have a long life, a trade-off that shapes the evolution of life span.

Other important trade-offs occur between the ability to eat one thing and the ability to eat many things, and between mating success and survival. There are many others. Whenever one analyzes the costs and benefits of changes in traits, trade-offs are usually found. They limit how much fitness can be improved by changing traits, for when traits cannot be changed independently of one another, the benefits gained by changes in one trait are often rapidly balanced by costs incurred in others, causing the response to selection to stop.

There are trade-offs between reproduction and survival, specialists and generalists, and mating success and survival.

Constraints

Organisms are not soft clay out of which adaptive evolution can sculpt arbitrary forms. Natural selection can only modify the variation currently present in the population, variation that is often strongly constrained by history, development, physiology, and the laws of physics and chemistry. Natural selection cannot anticipate future problems, nor can it redesign existing mechanisms and structures from the ground up. Evolution proceeds by tinkering with what is currently available, not by designing ideal solutions, and the variation currently available is often limited by constraints.

Natural selection can only modify the variation present. It cannot anticipate future problems.

The constraints imposed by physics and chemistry are straightforward. The diameter of an organism without a circulatory or respiratory system cannot be much greater than 1 mm, a limit set by the rate at which oxygen diffuses through water. Water-breathers would have great difficulty being endotherms, for they have to pass large volumes of water across large gill surfaces to extract oxygen, and moving water rapidly strips heat from warm bodies. The limbs of terrestrial animals must be thicker in heavier organisms, for the strength of a limb is determined by its cross-sectional area, whereas weight is determined by volume, which grows more rapidly (x^3) with the length of an organism (x) than does the cross-sectional area (x^2) of one of its limbs. That is why the legs of rhinoceroses are thicker, in proportion to their body lengths, than the legs of antelopes. Physics cannot be avoided.

Just as interesting are constraints that have evolved. Past adaptations can become future constraints, placing the imprint of history on a lineage. Here are four examples.

First, why are there no **parthenogenetic** (asexual) mammals? Early development in mammals requires one egg-derived and one sperm-derived haploid nucleus. The two types of nuclei are marked differently in the germ line of the parents by the attachment of methyl groups to the DNA molecules, a pattern known as **genetic imprinting**. Early development requires the expression of some genes derived from the father and some genes derived from the mother, which is determined by the sex-specific imprinting that occurs in the germ line of the parent. The parental patterns are erased later in development, allowing the offspring to imprint the genes that are appropriate to their own sex and making it possible to clone mammals from adult cells. If all the genes came from the mother, then some normally paternal genes would not be turned on at the right time, and early development would fail.

There are no parthenogenetic mammals because mammalian development requires one egg-derived and one sperm-derived nucleus.

Second, why are there no parthenogenetic frogs? In a freshly inseminated frog egg, the sperm donates the **centrosome** that replicates to form the poles of the first mitotic **spindle**. Activated eggs that lack the paternal contribution—the sperm-derived centrosome—divide abortively (Elinson 1989).

In both mammals and frogs, the constraints on parthenogenesis appear to result from a conflict between nuclear genes and mitochondrial genes, won by the nuclear genes. A genetic conflict exists whenever two types of genes have different patterns of transmission. **Mitochondria** are usually transmitted only through eggs, whereas nuclear genes are transmitted through both eggs and sperm. Because cytoplasmic genes have zero fitness in males, cytoplasmic mutants that induce parthenogenesis would be selected. The barriers to parthenogenesis in mammals and frogs may have evolved because sexual reproduction is advantageous to nuclear genes; they appear to protect the interests of nuclear genes against the invasion of cytoplasmic elements that could feminize their hosts (see Chapter 10).

The vertebrate eye is flawed. Nerves and blood vessels obscure the passage of light into the photoreceptors.

Third, the vertebrate eye, admired for its precision and complexity, contains a basic flaw. The nerves and blood vessels of vertebrate eyes lie between the photosensitive cells and the light source (Goldsmith 1990), a design that no engineer would recommend, for it obscures the passage of photons into the photosensitive cells (Fig. 1.7). Long ago, vertebrate ancestors had simple, cup-shaped eyes that were probably originally used only to detect light, not to resolve fine images. Those simple eyes

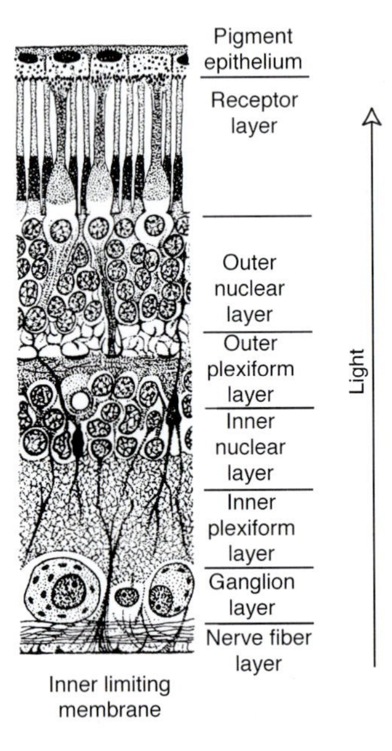

Fig. 1.7 The pigment epithelium of the vertebrate eye containing the light-sensitive cells develops inside the nerves and blood vessels through which light must pass, a design no engineer would approve. (From Romer 1962.)

developed as an out-pocketing of the brain, and the position of their tissue layers determined where the nerves and blood vessels lay in relation to the photosensitive cells. If the layers had not maintained their correct positions, relative to one another, then the mechanisms that control differentiation, in which an inducing substance produced in one layer diffuses into the neighboring layer, would not work. Once such a developmental mechanism evolved, it could not be changed without destroying sight in the intermediate forms that would have to be passed through on the way to a more 'rationally designed' eye.

During development the testes wrap the vas deferens around the ureters, like a person watering the lawn who gets the hose caught on a tree.

The fourth example concerns the length and location of the tubes connecting the testicles to the penis in mammals. In the adult, cold-blooded ancestors of mammals, and in present-day mammalian embryos, the testicles are located in the body cavity, near the kidneys, like ovaries in adult females. Because mammalian sperm develop better at temperatures lower than those found in the body core, there was selection, during the evolutionary transition from cold- to warm-bloodedness, to move the testicles out of the high-temperature body core into the lower-temperature periphery and eventually into the scrotum. This evolutionary progression in adults is replayed in the developmental progression of the testes from the embryo to adult, and as they move from the body cavity towards the scrotum, they wrap the vas deferens around the ureters, like a person watering the lawn who gets the hose caught on a tree (Fig. 1.8). If it were not for the constraints of history and development, a much

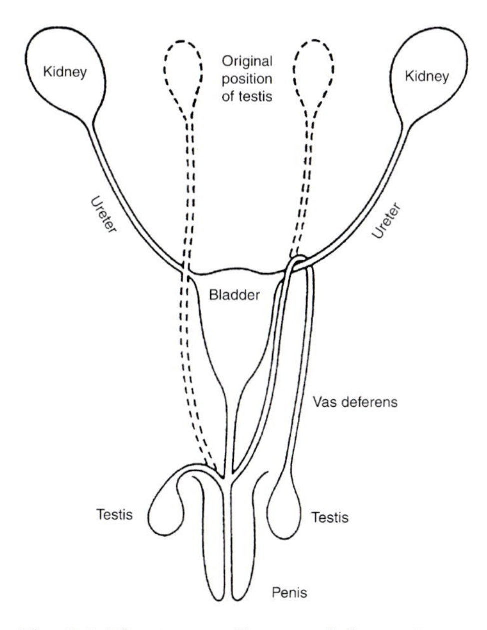

Fig. 1.8 The mammalian vas deferens is wrapped around the ureter like a hose wrapped around a tree by someone watering a garden. If the development of the testis were not constrained, the vas deferens could be much shorter (like the one on the left). (From Williams 1992.)

shorter vas deferens would have evolved, costing less to produce and perhaps doing a better job.

Speciation

Sexual organisms form clusters that we call species. Speciation helps to connect micro- to macroevolution.

The elementary processes in microevolution are genetic changes—first mutations, then changes in gene frequencies driven by selection, migration, and drift. Phenotypes with better reproductive success get more copies of the genes they carry into future generations, leading to inherited changes in the design of phenotypes. Adaptation occurs but perfection does not result because of constraints and the finite number of variants sorted by selection. Microevolution is, however, not the whole story, for the world is populated by organisms that form clusters of similar individuals, called **species**. Speciation, the splitting of populations into evolutionarily independent units, connects the microevolutionary processes occurring within populations to the macroevolutionary relationships and patterns of all living things. The species is the critical transition level between micro- and macroevolution.

Speciation is a by-product of several processes with diverse genetic consequences.

Speciation does not result from just one mechanism. It is a by-product of several processes—isolation from neighboring populations caused by geography, local habitat selection, and mate selection—with diverse genetic consequences. Populations do not speciate to become better adapted. Individuals choose mates either because they are locally available and are geographically isolated from other mates, or because choosing those mates (or mates found in those habitats and not others) results in fitter offspring. Speciation is a by-product of selection and drift operating on individuals.

Species can be defined either as potential mates or by phenotypic or genotypic similarity.

Species can be defined as sets of organisms that could mate with each other and produce viable grandchildren; or as sets of organisms resembling each other morphologically, or in their DNA sequences, that act as evolutionarily independent units. The first definition, the Biological Species Concept of Dobzhansky (1937) and Mayr (1942), was long thought to have decisive advantages, but some biologists now prefer the second definition, the Phylogenetic Species Concept, which also assumes that common ancestry causes the similarity. This approach leaves open the mechanisms leading to speciation, allowing one to consider alternatives without prejudice (Mallet 1995).

There are exceptions to every species definition.

No definition of species covers all known organisms. Species range from the relatively distinct and stable—as in birds and mammals—to the relatively indistinct and unstable, as in the hybrid complexes common in micro-organisms, plants, and some crustaceans, amphibians, and fish. Some bacteria exchange genetic information frequently enough to behave almost like sexual species (e.g. *Streptococcus*); others behave like asexual clones with nearly complete reproductive isolation (e.g. *Salmonella*). Speciation is discussed in more detail in Chapter 11.

Microevolution and macroevolution

Microevolution (Chapters 2–11, 15) describes processes occurring within species and populations. We can do experiments on microevolution, and we can study its causes directly. Macroevolution (Chapters 11–16) describes patterns perceived in the comparison of species and larger groups—families, orders, and phyla—that are described in systematics and paleontology.

The connections between micro- and macroevolution are to be sought in speciation and development.

How microevolution connects to macroevolution is an important question that has not yet been answered satisfactorily. A promising approach suggests that the microevolution of developmental mechanisms produces constraints on the further evolution of the organisms containing those mechanisms. Different lineages evolve different developmental mechanisms. All the species sharing developmental mechanisms may therefore also share similar body plans and evolve under similar constraints. This is probably why we can recognize major groups of organisms by their body plans, what Darwin called Unity of Type. Some progress in this direction is described in Chapters 6 and 15.

Biological causation

Biologists want to understand everything about organisms. Some study the immediate causes, seeking the answers to questions such as 'How does photosynthesis work?', 'What determines the sex of an organism?', and 'What causes disease?', in physiology, genetics, biochemistry, development, and related fields. Here the aim is to identify what causes the trait or process within the lifetime of a single organism. This is the study of **proximate** (mechanical) **causation**. It constitutes much of biology.

The causes of a trait during a single lifetime are called proximate; the evolutionary causes are called ultimate; all traits have both kinds of causes.

Evolutionary biologists ask different questions and investigate different kinds of cause, in search of answers to questions such as: 'Why does photosynthesis occur in the plastids and not in the cell nucleus?', 'Why do most species have approximately equal numbers of males and females, and why are there dramatic exceptions?', 'Why do many animals senesce, but many plants and fungi hardly at all?', 'Why do some small organisms have large relatives?'. These are also questions about causation, but on a time scale of many generations and at the level of populations and species rather than individuals. This is the study of **ultimate** (evolutionary) **causation**. Whereas in mechanical analysis the causes can be described as biochemical and physical processes, in evolutionary analysis one describes the causes as how natural selection, chance events, or constraints shaped the trait under study.

All traits have both mechanical and evolutionary causes; a complete explanation requires understanding of both kinds of cause. Isolating the two kinds of analysis from each other is a strategic error because it reduces the number of interesting questions that can be asked. A biologist should be able to see the world both ways—from the bottom up (from molecules to species) and from the top down (from selection to molecules).

Summary

- Evolution combines with physics and chemistry to explain biology. It is the only part of biology containing general principles not implicit in physics and chemistry. 'Nothing in biology makes sense except in the light of evolution' (Dobzhansky 1937).
- Evolutionary biologists have various specialties; each emphasizes different things. Some focus on mechanisms, others on patterns. No one approach can uncover everything about evolution that we want to know.
- Natural selection, the only mechanism known that can maintain and increase the complexity of organisms, is a non-zero correlation of reproductive success with a trait or gene. If the trait is heritable, the result is a change in the phenotypic design of the offspring of the more reproductively successful parents.
- If the variation of traits and genes is heritable and not correlated with the reproductive success of the organisms that carry them, they drift at random within limits set by constraints.
- How information is inherited determines how quickly gene frequencies can change and whether genetic conflicts exist. The major types of inheritance are sexual and asexual. Both preserve DNA sequences. Sexual reproduction also conserves gene frequencies and creates opportunities for conflict.
- Natural selection can rapidly produce states that would otherwise appear to be highly improbable.
- Adaptations do not evolve for the good of species, but species selection does produce large-scale patterns in the history of life.
- Speciation has produced an evolutionary tree, recording the history and describing the relationships of all life on Earth. The reconstruction of that history and those relationships combines the logic of phylogenetic analysis with the data of morphological and molecular evolution. Hypotheses about the relationships of taxonomic groups provide historical frameworks within which traits can be interpreted.
- There are two kinds of causes in biology, proximate (mechanical) and ultimate (evolutionary). Both shape all biological processes; both need to be understood; both are necessary; neither has precedence.

The process that distinguishes evolution from physics and chemistry is natural selection working on heritable variation to produce adaptations. It is discussed next.

Recommended reading

Darwin, C. (1859). *On the origin of species by means of natural selection or the preservation of favoured races in the struggle for life*. John Murray, London.

Futuyma, D. (1998). *Evolutionary biology*, (3rd edn). Sinauer, Sunderland, Massachusetts.

Questions

1.1 What aspects of biology can be explained by physics and chemistry without evolution having taken place?

1.2 What are the necessary conditions for natural selection? What are the sufficient conditions?

1.3 Which of the necessary conditions for natural selection must disappear for neutral evolution to occur?

1.4 You and your brother are walking through an African grassland. A lion attacks your brother. You come to his assistance. Describe your reactions in proximate and ultimate terms.

Landmarks in evolutionary biology

c. 340 BC	Aristotle invents scientific natural history while doing marine biology at Mytilene on the island of Lesbos; Theophrastus launches systematic botany.
1730–70	Linnaeus founds taxonomy; Buffon surveys all known life in the *Histoire naturelle*.
1795	Cuvier founds comparative anatomy, classifies the invertebrates in more detail but with the same principles as those used by Aristotle.
1815	Lamarck argues that a process of organic evolution can cause the differences observed between species, but gets the mechanism wrong. He produces a version of animal relations as good as any nineteenth-century phylogenetic tree, even those composed after Darwin.
1822	Geoffrey St. Hilaire speculates that arthropods are upside-down chordates.
1828	von Baer announces the first rigorous generalization in comparative developmental biology: the embryos of derived forms do not resemble the adults of more primitive relatives, but they pass through stages of development that resemble the stages of development of more primitive forms.
1830	The Cuvier–Geoffrey St. Hilaire debates over the existence and nature of homology: homology implies evolution from common ancestors. Cuvier resisted the notion of a common ancestor for mollusks and vertebrates, which Geoffrey defended. Goethe was among the spectators.

1831–36	Darwin circumnavigates the globe aboard the HMS *Beagle*, sees a sequence of fossil armadillos in Argentina and closely related, living finches in the Galapagos.
1838	Darwin's notebooks mention natural selection.
1844	Darwin drafts a manuscript laying out his ideas at length; it remains unpublished until Wallace's letter jolts him into action.
1858	Wallace sends Darwin a letter from Malaysia with the idea of natural selection. They publish brief accounts simultaneously.
1859	Darwin publishes *The origin of species by means of natural selection or the preservation of favoured races in the struggle for life*. Natural selection causes adaptive evolutionary change. Darwin later publishes other major books with key ideas: sexual selection, coevolution, the evolution of behavior.
1866	Mendel uses pea genetics to establish that genes behave like material particles present as two copies in diploid organisms; these copies segregate during gamete formation and assort independently among the gametes. Darwin either fails to read Mendel's paper or does not note its significance and continues to believe in blending inheritance.
1866	Haeckel publishes on the general morphology of organisms, states his famous biogenetic law, that ontogeny recapitulates phylogeny. Nineteenth-century evolutionary thought was fascinated by the idea that the development of individual organisms replays, in some sense, their evolutionary history. Darwin was no exception. Haeckel was a platonic thinker, one of a series of idealistic morphologists. Goethe was another.
1868–96	Weismann distinguishes germ line from soma, predicts that genes are material particles carried on chromosomes, predicts meiosis, speculates on the evolution of senescence and the evolutionary significance of sexual versus asexual reproduction, flirts with group selection, and ends up an uncompromising individual selectionist.
1900	Mendel's laws are rediscovered by Correns, Tschermak, and de Vries.
1915	Morgan, Sturtevant, and Bridges establish that genes are carried on chromosomes and that the behavior of chromosomes in meiosis and syngamy explains Mendel's laws. First genetic maps.

1918–32	Fisher, Wright, and Haldane establish the mathematical theory of population and quantitative genetics, explaining the roles and interactions of selection, mutation, inbreeding, migration, and genetic drift in large and small populations. Their success in solving problems in evolutionary theory that could not be solved with blending inheritance promotes genetics to a central position in evolutionary biology that it holds for more than 50 years.
1937–53	Dobzhansky, Mayr, Simpson, Stebbins, and Rensch establish that natural selection acting on genetic variation produced by Mendelian mechanisms is *consistent* with the phenomena of evolution from populations through species to higher taxa and the fossil record. They did not demonstrate that it was *sufficient*. Their body of work comes to be called The Modern or Neo-Darwinian Synthesis. It opposed the fundamental importance of variation to the nineteenth-century emphasis on platonic, ideal morphologies, or 'types'.
1946	H. J. Muller gets the Nobel prize for his work on mutations.
1947	Lack resurrects evolutionary ecology (Darwin founded it) with work on the evolution of optimal clutch size.
1950	Hennig proposes cladistic methods for the construction of phylogenetic trees.
1953	Watson and Crick discover the structure of DNA.
1952–57	Medawar and Williams predict the evolution of senescence and aging. Among the consequences: genes are the unit of selection, and phenotypes, from the gene's point of view, are disposable. We are phenotypes. We are disposable.
1963–64	Hamilton formalizes the theory of kin selection, explaining the evolution of apparently altruistic behavior and confirming the role of genes as units of selection and the disposability of phenotypes.
1964–83	Kimura develops the Neutral Model for molecular evolution.
1965	Lewontin and Hubby discover tremendous hidden genetic variation through electrophoretic analysis of enzymes from wild populations. The classical 'wild-type' model of evolutionary population genetics is destroyed. The debate on neutral versus adaptive evolution is joined.

1962–66	Maynard Smith and Williams expose the fallacies of group selection. Williams makes important contributions to life history theory, the evolution of sex, and sexual selection.
1975–78	Williams and Maynard Smith point out that sexual reproduction is a major problem for evolutionary theory.
1960–80	The expansion of evolutionary and behavioral ecology destroys the borders between evolution, ecology, and behavior.
1977	DNA sequencing becomes practical on a large scale; shortly thereafter molecular systematics expands rapidly. Arber's discovery of restriction endonucleases, enabling transgenic technology and the use of DNA microsatellites for genetic fingerprints, is honored with a Nobel prize.
1982	Maynard Smith publishes his book on evolutionary game theory.
1983–	Molecular developmental genetics reveals deep homologies in the genes controlling development, opening the way to understanding the mechanisms responsible for basic animal body plans.

Chapter 2
Adaptive evolution

Introduction

This chapter explores the causes of natural selection through examples. Organisms vary in their reproductive success; the correlation of this variation with traits is natural selection. Thus the causes of natural selection are all the reasons why organisms vary in reproductive success and why reproductive success is correlated with traits. In adaptive evolution, natural selection causes inherited traits correlated with reproductive success to increase in frequency and be maintained.

A trait responds to natural selection if it varies among individuals, is heritable, and is correlated with reproductive success.

To emphasize the subtlety and power of natural selection, we distinguish between its causes, which seem ordinary, and its consequences, which are extraordinary. Its causes are the reasons for individual variation in survival rates, in mating success, in offspring number, in offspring survival, in the survival and reproduction of relatives, and the reasons why variation in other traits is correlated with variation in reproductive success. Its consequences include the organization of the genome, the mechanisms of development and physiology, the structure and performance of brains, eyes, hearts, and other organs, much of behavior, much of population dynamics, and much of the ecological interactions of species: in short, much of biology. Natural selection can be underestimated because its ordinary causes have extraordinary consequences.

Natural selection has ordinary causes but extraordinary consequences.

Because natural selection is the correlation of a trait with reproductive success, survival itself is only important if it contributes to reproductive success. That is why male red-backed spiders commit suicide during copulation (see Prologue), why senescence evolves (Chapter 8), why selection for mating success can reduce survival (Chapter 9), and why worker bees will commit suicide to defend their nest (Chapter 14). Survival is only a means to achieve reproductive success. If death will bring the individual greater reproductive success than survival, then death will evolve.

Reproductive success, not survival, is what drives natural selection.

Natural selection is not limited to the differential reproductive success

Natural selection can act on several levels at once: in kin selection, sexual selection, and genetic conflict.

of individual organisms. In at least three situations it involves other units or levels: kin selection, sexual selection, and genetic conflict. To trace the overall outcome of selection acting on two or more individuals or levels at once, we follow changes in gene frequencies that reflect selection on all individuals and levels. Thus although natural selection is a phenotypic process, we often follow genotypic changes to understand its consequences.

Examples of natural selection and methods used to detect it

The large cactus finch

Biologists have gained great insight into natural selection by following known individuals from birth to death—noting how fast they grew, how old and large they were when they matured, how often they reproduced, how many offspring they had, how old they were when they died—and then doing the same for their offspring. This method can only be applied where organisms are large enough to be observed and can be relocated and individually identified with certainty. Examples include isolated populations of birds or mammals on islands, such as red deer on the Isle of Rhum in Scotland, song sparrows on Mandarte Island in Canada, collared flycatchers on Gotland in Sweden, and the large cactus finch on Isla Genovesa in the Galápagos.

Finches in the Galápagos have tremendous individual variation in reproductive success.

In studies of natural selection, a key question is 'How much does lifetime reproductive success vary in the population?' Grant and Grant (1989) gave a clear answer for the large cactus finch (Fig. 2.1). From 1978 to 1988 they caught and individually banded most of the birds on Isla Genovesa in the Galápagos, constructed detailed genealogies, and

Fig. 2.1 A pair of cactus finches foraging on *Opuntia* cactus in the Galápagos. (By Dafila K. Scott.)

followed the fates of offspring. During this period conditions were difficult and the population declined markedly; 79% of breeding males and 78% of breeding females produced no offspring that survived to breed. Most successful adults contributed only one offspring to the next generation, a few contributed two or three. The most successful parent was a male who bred from 1978 through 1982 and produced five recruits that, in turn, had produced recruits before the study ended. The distribution of females having 0, 1, 2, or 3 recruits per lifetime revealed the natural variation in female reproductive success (Fig. 2.2). The most successful female also lived the longest. Between 1978 and 1987 she laid 31 clutches in eight breeding years: her total of 110 eggs produced 58 fledglings, only three of which survived to breed themselves. Such tremendous individual variation in reproductive success, one of the two necessary conditions for natural selection, will translate into very strong selection on any traits correlated with reproductive success.

Cyanogenesis in *Lotus corniculatus*

Some plants defend themselves from slugs with hydrogen cyanide, the production of which is affected by two genes.

Slugs and snails pose a major threat to young plants soon after germination, when they are small and tender. Some plants defend themselves by producing hydrogen cyanide, a deadly poison. This ability occurs in at least 50 orders of flowering plants, in six ferns, and in several fungi (Ford 1975). In *Lotus corniculatus* this defense mechanism is controlled by two genes: one involved in the production of cyanide linked to a sugar that carries the cyanide in an internally harmless form until the plant is damaged; the other coding for an enzyme that catalyzes the production of hydrogen cyanide from the harmless form. The defense works without the enzyme, but it works better with it. Both genes are simple dominants, with presence of the defensive form dominant to absence.

Many natural populations are polymorphic for both genes, which are not linked, and many species of snails and slugs have been tested on noncyanogenic and cyanogenic leaves, with consistent results: they

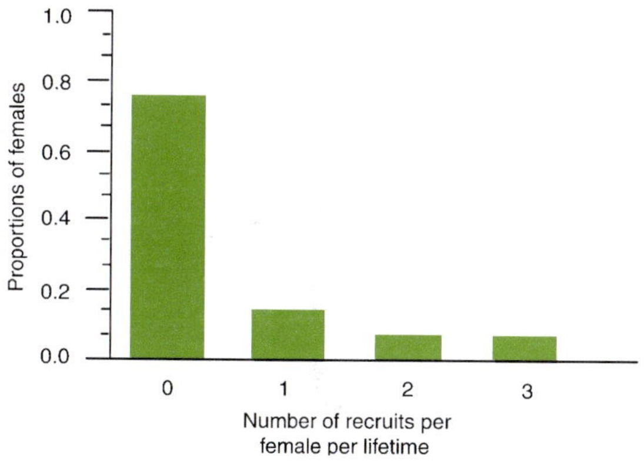

Fig. 2.2 Most female cactus finches have no surviving offspring; a few have up to three (n = 86). (From Grant and Grant 1989.)

Because some populations are polymorphic for both genes, cyanogenesis must be costly.

prefer to eat noncyanogenic leaves (Jones 1972). Thus the ability to produce hydrogen cyanide protects plants against defoliation by herbivores. However, cyanogenesis must also be costly, for if there were no cost, the genes would not be polymorphic—the cyanogenic forms would be fixed. The costs appear to be associated with temperature, for the frequency of cyanogenesis within species declines from 90–100% near the Mediterranean, through 60–80% in western Europe, to near absence in northern Russia.

Clutch size in kestrels

We can understand why a trait is adapted by manipulating it and seeing whether reproductive success decreases.

When a trait has evolved to a state of adaptation, the form found in nature has greater reproductive success than most alternatives against which selection could test it. To see whether a trait is adapted, we can manipulate it and note the consequences. The manipulated forms should have lower reproductive success than the natural state. Whereas

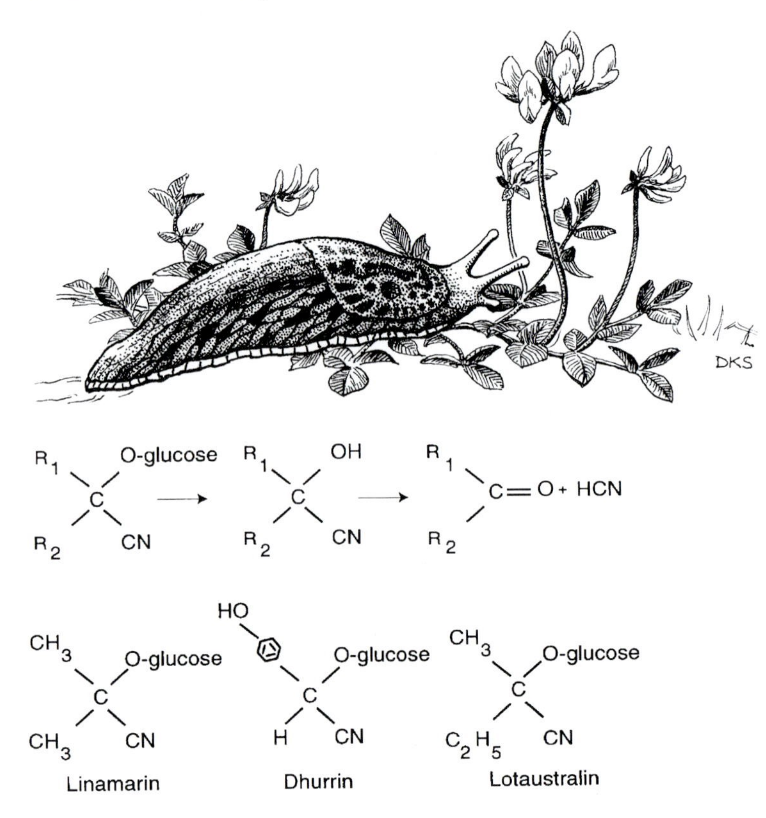

Fig. 2.3 Slugs are a major threat to plants, some of which defend themselves by producing poisonous hydrogen cyanide (sketch by Dafila K. Scott). The chemical compounds depicted in the lower row are those that carry the cyanide in an internally harmless form until the plant is damaged; the upper row shows the steps catalyzed by an enzyme that releases the hydrogen cyanide. R1 and R2 indicate chemical residues of two different forms. (From Harborne 1973.)

the previous studies used the naturally occurring variation in the populations, this and the following examples relied on artificial variation created by manipulations.

Biologists have long studied the reproduction of birds that naturally nest in holes and readily build nests in artificial boxes. One can take an egg or a nestling from one clutch and add it to another. Daan *et al.* (1990) did such a study on a falcon, the kestrel (Fig. 2.4), at a site in The Netherlands where survival was good and clutches were large. After their third year these kestrels had survival rates of about 70% per year, and after their first year their clutch sizes were about five eggs per clutch. However, survival to the third year, the size of the first clutch of the offspring, and the probability that a clutch succeeded varied with laying date and manipulation.

Kestrels with enlarged clutches fledged more offspring but had poorer survival. Kestrels with natural clutches had the highest reproductive success.

An interesting pattern emerged. The enlarged clutches fledged more offspring than the control or the reduced clutches, and the **reproductive value** of those clutches—the number that survived to reproduce, multiplied by the number of young they had—was also greater than the reproductive value of the control clutches. So far, it appears that it would be better for kestrels to lay more eggs. However, the parents of enlarged clutches had poorer survival than did the parents of control or reduced clutches. This effect on survival substantially reduced the number of offspring that the parents of enlarged clutches could expect to have during the rest of their life—their **residual reproductive value**. When both the reproductive value of the manipulated clutch and the subsequent reproductive performance of the parents were combined into a single measure, total reproductive value, it became clear that natural clutch sizes were adaptations. The controls, kestrels with clutches where young had been removed and then returned to the same nest, produced the most surviving offspring per lifetime: more than half a chick more than the

Fig. 2.4 A kestrel feeding a mouse to its nestlings. Increased reproduction in kestrels causes decreased adult survival. (By Dafila K. Scott.)

reduced clutches, and a full chick more than the enlarged clutches. Clutch size had evolved to maximize lifetime reproductive performance, taking adult mortality into account.

Long corolla tubes in Swedish orchids

Greater corolla depth selects for longer bee tongues, which in turn select for greater corolla depth.

Darwin was fascinated by the natural history of pollination. His study of the morphology of red clover and the bees that pollinate it convinced him that the length of the flower's corolla and the length of the bee's tongue had evolved so that the bee had to extend its tongue as far as possible while bending forward into the flower, bringing its head into proper position to transfer pollen. Thus greater corolla depth selects for longer bee tongues, and longer bees tongues select for greater corolla depth. If the length of the pollinator's tongue and the depth of the corolla were not limited by other factors, the process could result in very deep flowers and very long tongues. Later Darwin had the opportunity to examine the Madagascar star orchid, which has an extremely long floral tube (28–32 cm) (Fig. 2.5). Reasoning that the same selective forces would drive the **coevolution** of the orchid and its pollinator, Darwin predicted that it would be pollinated by a giant hawkmoth with a tongue about 30 cm long (Darwin 1859, p. 202). Forty years later, the hawkmoth (*Xanthopan morgani praedicta*) was discovered. It had a 30 cm tongue.

Orchids with artificially shortened corollas had reduced fitness.

Darwin's elegant explanation of floral tube depth remained untested until Nilsson (1988) manipulated the depth of the floral tubes of orchids on Baltic islands. He experimentally shortened the floral tubes by taping or tying them shut, squeezing the nectar above the constriction so that the pollinator's reward remained in place, then measuring the number of pollinia removed (male function) and the number of stigmas pollinated (female function) for each of the experimental treatments.

Fig. 2.5 The Madagascar star orchid (*Anagraecum sesquepedale*) with the hawkmoth (*Xanthopan morgani praedicta*). The length of the orchid's nectary and the length of the moth's tongue have coevolved. (By Dafila K. Scott.)

Orchids with artificially shortened corollas had reduced fitness through both male and female function (Fig. 2.6).

In both the kestrels and the orchids, much was learned about selection pressures by manipulating traits and noting the consequences. The clutches could be made both larger and smaller than the natural size; the floral tubes could only be shortened. We know that the kestrels produced clutches that led to greater reproductive success than either smaller or larger ones, but we only know that the orchids naturally produced floral tubes that led to greater reproductive success than did artificially shortened tubes. Thus the clutch size of kestrels is under **stabilizing selection** for an intermediate size. Whether the orchids' floral tubes are under stabilizing selection or **directional selection**—selection for ever greater length—remains an open question.

Aminopeptidase in mussels

There are at least three approaches to detecting selection acting on genes (Clarke 1975). First, one can document changes in allele frequencies and rule out other factors that can cause such changes, such as mutations, random events, or immigration. Secondly, if there is no selection or immigration, a population will move towards an equilibrium in which allele frequencies remain constant. Therefore, if the population remains out of genetic equilibrium for many generations, then either selection or immigration is occurring, and if immigration can be ruled out, selection is taking place. Thirdly, if the frequencies of particular alleles are correlated with environmental conditions, and if such correlations occur repeatedly, in different populations or in the same population every year, selection is indicated.

Selection is indicated if allele frequencies change and other causes are lacking, if a population remains out of genetic equilibrium and immigration can be ruled out, or if genes and environments are repeatedly correlated.

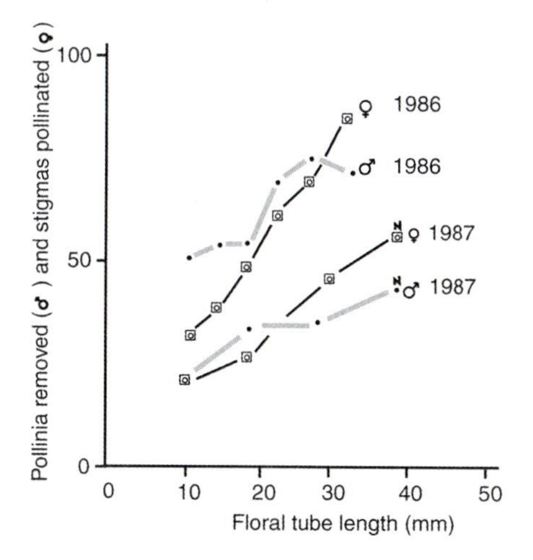

Fig. 2.6 The sexual performance of two species of Swedish orchids as a function of the length of their floral tubes. The longer the tubes, the more pollinia were removed and the more stigmas were pollinated. (From Nilsson 1988, reproduced by kind permission of the author and *Nature*.)

Mussels with the *Lap*^{94} allele acclimate faster to the hyperosmotic oceanic environment.

All these methods have been used to detect selection acting on genes in mussels. Mussels are bivalve mollusks, sessile as adults, that have external fertilization and planktonic larvae that disperse over large distances. A detailed study of the edible mussel, *Mytilus edulis,* revealed striking geographic and temporal variation in the frequency of the different naturally occurring alleles of the gene coding for an enzyme, aminopeptidase I, designated Lap (Koehn *et al.* 1983). One allele, *Lap*^{94}, had 20% greater specific activity than the other forms and was found at uniformly high frequency over large geographical areas in which populations experience oceanic salinity and relatively high temperatures. The enzyme cleaves proteins into free amino acids that affect the osmotic pressure of the cell. Mussels with the *Lap*^{94} allele therefore acclimate faster to the hyperosmotic oceanic environment than do mussels with the other forms of the gene.

But the *Lap*^{94} allele is detrimental in low-salinity environments, where it only occurs because of immigration.

However, the *Lap*^{94} allele is not advantageous in all environments. Every year, mussel larvae from populations encountering oceanic conditions invade estuaries and settle in environments where salinity fluctuates. Following settlement, individuals with *Lap*^{94} genes suffer dramatically higher mortality than do individuals without such genes, and those that do survive are in poorer condition, for they produce a pool of free amino acids that is too concentrated for the hypo-osmotic environments encountered in estuaries, where these metabolically expensive, nitrogen-containing, widely used molecules are leached out at a rate that cannot be compensated.

Antibiotic resistance

Bacteria rapidly acquire resistance to antibiotics through mutations, which can be transferred to other bacteria by transformation, transduction, and conjugation.

Staphylococcus aureus is a bacterium that causes dangerous infections in humans, including those recovering from operations, burn victims, and those suffering from wounds. In 1941, *S. aureus* could be treated with penicillin G, which inhibits the synthesis of the bacterial cell wall, but by 1944 some strains had already acquired resistance to penicillin, and by 1992 more than 95% of all *S. aureus* strains worldwide were resistant to penicillin and related drugs (Neu 1992). The pharmaceutical industry responded by synthesizing methicillin, but by the late 1980s resistance to methicillin was common, and strains of *S. aureus* resistant to methicillin were also resistant to all drugs structurally similar to penicillin, because a mutation in a chromosomal gene changes the structure of one element of the bacterial cell wall so that these drugs no longer bind to it.

Bacteria acquire resistance to antibiotics through several genetic mechanisms. The simplest is a new chromosomal mutation, which was responsible for the original resistance of *S. aureus* to penicillin. Some bacteria carry previously evolved resistance genes that are not normally expressed; they are only induced when the antibiotic is encountered. Resistance genes can be transferred to other bacteria through **transformation**, a process by which bacteria acquire genes by taking up free DNA molecules released by dying bacteria into the surrounding medium. The viruses that parasitize bacteria can also pick up a bacterial resistance gene and carry it with them into their next host, where it is inserted into the bacterial chromosome. If the bacterium survives the viral infection, it will have acquired antibiotic resistance. This mechanism is called **transduction**. Like transformation, it moves chromosomal genes between bacteria. Bacteria also contain small, circular pieces of extra-chromosomal DNA, called plasmids, some of which contain genes for antibiotic resistance, and they occasionally exchange plasmids and chromosomal genes in a third process, **conjugation**, in which they come into direct contact with each other. Furthermore, within the bacterial genome there are mobile genetic elements, transposons or 'jumping genes', that can pick up a resistance gene and move it along the main chromosome or into a plasmid. Transposons and plasmids may induce bacteria to conjugate, thereby creating opportunities for their horizontal transmission and that of the resistance gene.

Because of strong selection, transposition, and integration of resistance genes into the main chromosome of *S. aureus,* some strains have become resistant to many antibiotics, including erythromycin, fusidic acid, tetracycline, minocycline, streptomycin, spectinomycin, and sulfonamides and even to disinfectants and heavy metals, such as cadmium and mercury. In the mid-1980s, new drugs were discovered, the fluoroquinolones, that killed *S. aureus* at low concentrations and cured serious infections. Some thought the problem of antibiotic resistance had been solved. However, around 1990, strains of *S. aureus* that resisted the new drugs increased from less than 5% to more than 80% within 1 year in a hospital in New York, and resistance to fluoroquinolones spread around the world. Recently a drug, mupirocin, was found that killed strains of *S. aureus* resistant to fluoroquinolones. It was first used in the United Kingdom, where strains have already evolved resistance to it. The genes for resistance to mupirocin are carried on plasmids.

Staphylococcus aureus illustrates some properties of the evolution of antibiotic resistance that are true for many other dangerous bacteria (Fig. 2.7). Bacteria rapidly evolve resistance to virtually all antibiotics (Neu 1992). Resistance that evolves in one bacterium spreads horizontally to other bacteria through transduction, transformation, and conjugation. Hospitals are especially good breeding grounds for new forms of resistance because that is where antibiotics are used most intensively. Moreover, it costs much more to treat resistant forms. In the United States in 1992, it cost about $12 000 to treat one case of nonresistant tuberculosis, including drugs, procedures, and hospitalization; treatment for one case of multi-drug-resistant tuberculosis cost about $180 000 (Cohen 1992).

Many strains of infectious bacteria are now resistant to most antibiotics, and it costs a great deal more to treat resistant forms.

At human body temperatures, bacteria have generation times of about an hour, and they can multiply rapidly to form populations of billions of cells in which many mutations occur every generation. Strong selection applied by antibiotics to large populations with short generation times produces rapid evolution. Since the invention of antibiotics, about 5

About 5 million tons of antibiotics have exerted massive selection on the world's bacteria.

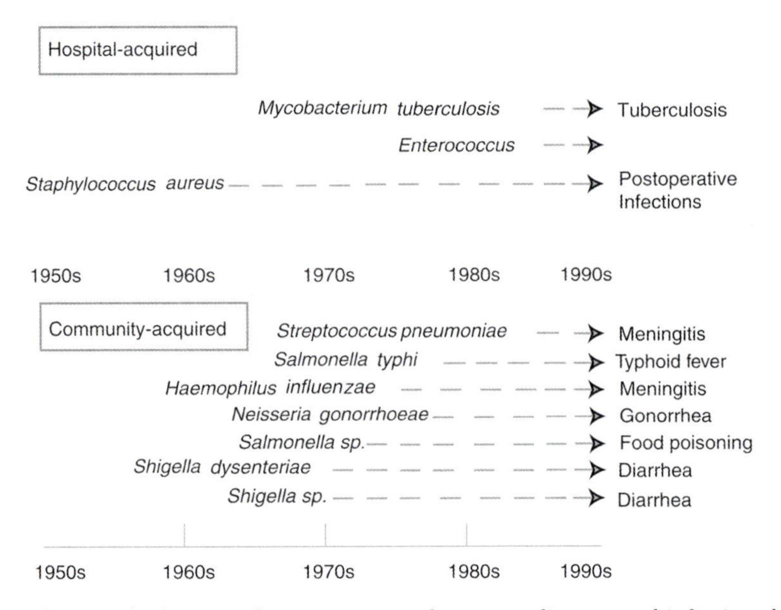

Fig. 2.7 The bacteria that cause many dangerous diseases and infections have rapidly acquired resistance to antibiotics during the past 20 years. (From Cohen 1992.)

million tons of these drugs have been used in humans and domestic animals, exerting massive selection on the world's bacteria. This cure contained the seeds of its own evolutionary destruction. The more people that doctors cure with antibiotics, the more rapidly new forms of resistance evolve. This open-ended coevolutionary arms race with bacteria is profitable for the pharmaceutical industry but risky for mankind. An expert said recently, '. . . the post-anti-microbial era may be rapidly approaching in which infectious disease wards housing untreatable conditions will again be seen' (Cohen 1992, p. 1055).

Methods for detecting adaptive evolution: summary

In all the cases discussed, natural selection was generated by variation in the reproductive success of individual organisms, but the response to selection was mediated by genes with quite different effects. The reproductive success of birds is influenced by many genes with complex interactions; antibiotic resistance in bacteria and defense against slugs in *Lotus corniculatus* are influenced by a few genes with relatively simple connections to the phenotype.

Many studies have demonstrated natural selection.

Several methods detect natural selection in action. The most direct and convincing method is to document variation in the lifetime reproductive success of parents and offspring for several generations. To study natural selection, we can use either natural variation in polygenic traits and correlate it with natural variation in reproductive success, or we can generate artificial variation in the trait and note the consequences. To study adaptive evolution of individual genes, we can measure changes in gene frequencies over time and rule out other factors that can cause such changes, or we can correlate particular genes with particular environmental conditions. If such correlations repeat themselves, either in different populations, or in the same population from year to year, selection is indicated.

Endler (1986) discusses several other, less direct methods of detecting natural selection. In at least 99 species of animals and at least 42 species of plants, adaptive evolution has been demonstrated in morphological, physiological, or biochemical traits. However, few of those studies measured lifetime reproductive success, few dealt with more than one or two traits, and in most one could not pinpoint the mechanism of selection.

The ways of classifying selection

Types of selection can be classified in several ways: as **natural selection** and **sexual selection**; as **stabilizing**, **directional**, and **disruptive selection**; as **density-dependent** and as **frequency-dependent selection**; as **individual**, **kin**, **group**, and species selection; and as direct and indirect selection. Each classification focuses attention on a different aspect or level of selection. Not all combinations of the different classifications are equally likely or even logically possible. For example, stabilizing, directional, or disruptive natural selection can operate on individuals, families, or groups, but sexual selection operates almost exclusively on individuals.

Natural selection and sexual selection

Sexual selection is a component of natural selection associated with mating success. People who work on sexual selection often refer to sexual and natural selection as though they were two different things. Selection for mating success does produce subtle and surprising effects. However, sexual selection is not something qualitatively different from natural selection. The same condition is necessary for both: variation in lifetime reproductive success. Sexual selection is a special aspect of natural selection.

Sexual selection is the component of natural selection associated with mating success.

When Darwin proposed sexual selection, it was not yet clear that natural selection is not just the correlation of a trait with survival but with survival combined with reproduction to yield reproductive success. Darwin distinguished between natural and sexual selection to explain the evolution of male traits, like the peacock's tail, that reduced male survival. Reasoning that natural selection could not produce traits that reduced survival, Darwin looked for some other process to explain them. The process he suggested was sexual selection through female choice. If peahens preferred peacocks with large, showy tails, then the greater mating success of such males might compensate for their reduced ability to escape from predators.

Darwin's distinction between natural and sexual selection was not logically necessary.

Now we know that natural selection results from the correlation of traits with lifetime reproductive success. Survival is only important if it contributes to future reproduction. Many traits besides sexually selected characters reduce survival, including reproduction itself. If Darwin had understood natural selection in the modern sense, he could have argued that the peacock's tail was the product of natural selection because the benefits that it brought the male in mating and reproduction were large enough to offset its survival costs. Thus his distinction between natural and sexual selection was not logically necessary, but its focus on a special aspect of selection important for the evolution of behavior has been useful.

Sexual selection consists of two different processes: intrasexual competition for mates and intersexual choice of mates.

Sexual selection itself can be further classified into two different processes. Competition for mates can select for fighting ability, resulting in larger body size and weapons in the sex that fights for mates. That sex is often the male; examples include the antlers of male deer and the greater body size of male than of female elephant seals, which is called sexual dimorphism. Competition for mates is straightforward and can be interpreted as a component of natural selection where the benefits of access to mates are balanced by the costs of producing the traits useful in gaining that access. Traits besides fighting ability may be selected, including the ability to manipulate and deceive competitors. In some fish (wrasses, bluegills, salmon), small, subordinate males mimic females to steal copulations from large, dominant males.

A female might choose a male mate because of direct benefits,

Active choice of mates, the other type of sexual selection, produces subtle and surprising effects. Both sexes may choose mates and both sexes may compete for mates, but often females choose males and males compete for females—in many species those are the dominant processes. Thus, much of sexual selection deals with the evolution of female

Fig. 2.8 A peacock displaying to a peahen, with a tiger lurking in the bushes. Darwin invented sexual selection for traits like the peacock's tail to explain traits that increased mating success while decreasing survival. (By Dafila K. Scott.)

preferences and their consequences, both for male traits and for the female preferences themselves. What criteria might a female use in choosing a male mate? She could look for traits indicating that the male will make a good parent, such as possession of a good feeding territory. She could avoid males that were obviously sick or carrying ectoparasites that might be transmitted to her or her offspring. Those two ideas are straightforward, but there are other, more subtle reasons to choose.

The female could choose a male because he was attractive to females in general, thus increasing the likelihood that her sons would also be attractive to females and thereby increasing their mating success. This is the sexy son hypothesis. It leads to a coevolution of male traits and female preferences that can, in principle, greatly exaggerate male traits.

. . . because he would give her sexy sons,

The female could also choose a male because he communicated to her an honest signal of his vigor. Honest signals must be costly. If they were not, a weak male could cheat and transmit a deceptive signal, indicating that he was vigorous when in fact he was not, and females evolve to ignore cheaters. Thus honest signals can be considered as costly handicaps. According to this hypothesis, males produce expensive structures or perform costly acts to signal their vigor, and females prefer males with such structures or behaviors because they have good genes. This is the good genes or handicap hypothesis, it depends on honest, expensive signals, and the male traits are often thought to indicate resistance to disease and parasites. It also can lead to the evolution of exaggerated male traits.

or because he communicated an honest and costly signal of his vigor.

Sexual selection is discussed in greater detail in Chapter 9.

Stabilizing, directional, and disruptive selection

Reproductive success is always under directional selection to increase (see Chapter 4). Most other traits only encounter directional selection when the environment changes. Then they begin to change in the direction of their correlation with reproductive success, but their change is soon halted by trade-offs. Traits involved in trade-offs encounter stabilizing selection as they change. Stabilizing selection favors an intermediate value of the trait over the extremes; it selects for an intermediate optimum. For example, in the kestrel study, stabilizing selection favored clutches of intermediate size, for both larger and smaller clutches resulted in lower lifetime reproductive success, and the reason for the lower lifetime reproductive success with larger clutches was a trade-off with adult survival.

Disruptive selection favors extreme values of the trait over intermediate values (Fig. 2.9); in the evolution of gametes of different sizes (**anisogamy**), very large and very small gametes (eggs and sperm) are favored over gametes of intermediate size.

Stabilizing selection favors an intermediate value of the trait over the extremes. Directional selection favors an extreme value. Disruptive selection favors extreme values over intermediate values.

Is selection usually stabilizing, usually directional, or stabilizing for some kinds of traits and directional for others? Do organisms live in an equilibrium world where they are as well adapted to their environment as they can be, where a further response to selection will only occur if

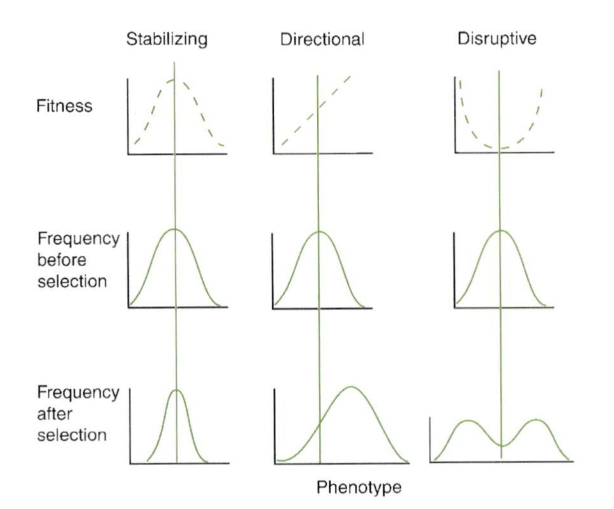

Fig. 2.9 Stabilizing, directional, and disruptive selection. Top row: relationship of phenotypes to fitness. Middle row: frequency of phenotypes before selection (same in all cases). Bottom row: frequencies of phenotypes after selection. (From Futuyma 1998.)

Stabilizing selection optimizes traits and reduces their genetic variation. Directional selection may increase the genetic variation of the traits it acts on.

the environment changes, which would be the case if selection is usually stabilizing? Or do they live in a nonequilibrium world that never attains stability, where selection is usually directional? The distinction is important because biologists that make one assumption have problems communicating with biologists that make the other. If selection is usually stabilizing, then traits are optimized and we expect their genetic variation to decrease as selection gets stronger (Chapter 5). If selection is usually directional, then traits may display more genetic variation as selection gets stronger. People working on traits such as clutch size usually assume stabilizing selection; people working on sexually selected traits usually assume directional selection.

Density-independent and density-dependent selection

Density-independent and density-dependent selection refer to the population density under which selection occurs.

These terms apply to the ecological conditions under which selection occurs. Density-independent selection refers to selection that does not change when population density changes, and density-dependent selection refers to selection that changes from low to high population density. Again, these kinds of selection correspond to different assumptions about the natural world. If one thinks that a population is rarely limited by resources and spends much of its time growing rapidly as it recovers from intermittent catastrophes, then traits favoring rapid growth and reproduction will be favored at the expense of others. If one thinks that a population is normally limited by its resources and at equilibrium, neither growing nor declining, then traits favoring reproductive success under food stress will be favored at the expense of others.

Frequency-dependent and frequency-independent selection

If a trait or gene is under frequency-dependent selection, then the selection it experiences depends on whether it is rare or common in the population. Positive frequency-dependent selection favors common types; negative frequency-dependent selection favors rare types. Selection favoring rare traits or rare genes acts to increase and maintain the diversity of the population, for the more common the trait, the greater its disadvantage, and the rarer the trait, the greater its advantage. Frequency-dependent selection is important in the evolution of behavior, the evolution of sex and of sex ratios, and the maintenance of genetic variation.

Frequency-dependent selection is important in the evolution of behavior and of sex, and in maintaining genetic variation.

The classic example of frequency-dependent selection is Fisher's (1930) explanation of the stability of a 50:50 sex ratio. If mating is random and more males are present than females, then it pays parents to produce female offspring because they will get more grandchildren through daughters than through sons, which would have to compete with the more numerous males for access to mates. If more females are present than males, then it pays parents to produce male offspring, for, by the same reasoning, they will get more grandchildren through sons than through daughters. The result, given Fisher's assumptions, is that the population will evolve from any starting sex ratio to a 50:50 sex ratio and then remain in that state.

Individual selection, kin selection, group, and species selection

Most features of organisms—their development, biochemistry, physiology, and life history—are the product of individual selection. When we talk about individual selection, we think of individuals acting as agents for their genes, genes acting on traits that contribute to the reproductive success of the individuals that carry them. Individual selection produces adaptations.

Individual selection refers to processes involving only the reproductive success of single individuals.

Species selection differs in important ways, including the time scale on which it operates and the kinds of patterns that result from it. It refers to the process of sorting among species that occurs when different lineages have different rates of speciation and extinction. Whereas individual selection happens on a time scale of generations—usually days, months, or years—species selection acts on a time scale of tens of thousands to millions of years. The elements of individual selection, the birth and death of individual organisms, happen frequently. The elements of species selection, speciation and extinction, happen much less frequently, too seldom to produce adaptations. Instead, they produce patterns in the abundance of families, orders, classes, and phyla. Such patterns include radiations—the rapid expansion through speciation of a particular group—and extinctions of large groups.

Species selection refers to sorting among species caused by differential rates of speciation and extinction.

Kin selection refers to selection processes involving interactions

Kin selection refers to selection involving interactions between relatives.

between related organisms. For example, consider two baboons who are brothers. If one is attacked by a leopard, it is likely to die, but if the other comes to help and together they chase the leopard away, the risk to either of them may be small. Thus at little risk to oneself, one can sometimes increase the reproductive success of a relative. Such conditions favor the evolution of helping and cooperation among relatives. In special cases, complete reproductive self-sacrifice can evolve, as in the sterile worker castes of some bees and all ants and termites. Kin selection is discussed in Chapter 14.

Group selection refers to the selection of properties of groups by processes acting on groups. Group selection is rarely effective because individuals are much more numerous and have shorter generations and more genetic variation.

Group selection has been controversial. To see why, consider one of the examples that set off the controversy. Wynn-Edwards (1962) suggested that animals breed in colonies or groups to get information on population density. If overpopulation threatened resources for the next generation, they could then reduce reproduction so that there would be enough food for all the offspring. The problem with this explanation is that selection for prudent behavior to benefit the group is opposed by more efficient selection to improve the reproductive performance of individuals within each group. Selection on individuals is usually stronger than selection on groups, and the response to individual selection is faster and more precise than the response to group selection, for three reasons:

(1) individuals are born, reproduce, and die much more frequently than do groups;

(2) there is usually much more genetic variation among individuals than among groups; and

(3) there is normally a much stronger correlation between the reproductive performance of individuals and individual traits than there is between the reproductive performance of groups and group traits.

These three factors combine to generate stronger selection and a stronger response to selection for individuals than for groups.

Traits evolve to increase the reproductive success of individuals and kin groups.

It is safe to assume that traits usually evolve to increase the reproductive success not of groups or species but of individuals and families. Most of the characteristics of groups and species are not produced by selection acting on groups or species. They are by-products of the individuals out of which groups and species are composed. Those individuals have been produced primarily by selection acting on individuals. Exceptions are logically possible but are probably not frequent.

Direct and indirect selection

Indirect selection occurs whenever pleiotropy or linkage create connections between traits under selection and other traits or genes.

Direct selection is straightforward and has been illustrated in the examples above. Sometimes, however, traits evolve although they do not directly affect the reproductive success of their carriers. Many genes affect more than one trait; they are called **pleiotropic** genes. If a trait is affected by pleiotropic genes, then it may evolve because it is genetically connected to other traits that do affect reproductive success. If a neutral gene is closely linked to a gene under selection, it will change in frequency although it produces no phenotypic effects that are correlated with reproductive success. **Hitch-hiking** describes changes in indirectly selected traits and genes (see below).

The strength of selection and the rate of evolutionary response

The strength of selection in nature

Darwin and the early population geneticists thought that selection was weak and that evolution was slow and gradual. This view lasted into the 1970s, when studies began to accumulate demonstrating strong selection and rapid evolution. Endler's (1986) summary of studies of selection in natural populations made clear that selection was quite strong on some species, traits, and genes and weak or absent on others. He also compiled the frequencies of directional, stabilizing, and disruptive selection.

Selection in natural populations can be strong. Directional, stabilizing, and disruptive selection all occur in natural populations.

From studies on 25 undisturbed species, Endler calculated 262 estimates of the strength of directional selection (Fig. 2.10a): 102 of them were significantly greater than zero, and more than half were impressively large. From studies on 32 species, including five fossil species, he calculated 330 estimates of the strength of stabilizing or disruptive selection (Fig. 2.10b): 78 of them were significantly different from zero. Most of those indicated stabilizing selection, but a minority documented disruptive selection, sometimes quite strong.

Thus selection in natural populations can be strong; and directional,

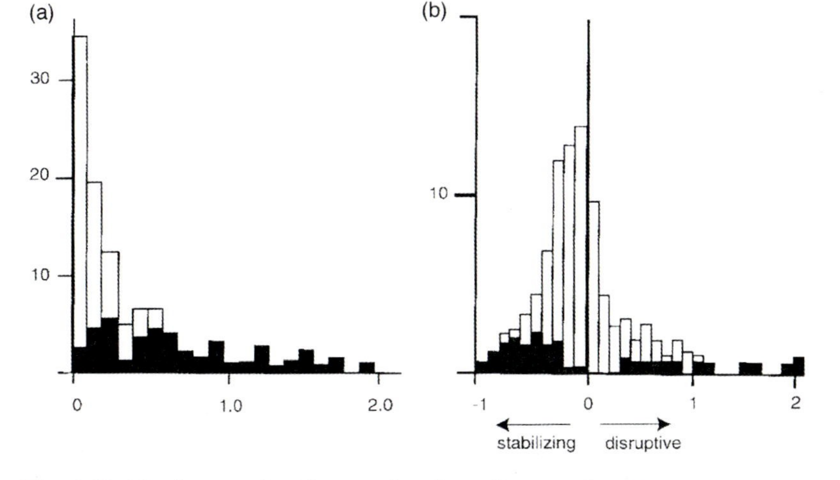

Fig. 2.10 The frequencies of examples of (a) directional and (b) stabilizing or disruptive selection in the literature. The *x*-axis indicates the selection differential, significant differences from zero are indicated in black. (From Endler 1986.)

stabilizing, and disruptive selection all occur in nature. Because the strength of selection varies from strong to zero, we cannot assume it a priori: if we need to know it, we have to measure it. Endler's survey may have been biased towards strong selection, as he acknowledged, for such results are more likely to be published, but that criticism does not affect the main conclusions: selection can be strong, and the major types of selection all occur in nature.

The rate of evolutionary response to selection

Microevolution is fast in large populations with lots of genetic variation under strong selection.

It is a common misconception that evolution is slow and gradual. In fact, the rate of evolution can be either slow or fast. The most rapid change has been measured in guppies in Trinidad and finches in the Galapagos, in both of which the rate of change of some traits approached one standard deviation per generation. Microevolution can be very fast, much faster than one would infer from fossils, and it is fastest in large populations for traits with a great deal of genetic variability under strong selection.

The context-dependence of selection

Selection always involves interactions between the thing selected and its environment: for genes, the other genes in the genome; for traits, the other traits in the organism; for organisms, the other organisms in the population.

The type of selection experienced by a gene in a genome, or a trait in an organism, or an organism in a group, results from an interaction between the environment, the focal gene or trait or organism, and the other genes, traits, and organisms contributing to survival and reproduction. Selection thus has both external and internal causes. It always involves interactions between the thing selected and its environment. For a gene, the other genes in the genome are important; for a trait in an organism, the other traits in the organism are important; and for an organism in a population, the other organisms, especially its potential mates and competitors, are important parts of the selective environment. The external environment is not an absolute but a relative agent of selection, for even the selective impact of a physical factor like temperature can depend on such interactions; for example, the impact of cold or heat may depend on competition for refugia.

Consider selection in an asexual clone. The entire genome is transmitted like a single gene, and any advantages or disadvantages created by a new mutation at one location may be canceled or reinforced by mutations occurring at other locations. Or consider the selection experienced by a gene in a sexually reproducing organism on a chromosome containing many genes. At any one time, the chances are good that several of the genes will be subject to selection, and they will be causing changes in the frequencies of all genes close to them on the chromosome. The action of indirect selection (or hitch-hiking, see above) is described as a selective sweep, by analogy to a broom that sweeps up whatever it encounters, including much besides the object of attention. In asexual organisms, all genes are subject to selective sweeps, caused by selection acting anywhere in the genome. In sexual organisms, genes encounter selective sweeps caused by selection acting on linked genes on the same chromosome. This is just one of several reasons why the fate of an allele depends on what alleles are present at other loci. What happens to one gene depends on what happens to many other genes.

Organisms can also choose the environmental context in which they will live and thus the kinds of selection pressures that they will encounter. For example, phoronids are marine worms that live in sand. Their planktonic larvae actively choose sand with phoronids living in it as a substrate to settle on. By selecting an environment in which other phoronids have already survived, they greatly increase their chances of surviving. Phoronids are living fossils that have not changed much for hundreds of millions of years. One reason for the lack of change may be that their larvae have always chosen the same environment in which the adults live, an environment that has always been available and has not changed very much. A more familiar example is provided by house sparrows. House sparrows prefer to live in and around human settlements; it is hard to find one just 50 meters into a forest, where selection pressures are quite different. The presence of humans protects house sparrows from many predators, and house sparrows choose a safe habitat. If performance in human settlements trades off with performance in wild habitats, then the habitat choice of house sparrows is leading to a type of adaptive evolution that will cause them to suffer if human settlements disappear. They are domesticating themselves.

Larvae can choose the environment in which the adults will encounter selection.

Cultural evolution

If things that can reproduce vary in their reproductive success, and if some of their success can be inherited by their offspring, they will undergo adaptive evolution and experience heritable change that improves their reproductive performance. The logic of natural selection is clear. If a response to selection does not occur, then one of the assumptions just listed must be false: either the things do not reproduce themselves, or they do reproduce but do not vary in their reproductive success, or they do vary in their reproductive success but that success is not correlated with anything heritable.

The logic of natural selection is clear; it must be true, . . .

What types of things are subject to natural selection and evolve? In what sense do ideas, religions, agricultural practices, or computers evolve in response to natural selection? What corresponds to reproductive success and to inheritance? How does cultural change connect to biological evolution? Neither reproduction nor inheritance can be as precisely defined for cultural evolution as they can in biological evolution, but something analogous appears to be going on. It is an analogy worth exploring (see the list of recommended reading for Chapter 17).

and its scope of application may be surprisingly broad.

Summary

This chapter described how adaptive evolution works, what it works on, how we detect selection, how different types of selection are classified, and how strong selection actually is.

- Adaptive evolution is caused by variation in the number of copies of their genes that individuals contribute to future generations. Much of that variation results from variation in the lifetime reproductive success of individuals themselves. Some of it results from the effects that individuals have on the reproductive success of their relatives, through which copies of their genes also are contributed to future generations. While selection acts on organisms, it has consequences for all levels in the biological hierarchy, from genes to species. Those consequences only endure if selection changes gene frequencies.
- Different methods are used to detect natural selection of organisms, traits, and genes. To detect selection of organisms, we can measure variation in lifetime reproductive success. To detect selection of traits, we can manipulate the trait and note the consequences for survival and reproduction. To detect selection of genes, we can measure changes in gene frequencies and rule out other causes of such changes.
- Selection can be classified at least four ways:
  (1) as natural selection and sexual selection;
  (2) as individual, kin, and group selection;
  (3) as directional, stabilizing, and disruptive selection; and
  (4) as density- or frequency-dependent selection.
  The strength of selection in nature varies from strong to zero.
- Selection is a context-dependent product of interactions between things and their environments. The relevant environment for traits includes the state of the other traits in the population; the relevant environment for genes includes the other genes in the population. Organisms can interact with environments to determine the selection they experience.
- Adaptive evolution occurs in populations of any things that vary, reproduce themselves, have some form of inheritance, and some non-zero correlation between heritable variation and reproductive success. Those things do not have to be organisms.

This chapter described adaptive evolution. The important consequences of neutral evolution are the contrasting focus of the next chapter.

Recommended reading

Bell, G. (1996). *The basics of selection.* Chapman & Hall, New York.

Dawkins, R. (1982). *The extended phenotype. The gene as the unit of selection.* W.H. Freeman, New York.

Grant, B. R. and Grant, P. R. (1989). *Evolutionary dynamics of a natural population. The Large Cactus Finch of the Galápagos.* University of Chicago Press, Chicago.

Questions

2.1 In the nineteenth century, people thought that senescence evolved so that old organisms would die, releasing resources for use by younger organisms. Discuss this in terms of individual versus group selection. If this explanation is going to work, what must be the interaction between parents, offspring, resources, and other unrelated individuals in the population?

2.2 In one type of mimicry, one species, the mimic, which is not poisonous or otherwise dangerous, evolves to look like another species, the model, which is poisonous or otherwise dangerous. Mimicry is common among butterflies, which are eaten by birds. Birds learn from what they eat. Is the selection pressure on mimics a function of the frequency with which birds encounter models?

Chapter 3
Neutral evolution

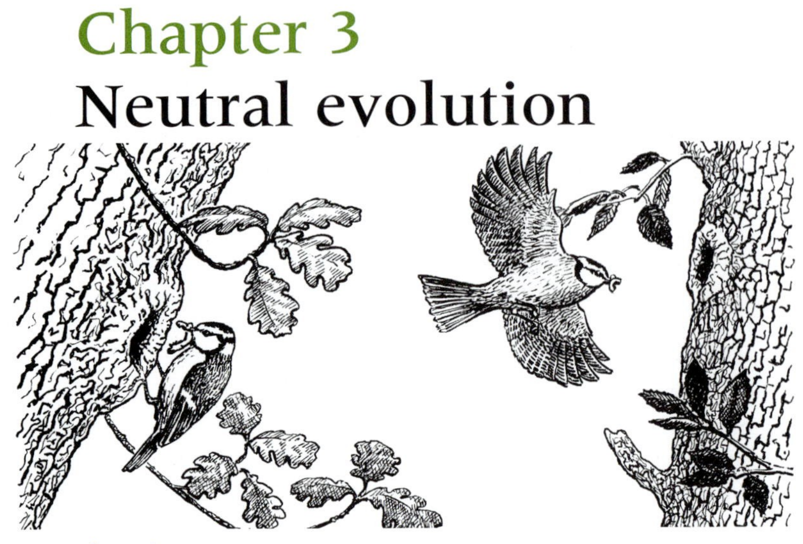

Introduction

Chapter 2 pointed out that adaptive evolution requires a correlation between genetic variation and variation in reproductive success. We will now use **fitness** as a synonym for reproductive success. Only when some of the variation in fitness is heritable will changes in the genetic composition of the population improve adaptation. From this follow important questions: 'What is the quantitative relationship between fitness variation and genetic variation?', 'How much genetic variation is there?', and 'How often do genetic differences imply fitness differences?' The answers to these questions determine whether adaptive evolution can occur; they touch the core of evolutionary biology. We address them in Chapter 5.

How do gene frequencies change when there is no relationship between genetic variation and reproductive success?

In contrast, what is the evolutionary significance of *non*-heritable variation in reproductive success? Will variation in fitness that is *not* correlated with genetic variation *not* change the genetic composition of the population? Is such variation therefore unimportant in evolution? This chapter discusses how gene frequencies change when there is no relationship between genetic variation and reproductive success.

The relationship between genetic variation and fitness

Variation relevant for evolution occurs at the level of the genotype, the phenotype, and fitness.

A closer look at the logical relation between genetic variation and fitness shows us where that relationship could be disconnected. Variation relevant for evolution occurs at three levels: the genotypic level, the phenotypic level, and the fitness level. Figure 3.1 sketches the connections between the three levels. The different **genotypes** in a population can be represented by points on the genotype plane (a realistic genotypic space would have many dimensions; this one has just two for convenience).

Similarly, points in the phenotypic plane represent different phenotypes, and points in the fitness plane represent different fitness values. Because the connections between genotype, phenotype, and fitness depend on the environment, such a scheme is only valid in a particular environment.

Figure 3.1 illustrates 'genotypic redundancy' with respect to phenotype—different genotypes may specify the same phenotype (G_1 and G_2 both correspond to P_1)—and 'phenotypic redundancy' with respect to fitness (P_1 and P_2 both have fitness F_1). Because the three genotypes G_1, G_2, and G_3 all have the same fitness, their differences are *neutral* with respect to fitness: the variation among them is not correlated with variation in reproductive success. On the other hand, the difference between any one of them and G_4 causes a fitness difference and evokes a response in the population leading to adaptive evolutionary change.

Neutrality arises because genotypes are redundant with respect to phenotypes and phenotypes are redundant with respect to fitness.

The example in Fig. 3.1 is theoretical. How likely is selectively neutral genetic variation in reality? We describe next an experiment that illustrates neutral evolution.

Experimental evolution in *Escherichia coli*

Conclusions drawn from an experiment are considerably strengthened if the same outcome is observed every time the experiment is repeated. At first sight you might think that this would make experimental evolution problematic. Evolution is, after all, an historical process, and an evolving population changes. Repeating the process requires that conditions be recreated precisely as they were in the past, which is not possible. However, experiments done by Lenski and his colleagues come close to

Experimental evolution in bacteria allows us to assess the repeatability of evolution.

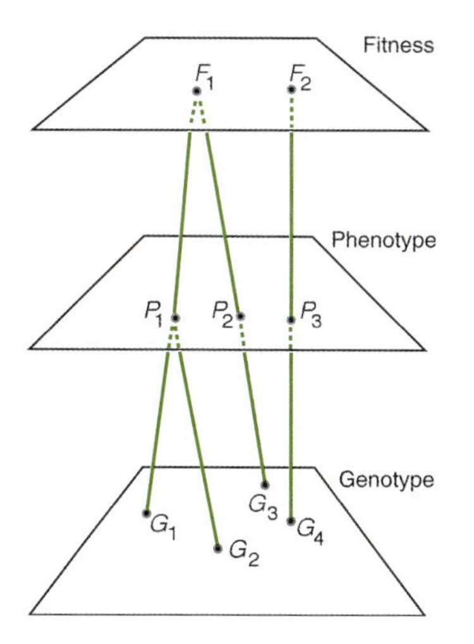

Fig. 3.1 Schematic relations between variation at the levels of the genotype, the phenotype, and fitness. G^i, i = 1–4 denote different genotypes; P^i, i = 1–3 denote different phenotypes; and F^i, i = 1,2 denote different fitnesses. The variation between the genotypes G_1, G_2, and G_3 is neutral with respect to fitness.

replicating evolution—not by repeating an evolutionary process exactly but by studying evolution in replicated populations evolving simultaneously in identical environments. They work with the bacterium *Escherichia coli,* whose short generation time allows experiments to extend over thousands of generations. The replicated populations were initially identical, consisting of a single genotype, and can be revived after freezing at –80 °C. Thus derived and ancestral populations can be compared directly, for example by measuring fitness differences in competition. We consider one experiment that reveals much about the repeatability of evolution.

At one level evolution was repeatable: identical starting conditions and identical environmental conditions produced similar outcomes.

Twelve initially identical populations were allowed to evolve independently in identical glucose-limited environments (Travisano *et al.* 1995). After 2000 generations the fitness of the bacteria in the derived populations had improved relative to their common ancestor by 35%. The 12 replicate populations differed in fitness from one another by only a few per cent and resembled each other closely in several respects: they all had higher maximal growth rates, larger individual cells, and fewer cells at stationary phase than their common ancestor. So far evolution appears to be repeatable and largely deterministic: identical starting conditions and identical environmental conditions produce an almost identical evolutionary outcome in 12 replicate populations.

The result, however, could mean either of two things, as the authors realized. The 12 populations might have achieved the observed adaptations by an identical evolution process involving the same genetic and physiological changes, or they might have reached the same end result through different changes at the genetic and physiological level. These two possibilities are illustrated in Fig. 3.2.

To discriminate between these possibilities, the authors introduced all 12 evolved populations to novel environments, substituting other sugars

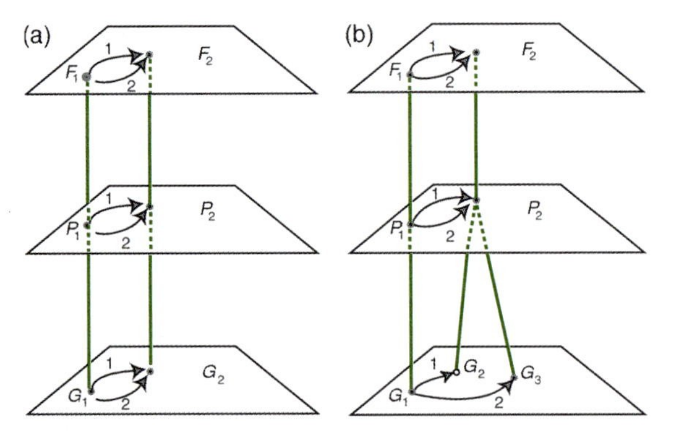

Fig. 3.2 Experimental evolution in two replicate populations 1 and 2. In case (a) the same genotype has evolved in both populations; in case (b) different genotypes G_2 and G_3 have been selected in the two populations, but both these genotypes correspond to the same phenotype P_2 (at least in the environment in which selection occurred).

(maltose or lactose) for glucose in the culture medium. When they measured the fitness of the bacteria relative to the ancestral genotype in the novel environments, they observed 100-fold greater genetic variation for fitness among the 12 populations than they had in the glucose environment in which the populations had evolved. Some populations were able to grow much better in a novel environment than others. Although the replicate populations had acquired identical modifications at the *phenotypic* level during the 2000 generations of evolution, they nevertheless differed at the *genotypic* level.

Although evolution was repeatable at the phenotypic level, it was not repeated at the genotypic level.

The authors' interpretation was that the mutations that distinguish the different lines produce the same phenotype in the glucose environment but differ in their side-effects in the novel environments, suggesting that Fig. 3.2b was more nearly correct than Fig. 3.2a. Whatever the interpretation, these experiments demonstrate that chance events, different in each of the 12 independently evolving populations, were an essential ingredient of the evolution observed.

Chance events in the form of genetic mutations were an essential part of the outcome, . . .

In these experiments both adaptive and neutral evolution occurred simultaneously. The adaptive evolution in the 12 populations resulted from a response to heritable variation in fitness in the glucose environment. The 12 populations diverged genetically in part because they experienced different mutations that were selected. However, the differences in their eventual genetic composition were close to *neutral* in the glucose environment because they all had about the same fitness. A mixture of the 12 evolved populations tested in the glucose environment would have contained genetic variation uncorrelated, or very weakly correlated, with fitness.

in which the genetic differences were apparently neutral.

Note that the term 'neutral' always refers to a comparison. In this experiment, neutrality refers to the absence of differences in fitness, measured by competition with the ancestor, between the 12 evolved populations. When a new mutation occurs that has the same fitness effects as the allele from which it mutated, the mutation is called 'neutral'. A neutral mutation may affect the reproductive success of its carrier. The characteristic that makes it neutral is that its effect does not differ from that of the resident allele from which it originated.

The term 'neutral' compares the mutation to the allele from which it originated.

The rest of this chapter explores the causes and consequences of neutral evolution. Next we consider what types of mutations are likely to experience neutral evolution. Because neutral alleles are, by definition, not correlated with variation in fitness, their frequencies in populations undergo undirected change caused by chance effects. We then discuss the processes that cause such random change, the role of neutral genetic change in molecular evolution, and end with some applications.

Reasons for no correlation between genetic variation and fitness

Genetic variation not expressed in the phenotype

Synonymous substitutions do not change the coded amino acid; they are neutral or nearly neutral.

Some nucleotide changes, called **synonymous substitutions**, do not change the coded amino acid because the genetic code is redundant, mostly in the third position of the nucleotide triplets that code for amino acids. For example, AAG (adenine–adenine–guanine) and AAA both code for the amino acid phenylalanine. The redundancy for the amino acid leucine is even more impressive; six codons—AAT, AAC, GAA, GAG, GAT, and GAC—code for leucine. If all possible nucleotide substitutions in all triplet codons are equally likely, then about 25% of all substitutions are synonymous—at the third position, up to 70% (Li and Graur 1991). Because synonymous substitutions do not change the phenotype, they are neutral or nearly neutral.

The reason we say 'nearly neutral' is **codon bias**, which means that synonymous codons do not always occur in equal frequency. In many species usage of synonymous codons is nonrandom, indicating selection favoring particular codons, possibly caused by differences in abundance of the corresponding transfer RNAs. Because the selection involved is believed to be very weak, the assumption of neutrality is approximately true.

Mutations in introns and in pseudogenes are neutral—if they do not interact with genes that are expressed.

Other mutations not likely to have phenotypic effects occur in **introns** and in **pseudogenes**. Introns are sequences within eucaryotic genes that are removed by gene splicing after transcription into RNA. Because introns do not code for proteins and are not expressed in the phenotype, mutations in introns are likely to be neutral. (Some caution is appropriate, for a few cases are known in which a mutation in an intron interferes with the expression of its gene.) Pseudogenes are DNA sequences derived from functional genes by gene duplication; they are no longer expressed and therefore nonfunctional. Mutations in pseudogenes have no phenotypic effect and are neutral. Selective neutrality of mutations in introns and pseudogenes is supported by molecular evidence that introns evolve faster than the translated parts (**exons**) and that pseudogenes evolve faster than functional genes.

Neutral amino acid variation

Changes in amino acids that do not affect protein function are nearly neutral.

Some changes in the sequence of amino acids of a protein do not affect its function; such regions show higher rates of amino acid substitutions when proteins of related species are compared. For example, apolipoprotein molecules carry lipids in the blood of vertebrates and have a lipid-binding site consisting of hydrophobic amino acids. Comparisons of apolipoprotein sequences from several groups of mammals suggest that in these domains one hydrophobic amino acid can be replaced by another without affecting function. If such amino acid substitutions have an effect on reproductive success, it is probably very small: they are approximately neutral.

The canalization of development

Some characters show no, or very little, phenotypic variation, despite considerable environmental and genetic variation. These are called canalized characters because the final phenotypic outcome is kept constant, as though development were confined within a canal that did not allow deviations from its course. When environmental or genetic variation is extreme, the canalization may break down, revealing genetic variation for the trait that had been hidden and demonstrating that the normal state was canalized. For example, *Drosophila melanogaster* normally has four scutellar bristles, but in flies homozygous for the mutation *scute* the number of bristles is reduced to an average of two, with some variation. This variation can be used to select for lower or higher bristle numbers; it is genetic. Extreme temperature treatments also reveal the underlying genetic variation that is normally 'invisible' due to canalization. Because of developmental canalization, the phenotypic effect of mutations in genes affecting a canalized trait is usually suppressed; such mutations are neutral so long as they are not expressed. In this example, genes causing variation in bristle number would be neutral in normal flies and potentially non-neutral in flies homozygous for the mutation *scute*.

Genes affecting a canalized trait will be neutral when not expressed.

Mechanisms that cause random evolutionary change

Several mechanisms cause random changes in the genetic composition of a population. They affect both neutral genetic variations and genetic variation that is responding to selection—adaptive genetic variation is not immune to random change. Random change in the genetic composition of a population is called **genetic drift**. Neutral genetic variation is subject to chance processes unless the neutral gene is located close to a gene undergoing selection, in which case it 'hitch-hikes' with the selected gene. Genetic variation correlated with fitness is also subjected to change directed both by natural selection and by genetic drift. Which of the two forces determines the outcome depends on their relative strength. Thus the two types of genes, neutral and selected, actually lie along a continuum. Frequency change of neutral genes is dominated by drift but perhaps contaminated by selection. Frequency change of selected genes is dominated by selection but perhaps contaminated by drift.

Neutral genetic variation only experiences chance processes; adaptive genetic variation experiences both directed and undirected change.

The strength of selection on a trait decreases with the correlation of the trait with reproductive success and with the amount of variation in reproductive success, and genetic drift is stronger in small than in large populations. The mechanisms that cause random change include mutation, the Mendelian lottery, founder effects, and genetic bottlenecks. Below we discuss these processes, then model genetic drift of gene frequencies as a statistical sampling process.

Mutation

Evolutionary change is based on germ-line mutations.

All genetic variation originates from mutation. A mutation is a hereditary change in the DNA sequence or in chromosome number, form, or structure. Because it is a chance process, mutation contributes to genetic divergence between populations, for different populations receive different mutations, as in the experiments described above on *E. coli*. Most mutations arise from errors during DNA replication. Mutations can occur in somatic cells as well as in the germ line (cells that end up as eggs and sperm). Somatic mutations can affect the function of individual organisms, both positively and negatively. For example, somatic mutation helps to generate antibody diversity in the immune system and to defend against pathogens; some somatic mutations cause cancer. Germ-line mutations are more important for evolution because, unlike somatic mutations, they are transmitted to future generations.

Types of mutations

Mutations can be classified into many types, depending on genetic details. We discuss only the types most important for evolutionary biology.

Mutations can be point mutations,

The most common type of mutation is a single nucleotide (DNA base pair) change, or **point mutation**. Often point mutations have little effect on the phenotype and on fitness. Because of the redundancy of the genetic code, many nucleotide changes do not change the coded amino acid. When point mutations occur in noncoding DNA, they are also effectively neutral. However, occasionally point mutations cause large fitness effects. For example, a point mutation causes the substitution of valine for glutamic acid in the sixth position of the β-chain of human hemoglobin. The resulting *HbS* allele (Hb for hemoglobin, S for sickle) is responsible for an abnormal sickle shape in red blood cells, causing anemia in people homozygous for the mutation but protecting heterozygotes for the mutation from malaria. The consequences of this point mutation are discussed in Chapter 5.

deletions and insertions,

Other types of mutation are deletions and insertions. A deletion is the loss of a chromosomal segment. The effect of a deletion depends on its size. A large deletion, involving tens to thousands of genes, will have severe effects and will often be lethal. A small deletion within a single gene inactivates the gene. For example, a 32 bp (base pair) deletion in the *CCR5* gene on chromosome 3 in humans has striking effects. This gene codes for a chemokine receptor involved in infection by HIV-1, the virus causing AIDS. People homozygous for this deletion appear to be immune to AIDS, and heterozygotes exhibit slower progression into the clinical stages of AIDS.

or changes in the total amount of DNA and number of genes.

Mutations that change the amount of DNA or the number of genes are key events in evolution. Among these are **polyploidization**, doubling of the complete set of chromosomes, and **duplication** of DNA sequences. Polyploidization–together with other mutations affecting the number and structure of chromosomes–is the main process responsible for differences in karyotype among species. It increases the total amount of DNA in the genome, providing material for the evolution of new functions. Duplicate copies of genes will accumulate mutations independently and may diverge to acquire a new function, as will duplications of small parts of chromosomes if the duplicated regions are at least as large as a gene.

Repeated gene duplication produces multigene families.

There are many **multigene families**, consisting of genes that have arisen by duplication from a common ancestral gene and have retained similar function. Examples in mammals include genes coding for heat-shock proteins (involved in protection of cells against environmental stress), for globin proteins (involved in oxygen transport), for apolipoproteins (involved in lipid metabolism), oncogenes (implicated in cancer), and genes involved in the immune system. Figure 3.3 shows the evolutionary history of the human globin genes. A very ancient duplication allowed divergence into two types of functional globin proteins: myoglobin, for oxygen storage in muscles, and hemoglobin, for oxygen transport in blood. Further duplications and divergence have produced the α and β families of hemoglobin, which consist of functional genes such as α1, θ1, and ζ in the α family, and ε, γ, δ, and β in the β family, and pseudogenes (nonfunctional remnants of once functional genes), such as ψα1 in the α family.

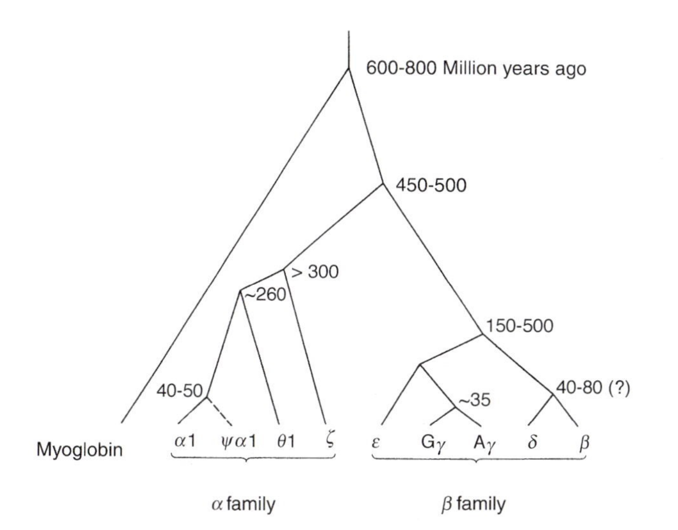

Fig. 3.3 Phylogenetic tree of human globin genes, illustrating a series of gene duplications. The human globin gene family consists of three groups, the myoglobin gene on chromosome 22, the α-globin genes on chromosome 16 and the β-globin genes on chromosome 11. Several pseudogenes also belong to this gene family. Pseudogenes are remnants of duplicated genes that have become nonfunctional due to mutations. Hemoglobin is made up of two protein chains, one coded by a gene from the α group and one by a gene from the β group. The various combinations differ in oxygen-binding affinities and appear at different developmental stages (embryo, fetus, adult). (From Li and Graur 1991.)

How random are mutations?

Mutations are not random, in the sense that they occur more frequently in some locations, and under some circumstances, than others.

It is often stated that adaptations are produced by natural selection acting on variation resulting from random mutations. What does the word **random** mean in this context? Because some parts of a genome experience much higher rates of mutations than other parts, mutation is not random with respect to where it occurs. Mutations can also be triggered by a specific signal, for example, in the fungus *Neurospora crassa,* where newly duplicated sequences trigger a specific mutational response (called RIP) that deactivates the repeated sequence (Selker 1990). RIP is an adaptive mutation, for it prevents the harmful accumulation of non-functional repeated sequences. Enhanced mutation rates at places in the genome where a high level of genetic variability is advantageous are also adaptive. Examples include the high level of somatic mutation generating antibody diversity in the vertebrate immune system, and the highly mutable bacterial genes involved in the interactions of pathogenic bacteria with their hosts (Moxon *et al.* 1994). Mutations do not occur at random with respect to their location in the genome. Some genes mutate more frequently than others.

However, the critical question is: 'Do mutations with a specific phenotypic effect occur more often when they are advantageous than when they are not?' If so, adaptations could be produced by mutation alone, and natural selection would be less important. Such a 'directed' mutational process is called Lamarckian because it resembles the idea expressed by Lamarck (1744–1829) that an adaptation acquired by an organism during its lifetime can be transmitted to its offspring. This would be the case, for example, if an animal could transmit to its offspring the immunity to a disease that it had developed through an immune response; but it cannot. Darwin adopted the idea of Lamarckian inheritance, for he thought that the use and disuse of parts could produce heritable modification; but he was mistaken. For example, he thought that winglessness in ostriches arose in part because their ancestors did not use wings and in part as a response to natural selection. Our present knowledge of genetics does not rule out Lamarckian mutations completely, but there is no indication that they occur at the level of genetic mutations (changes in DNA sequence), and there is no evidence that they are very important.

Mutations are random in the sense that there is no systematic relationship between their phenotypic effect and the needs of the organism in which they occur.

Mutations are certainly random in the sense that there is no systematic relationship between their phenotypic effect and the actual needs of the organism in which they occur. Note that it is the *specific* phenotypic effect of a mutation that matters here. Vertebrates 'need' antibody diversity to produce an effective immune response, and a mutational process helps generate this diversity. However, this is not a Lamarckian process because the presence, for example, of influenza virus does not affect the probability that a somatic mutation yields a lymphocyte clone effective specifically against influenza virus.

Thus mutations are one source of randomness in evolution. Another source is sexual reproduction.

Sexual transmission

Chance also plays an essential role in the sexual transmission of genes from parents to their offspring, as reflected in the phrase 'the **Mendelian lottery**'. Consider an individual heterozygous at some locus. Although it produces equal numbers of gametes that carry one or the other allele at this locus, how many copies of each allele it transmits to its offspring is subject to chance. In a large population individual variations in transmission tend to cancel each other out, but in small populations this does not always occur, and allele frequencies may change as a consequence. The Mendelian lottery can be easily seen in the distribution of sons and daughters within families. When many families are taken together, roughly equal numbers of sons and daughters are observed as expected, but within individual families striking deviations from the 1:1 expectation are common. What holds for sex chromosomes holds for all chromosomes: small samples of a random process often deviate strikingly from average expectation.

The Mendelian lottery generates chance variation in the alleles that get transmitted to offspring.

The founder effect and genetic bottlenecks

New populations are sometimes founded by a small group of individuals in which gene frequencies differ considerably from the frequencies in the parent population, simply by chance. This is called the **founder effect**. Some alleles may be completely absent; others rare in the parental population can reach high frequency in the new population simply because they happened to be present in a founder. For example, in the sixteenth and seventeenth centuries small groups of European colonists gave rise to the Afrikaans-speaking population in South Africa, derived from Dutch founders, and to the French-speaking population of Québec, derived from French founders. Some genetic diseases rare in Europe occur at relatively high frequency in these populations. The disease porphyria variegata, an autosomally inherited dominant disorder of heme metabolism, is very rare in most populations but occurs in about 1 in 300 Afrikaners. Most of the estimated 10 000–20 000 carriers in South Africa are descendants of a single Dutch couple who arrived and married in Cape Town in 1688. At the time the settlement numbered a few hundred people.

Populations founded by a few individuals do not contain a representative sample of the genes in the parent population.

Similarly, if because of some catastrophe only a few individuals survive to breed, the genetic composition of the population changes dramatically as it passes through a **genetic bottleneck**. Many alleles are lost and others rise to high frequency. Even if the few surviving individuals do so because of a selectively superior genotype, adaptive evolution is likely to affect the allele frequencies directly only at a few loci. Changes at most loci will be random.

When a population passes through a bottleneck, many alleles are lost and others rise to high frequency.

Genetic drift

The fate of neutral genetic variation in a population: the gene-pool model

The gene-pool model mirrors the genetics of a population in which individuals mate randomly.

Random change in allele frequencies due to chance factors is called **genetic drift**. Here we consider a simple model of genetic drift, limited to the random effects of the Mendelian lottery and variation in family size. Both factors are described by the same process in the *gene-pool model*. This model approximates the population genetics of a population in which the individuals mate randomly (including selfing). We characterize the individuals in the population by their genotype at a locus A, with two alleles A and a. Thus individuals are either AA, Aa, or aa. The formation of a new generation is represented as follows. Each parent produces an equal and large number of gametes; the gametes from all parents collectively form a gene pool from which an offspring is formed by drawing two gametes (genes) at random. This is repeated N times to produce an offspring generation of N individuals (Fig. 3.4). The frequency of the A allele in the offspring is given by a binomial distribution because the repeated drawing of a gene from the gene pool is a repeated chance event with two possible outcomes, like tossing a coin.

What can we conclude from this model? Imagine we start with many identical parental populations, and for each an offspring generation of N individuals is formed as described above, independently. Then we expect that among the populations the **gene frequency** p' of A in the new gen-

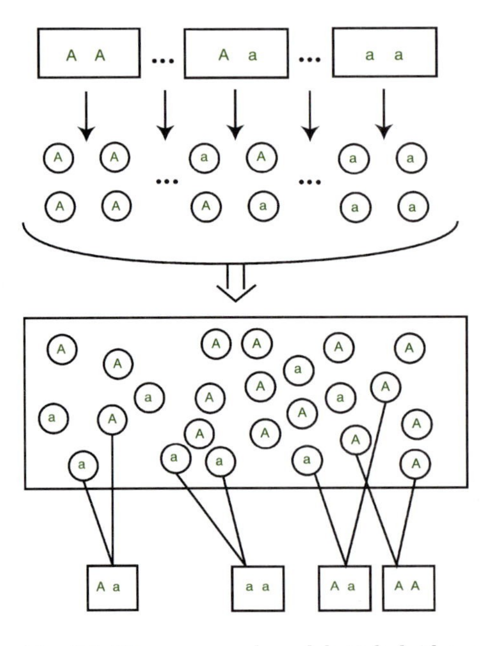

Fig. 3.4 The gene-pool model. N diploid parents contribute haploid gametes to a gene pool, from which N diploid offspring are drawn at random.

eration will vary according to a binomial distribution. We do not expect a change in gene frequency when we average over all populations: upward and downward deviations from the original gene frequency p will be equally likely. If we repeat the same procedure over many generations, results like those in Figure 3.5 are expected: increasing dispersion of the gene frequency among the replicate populations. Sooner (when N is small) or later (when N is larger) either the A or the a allele will be lost. The dispersion of gene frequencies due to chance is called genetic drift. In statistics it would be called the propagation of sampling error. It is stronger in small populations than in large populations, because a small random sample from a population is likely to deviate more from the population composition than a large sample.

Genetic drift is the dispersion of the frequency of a neutral gene in replicate populations, stronger in small than in large populations.

The significance of genetic drift in molecular evolution

In the 1960s new technology revealed the amino acid sequences of proteins. By comparing the sequences of proteins such as hemoglobin and cytochrome *c* from different species, and using paleontological estimates of times to last common ancestors, biologists could estimate the rate of evolutionary change in the protein sequences. For example, dogs and humans, which split from a common ancestor about 100 million years ago, have evolved independently for 200 million years, and the α hemo-

Proteins have characteristic and roughly constant rates of change that differ considerably from protein to protein.

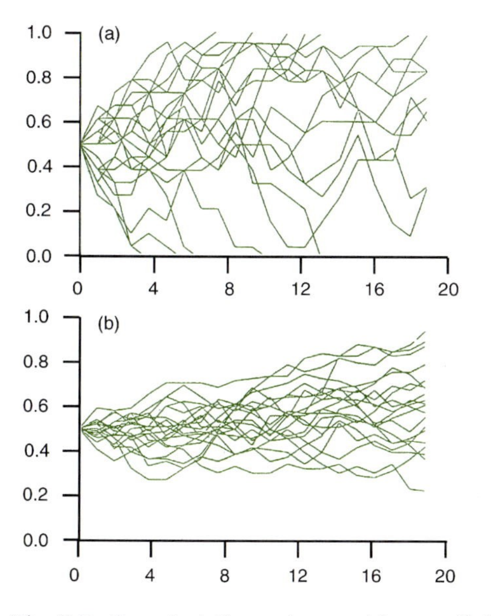

Fig. 3.5 Genetic drift at a locus with two alleles with initial allele frequencies of 0.5 has been simulated in (a) 20 small populations (nine diploid individuals) and (b) 20 larger populations (50 diploid individuals). In the absence of mutation, fixation of one of the alleles occurs inevitably, but more rapidly (on average) in the smaller populations. (From Ridley 1996.)

globin protein sequences in dogs and humans differ at 23 of the 141 amino acids, or 16.3%. That suggests that the mean number of substitutions per amino acid is about 1 per 1.23 billion years. A strikingly similar figure is obtained from comparisons of α hemoglobin evolution between other pairs of vertebrates. Other proteins also have characteristic and roughly constant rates of change, rates that differ considerably among proteins. For example, histone proteins evolve very slowly, while fibrinopeptides change about 80 times faster. The differences in rates are caused by chemical and functional constraints. Histones interact very closely with the DNA, every amino acid having a precise role and being difficult to replace without loss of function; whereas amino acid changes in many parts of fibrinopeptide molecules, which are involved in blood clotting, have little effect on their function.

The rate of nucleotide substitution in a DNA sequence is also roughly constant.

The first data on molecular evolution were on amino acid sequences, but now molecular evolution is usually described as nucleotide substitutions in DNA sequences. Just as there is a fairly constant and characteristic rate of change per amino acid in a protein or class of proteins, there is also a roughly constant and typical rate of nucleotide substitution in a DNA sequence or class of DNA sequences. Not surprisingly, the rate of synonymous substitutions, which has a large random component, varies much less between different proteins than the rate of non-synonymous substitutions, some of which are neutral and some of which experience strong selection.

Molecular clocks

Each protein appears to have its own molecular clock.

The approximately constant rates of amino acid substitutions in all lines of descent of protein sequences suggest that each protein has its own **molecular clock** that sets the pace of evolutionary change. This can be used as a method to estimate the time elapsed since the divergence of independent evolutionary lines. Molecular clocks are further discussed in Chapter 12. There has been much discussion and controversy about the constancy of the molecular clocks, and several instances have been found where the clock appears to tick faster in one lineage than in another. Nevertheless, evolution in the parts of the DNA that we can expect to be neutral appears to proceed at a fairly constant rate. The constancy of the rate indicates the randomness of the mechanism driving the change. This is in sharp contrast to morphological evolution. Many cases have been found in which evolutionary rates in genes and morphology seem to be uncoupled. Near constancy of morphology over hundreds of millions of years may go together with large changes at the genetic level. Conversely, large changes in morphology may occur in short time periods during which most genes change little (see Chapter 6), probably because such morphological change is driven by changes in a few genes with large effects.

The neutral theory of molecular evolution

In an influential paper Kimura (1968) proposed that most of the evolutionary change at the molecular level occurs as a consequence of random genetic drift of mutant alleles that are selectively neutral or nearly neutral. A heated controversy followed. At the time, many evolutionary biologists could not accept the idea that evolutionary change could be a random process; they maintained that changes in allelic frequencies in populations are adaptive and largely determined by natural selection. Note, however, that Kimura did not suggest that all evolutionary change is driven by genetic drift, only most of the nucleotide changes observed at the molecular level. There was never disagreement on the adaptive significance of most morphological, life history, and behavioral evolution.

Kimura proposed that most evolutionary change in DNA occurs through random drift of neutral alleles.

Despite much progress, how much of the genetic variability measured by molecular methods is produced by random genetic drift and how much by adaptive evolution is still not clear. On the one hand, it is clear that the selection forces driving the evolution of DNA sequences that are not expressed and have no direct function must be weak. On the other hand, it is also clear that many a priori neutral genes are hitch-hiking on selected genes. The controversy over genetic drift versus natural selection in molecular evolution is further discussed in Chapter 5.

How much molecular variation is produced by random drift and how much by adaptive evolution is still not clear.

Summary

Adaptive evolution requires a correlation between genetic variation and variation in reproductive success. This chapter considers the evolutionary consequences of genetic variation that is uncorrelated with reproductive success.

- Different alleles with the same effects on fitness are said to be neutral. Genetic variation that is not expressed in the phenotype is likely to be neutral, as is variation that causes functionally unimportant phenotypic variation.
- Neutral genetic variation is subject to undirected evolutionary change. Several processes cause undirected, or random, changes in allele frequencies. The most important are mutation, sexual transmission, random variation in family size, and founder effects. The random change in allele frequencies caused by these processes is called genetic drift; it has stronger effects in small populations than in large populations.
- An influential theory claims that much of the evolutionary change measured by molecular methods is determined by genetic drift.
- One argument for the 'neutral theory' is the relatively constant rate of change of the amino acid composition and the DNA sequence of a protein in all lines of descent from a common ancestor. Each protein is characterized by a characteristic rate of change: it evolves according to its own 'molecular clock'.

The neutral theory has generated much controversy and stimulated intensive research on the relationship between molecular variation and fitness variation. In the next chapter, we examine the contrasting consequences of different types of inheritance—sexual versus asexual, Mendelian versus quantitative—for directed genetic change under selection.

Recommended reading

Kimura, M. (1983). *The neutral theory of molecular evolution.* Cambridge University Press, Cambridge.

Li, W.-H. and Graur, D. (1991). *Fundamentals of molecular evolution.* Sinauer, Sunderland, Massachusetts.

Questions

3.1 Some genetic differences are subject to natural selection in some environments but are selectively neutral in others. Was some human genetic variation that is now neutral formerly under selection? Which traits were involved? What consequences do you expect from the change in selection regime?

3.2 We have no evidence that an infection by a particular pathogen can induce mutations in cells of the immune system causing immunity to this specific pathogen. But suppose such a mechanism did exist—would it be an example of Lamarckian evolution?

3.3 Does genetic drift only affect the frequencies of neutral alleles, or does it also affect the frequencies of alleles that are subject to natural selection?

Chapter 4
Evolution as changes in the genetic composition of populations

Introduction

Chapter 2 discussed several examples of evolutionary change produced by natural selection acting on phenotypes, including cyanogenesis in plants, clutch size in kestrels, and antibiotic resistance in bacteria. Because the changed phenotypes were inherited, there must have been genetic change as well. Knowing how genetic change occurs is essential for understanding and predicting an evolutionary process, because the direction and rate of phenotypic change under selection depends on the relation between genotype and phenotype and on the genetic composition of the population. This chapter discusses the genetic response to selection.

Understanding genetic change is one key to understanding evolution.

Two approaches have been developed to understand genetic change under selection, population genetics and quantitative genetics.

The two main approaches are population genetics, for simple genotype–phenotype relations, . . .

Population genetics can be used when the genotype–phenotype relation is relatively simple, when genetic differences between two alleles at one locus have phenotypic effects large enough to be detected unambiguously. Many traits whose variants differ qualitatively have simple genotype–phenotype relations, including those of the garden pea (*Pisum sativum*) used by Mendel in the experiments from which he deduced the rules of genetic transmission. His plants differed in flower color (purple and white), in seed color (green and yellow), and in seed shape (round and wrinkled). In all three traits the phenotypic differences correspond to simple differences between alleles at one locus, with each locus on a different chromosome. Other examples include some coat color differences in mammals, many rare genetic diseases in humans, and the DNA sequence variation important in molecular evolutionary genetics. When the genotype can be inferred from the phenotype, population genetics can describe a population in terms of genotype frequencies and can model evolution as changes in these frequencies.

and quantitative genetics, for complex ones.

Quantitative genetics is used when unknown or complex genotype–phenotype relations result in a continuous distribution of phenotypic variation instead of discrete phenotypic classes. For example, height, weight, longevity, milk yield in cows, and oil content in seeds vary continuously and quantitatively among individuals. Such traits are determined by several to many genes, none of which has effects large enough to create a recognizable phenotype that can be unambiguously assigned to a specific gene. Because quantitative traits do not segregate in genetic crosses like qualitative traits, the genotype cannot be inferred from the phenotype, and quantitative genetics cannot describe populations in terms of genotype frequencies. Instead it focuses on the means and variances of traits, estimates how much of the phenotypic variation is due to genetic differences between individuals, and uses that estimate to predict how fast the mean of a trait will change under selection.

We now consider simple models of evolutionary change in populations, taking first the population and then the quantitative genetic approach. Then we discuss the connections between the two.

Genetic systems: sexual and asexual, haploid and diploid

Four genetic systems: sexual or asexual, haploid or diploid.

The impact of selection on a population depends strongly on the genetic system of the organisms. The two key features of a genetic system are whether reproduction is sexual or asexual and whether the adult organisms are haploid or diploid. The genetic system is especially important in population genetics, whose models must specify its details. We describe four genetic systems (Fig. 4.1), then analyze the population genetics of two of them to see how selection produces genetic change and what difference the genetic system makes.

Who does what: a survey of genetic systems

Asexual haploids: procaryotes, some fungi, and cellular slime molds.

Asexual organisms with predominantly haploid life cycles are mostly procaryotes, bacteria, and cyanobacteria (blue-green algae) (Fig. 4.2). This genetic system also occurs among the Fungi Imperfecti, such as *Penicillium*, and in the cellular slime molds. Including the bacteria, these amount to about 20 major groups of organisms, including most of those that cause infectious diseases. Individuals with this genetic system outnumber by far all the other organisms on the planet. The asexual haploids are very common.

Sexual haploids: some algae, most fungi, the mosses.

Sexual organisms with predominantly haploid life cycles include some red algae, some conjugating green algae, most fungi, including the

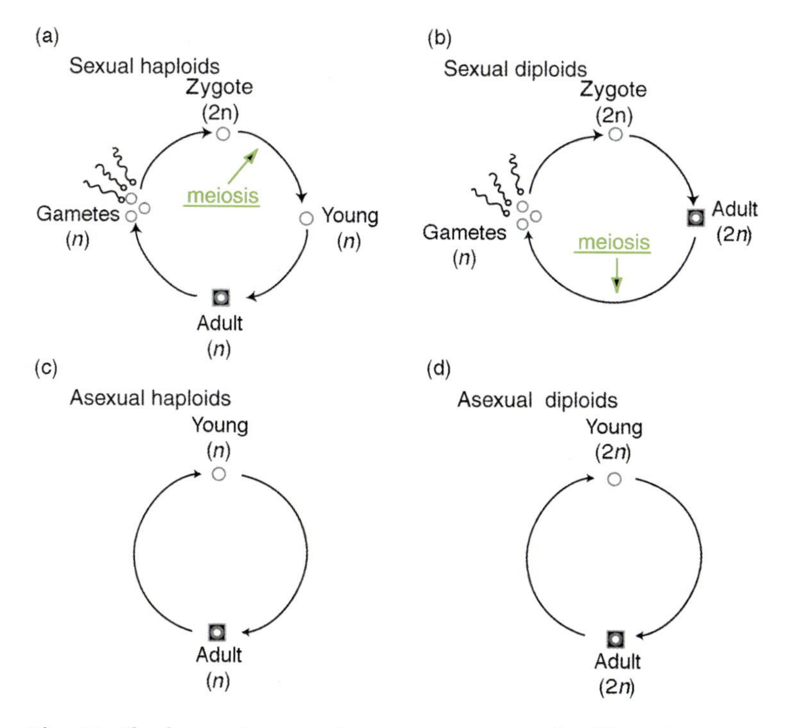

Fig. 4.1 The four major genetic systems represented as life cycles. (a) Sexual haploids; (b) sexual diploids; (c) asexual haploids; (d) asexual diploids. Selection can occur at every stage in the life cycle, but is assumed here to act only during growth from zygote to adult.

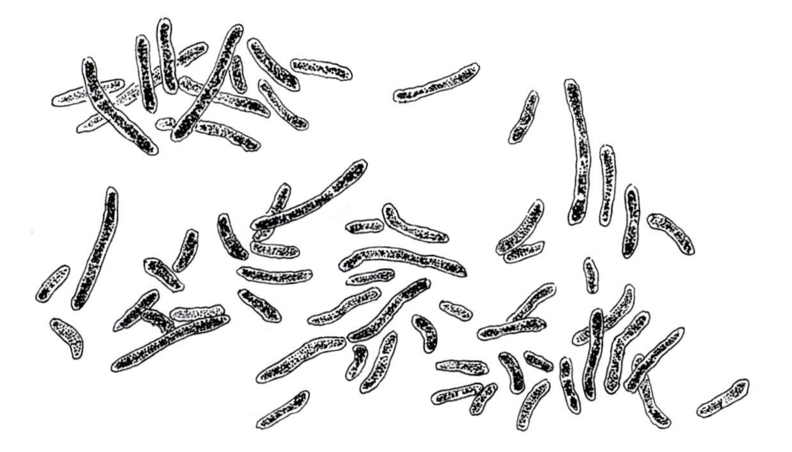

Fig. 4.2 A haploid asexual bacterium *Mycobacterium tuberculosis*, the pathogen that causes tuberculosis. (By Dafila K. Scott.)

ascomycetes (e.g. *Neurospora*), and the mosses (Fig. 4.3).

There are fewer than 10 major groups with this genetic system, but the system is not rare—these algae, fungi, and mosses are represented by many species that can occur in large populations.

Fig. 4.3 A haploid sexual moss, a sexually reproducing organism in which the dominant phenotype in the life cycle is haploid. In this figure the fruiting bodies, the sporophytes, are diploid, but the gametophyte, the plant we refer to as moss, is haploid. (By Dafila K. Scott.)

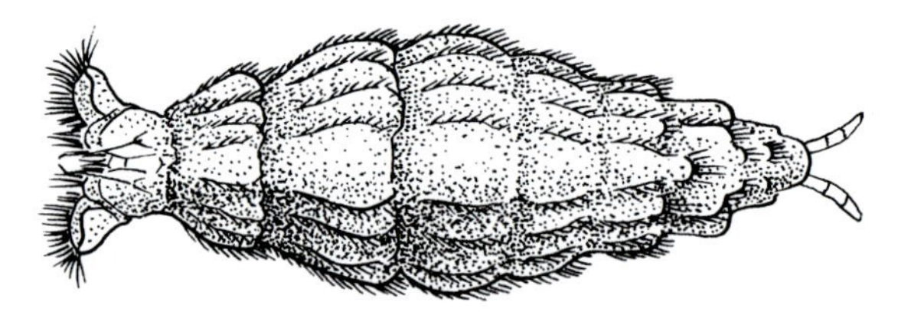

Fig. 4.4 A diploid asexual bdelloid rotifer, one of the scandalous ancient asexuals that is not supposed to have survived (see Chapter 7).

Asexual diploids: dinoflagellates, some protoctists.

Asexual organisms with predominantly diploid life cycles include the dinoflagellates, more than 10 groups of protoctists (unicellular algae, protozoa, and unicellular groups resembling fungi), and several groups of multicellular animals (Fig. 4.4). It is not yet known whether some of the species in these groups are haploid or diploid.

Sexual diploids: most multicellular animals and plants, some algae, protozoa, and fungi.

Sexual organisms with predominantly diploid life cycles include about 20 animal phyla, the multicellular plants with alternating haploid and diploid stages whose diploid stage is larger and longer lived (ferns, cycads, conifers, flowering plants), and several groups of algae, protozoa, and fungi. Sexual diploids tend to be large, long-lived, and familiar (Fig. 4.5).

Thus the two most common types of life cycles are asexual haploids and sexual diploids. The former often parasitizes the latter. Sexual haploids and asexual diploids are less common but important.

Haploid–diploid alternation: many protozoa, mushrooms, rotifers, cnidarians, some annelids, many arthropods.

As often happens in biology, many groups cannot be so simply classified. Among these are about six groups of sexually reproducing organisms that alternate between haploid and diploid phases, with neither dominating. These include the Foraminifera, important marine protozoa with a long fossil record; the Basidiomycetes, of which mushrooms are

Fig. 4.5 A sexual diploid animal and plant: a hawkmoth pollinating a wild carnation. (By Dafila K. Scott.)

well-known representatives; the Microsporidians, an important group of parasites; and the Apicomplexa group of protozoans, which includes *Plasmodium*, the cause of malaria. In addition, there are eight predominantly diploid animal phyla with species that alternate sexual and asexual reproduction, among them some rotifers, cnidarians, annelids, and arthropods, including aphids and water fleas (*Daphnia*).

Thus each of the four major systems has numerous and important representatives. Some organisms do not fit into this binary classification. Their evolutionary genetics are more complicated than the two most common genetic systems, asexual haploid and sexual diploid, analyzed next.

Population genetic change under selection

To compare the effects of asexual haploidy with those of sexual diploidy, we imagine a standard situation, a population into which a new, advantageous allele has entered by mutation or immigration. Then we work out the fate of this new allele in both genetic systems.

Consider the fate of a new advantageous mutation that is initially rare. It has a fitness advantage of $1+s$, where s is the selection coefficient.

We start with a population where most individuals have the same genotype at a locus A, A_1 in the haploids and A_1A_1 in the diploids, and where, due to mutation or immigration, a few individuals carry an A_2 allele. We assume that the population is so large that chance events (genetic drift, discussed in Chapter 3) can be neglected. Organisms with the allele A_2 are assumed to have greater reproductive success—higher fitness—than organisms with A_1. Here we restrict differences in fitness to differences in survival to maturity, thus assuming that the different genotypes do not differ in fertility or in survival later in life. In the haploid case, the fitness benefit of the A_2 allele is expressed as $1 + s$ times the fitness of an A_1 individual. In the diploid case the fitness of an individual with genotype A_2A_2 is $1 + s$ times the fitness of an A_1A_1 individual. The parameter s is called the *selection coefficient*. Because s increases fitness, the

A_2 allele should increase in frequency and eventually replace the A_1 allele. The models tell us how fast this happens.

Model its spread over many generations by breaking the process down into little steps, from one generation to the next.

Let the frequency of A_1 in the present generation be p (so that $0 \leq p \leq 1$) and that of A_2 be $q = 1 - p$. In a haploid population of 1000 individuals of which 990 are A_1 and 10 are A_2, $p = 0.99$ and $q = 0.01$. These frequencies are measured in the newborn of the asexual life cycle or in the gametes of the sexual life cycle. In the next generation the allele frequencies of A_1 and A_2 are written as p' and q'. If we can determine how p changes to p', and repeat that process over many generations, then we have understood genetic change under selection.

Assume that mating is random with respect to this gene—that it is not involved in mate choice—and that generations are separate.

We assume that mating occurs at random in the sense that the probability of a mating does not depend on the genotype of the individuals involved. This is realistic for genes not involved in sexual selection. Thus the probability that individuals of given genotypes will mate is just the product of their frequencies in the population. Finally, it is convenient to assume that generations are separate: the parents reproduce and die before the offspring reach reproductive age. This assumption is realistic for annual plants; it is only approximate for others.

Asexual haploid populations

This life cycle (see Fig. 4.1c) dominates among bacteria and is common in algae and fungi. While recombination may occur occasionally, we neglect it to study genetic change under pure asexual reproduction. Table 4.1 shows the derivation of a formula (Equation 4.1) describing the change over one generation in the frequency of the selectively superior allele A_2:

$$q' = \frac{q(1+s)}{1+sq} \qquad [4.1]$$

You can follow the derivation of Equation 4.1 step by step by going through Table 4.1 from the top to the bottom row. Thus in a given

Table 4.1 Calculation of gene frequency change in asexual haploids

Stage in life cycle	**Relative genotype frequencies ($p + q = 1$)**		**Relative fitnesses**	
Young in present generation	A_1	A_2	A_1	A_2
	p	q	1	1 + s
Adults	A_1	A_2	A_1	A_2
	$\frac{p}{p+q(1+s)}$	$\frac{q(1+s)}{p+q(1+s)}$	1	1
	$= \frac{p}{1+sq}$	$= \frac{q(1+s)}{1+sq}$		
Young in next generation	A_1	A_2		
	$p' = \frac{p}{1+sq}$	$q' = \frac{q(1+s)}{1+sq}$		

generation, offspring of the two genotypes A_1 and A_2 are formed with frequencies p and q. Then selection acts during their growth to adults, and the probability of survival of A_2 individuals relative to that of A_1 is $1 + s$. As a result, the frequencies of A_1 and A_2 among the adults are no longer p and q, but have changed as calculated in the fourth row of the table. (This calculation is necessary to keep $p + q = 1$ as required by the definition of frequencies.) That the surviving adults do not differ further in fitness is indicated by the identical fitnesses of the two genotypes. This implies that all adults have the same number of offspring irrespective of genotype, and the frequencies of the two genotypes among the newborn of the next generation (shown in the last row of the table) are the same as those among the adult parents.

Asexual haploids: the formula for genetic change from one generation to the next is built in simple steps.

The change in allele frequency given by Equation 4.1 is shown in Figure 4.6 (curve a).

Sexual diploid populations

Most familiar animals and plants are sexual diploids (see Fig. 4.1b). Many plants can reproduce sexually or asexually, but most animals can only reproduce sexually. Equation 4.2, derived in Table 4.2, describes the change in frequency of an advantageous mutation in a sexual diploid population. Because in diploids there can be heterozygotes, we need a new parameter, h, to describe the fitness of heterozygotes. We define the fitness of A_1A_2 heterozygotes to be $1 + hs$ times that of A_1A_1 homozygotes. Here h determines the dominance relationship between the two alleles: $h = 0$ means that A_2 is recessive and $h = 1$ means that A_2 is dominant. For $0 < h < 1$ dominance is incomplete or partial, which often occurs.

Sexual diploids: the formula for genetic change contains a new parameter, h, to describe the degree of dominance.

$$q' = q\left(\frac{1 + s(q + hp)}{W}\right) \qquad [4.2]$$

where $W = p^2 + 2pq(1 + hs) + q^2(1 + s)$.

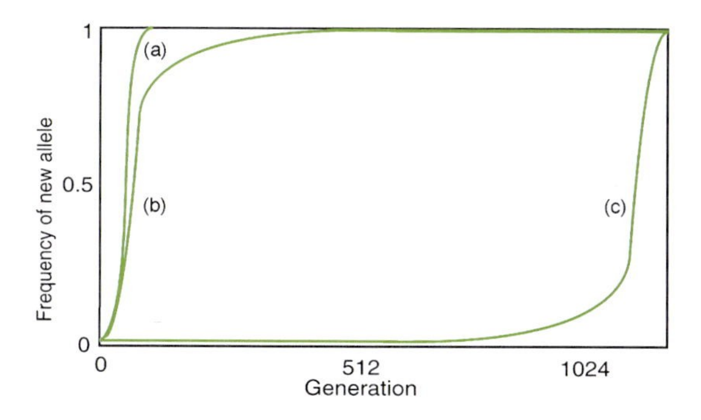

Fig. 4.6 Allele frequency change by selection. The curves show the increase of an initially rare allele (a) in an asexual population; (b) a dominant gene in a sexual diploid population, and (c) a recessive gene in a sexual diploid population.

Table 4.2 Calculation of gene frequency change in sexual diploids

Stage in life cycle	**Relative genotype frequencies ($p + q = 1$)**			**Relative fitnesses**		
Gametes forming present generation	A_1 p		A_2 q	A_1 1		A_2 1
Zygotes	A_1A_1 p^2	A_1A_2 $2pq$	A_2A_2 q^2	A_1A_1 1	A_1A_2 $1 + hs$	A_2A_2 $1 + s$
Adults	A_1A_1 $\frac{p^2}{W}$	A_1A_2 $\frac{2pq(1 + hs)}{W}$	A_2A_2 $\frac{q^2(1 + s)}{W}$	A_1A_1 1	A_1A_2 1	A_2A_2 1
	$W = p^2 + 2pq(1 + hs) + q^2(1 + s)$					
Gametes forming next generation	A_1 $p' = \frac{p(1 + shq)}{W}$		A_2 $q' = \frac{q[1 + s(q + hp)]}{W}$			

Hardy–Weinberg frequencies

The derivation of Equation 4.2 can be followed step by step by going through Table 4.2 from top to bottom. The transition from the second row to the fourth row deserves special attention. There we see that if the alleles A_1 and A_2 have frequencies of p and q among the gametes, and all gametes have equal chance of fertilization, the three types of diploid genotypes formed by random mating occur among the zygotes in proportions p^2, $2pq$, and q^2. This can be understood by noting that the formation of zygotes by random mating is equivalent to random pairing of gametes, shown in Table 4.3.

Hardy–Weinberg: 'if nothing happens in a large population, gene frequencies do not change'.

The diploid genotype frequencies thus obtained—p^2, $2pq$, q^2—are called the *Hardy–Weinberg frequencies* after Hardy and Weinberg. They showed independently, in 1908, how diploid genotype frequencies are related to haploid allele frequencies in a large Mendelian population with random mating and no mutation, selection, or gene flow. They thus proved that Mendelian genetic transmission itself causes no change in these frequencies. You can check this by setting $s = 0$ (no selection) in Equation 4.2. Their result justifies the description of the genetic composition of a population in terms of allele frequencies rather than a much more complicated description in terms of genotype frequencies. If you know the allele frequencies, you can calculate the genotype frequencies.

Table 4.3 A Punnett diagram for Hardy–Weinberg frequencies

	A_1 p	A_2 q
A_1 p	A_1A_1 p^2	A_1A_2 pq
A_2 q	A_1A_2 pq	A_2A_2 q^2

The Hardy–Weinberg result, 'if nothing happens in a large population, gene frequencies do not change,' is the starting point of population genetics. The result is both important and remarkable. Why does meiosis have this consequence at the level of the population? Why is it normally so precisely fair to the different segregating alleles, allowing all of them to be maintained in the population? We return to this question in Chapter 11.

Why is meiosis so fair at the population level?

Another feature of Table 4.2 is the expression W in the sixth row of the table. It is called the *mean fitness* of the population, because it is the average fitness of the individuals. As you can check, it is the average of the genotype fitnesses, weighted by the genotype frequencies. The mean fitness, W, appears in the denominator of the derived allele frequencies to scale them as proper frequencies with values between 0 and 1, just as the expression $p + q(1 + s)$ did in the fourth row of Table 4.1.

Iteration of Equation 4.2 produces curves (b) (A_2 dominant) and (c) (A_2 recessive) in Fig. 4.6.

Implications of population genetics for evolutionary biology

Comparing genetic change in these two systems suggests several conclusions.

Allele substitution in natural populations is slow–fast–slow

In all cases the curve describing genetic change under selection is S-shaped (Fig. 4.6). The models predict slow change at low and high allele frequency and fast change at intermediate allele frequency. Therefore rare phenotypes are expected to increase or decrease very slowly (depending on whether they are selected for or against). When alleles are at intermediate frequencies, evolutionary change can be much faster.

Genetic change under selection is S-shaped; rare alleles increase or decrease very slowly. Change can be fast at intermediate frequencies.

There are not many examples of well-documented gene substitutions in natural populations. One reason is that in most cases where an advantageous allele is replacing an existing prevalent allele, the process is too slow to be witnessed within a few human generations. It is therefore not surprising that the few examples we have are all related to gene substitutions in species that have been confronted with drastic environmental changes caused by humans. These changes produced very strong selection for genetic variants that could cope with the new conditions.

A dramatic example of recent, rapid genetic change is the spread of resistance to pesticides. Since the Second World War large amounts of insecticides, particularly DDT and malathion, have been used to control insects that attack crops and spread human diseases. Initially, most insects were sensitive to the pesticides, chemical control was effective, and by the early 1960s the incidence of diseases transmitted by insects was much reduced. But by 1970, many insects had evolved genetic resistance to the main insecticides. For example, malaria declined to

An example of recent and rapid gene substitutions: the spread of resistance to pesticides.

about 70 million cases per year in 1960 but increased again to 200 million by 1970 due to the rapid spread of resistance to DDT among the mosquitoes that transmit the disease. By 1980 over 400 species of arthropods were resistant to the main pesticides.

In hundreds of species similar evolutionary processes occurred independently: rare alleles conferring resistance increased in frequency after the massive applications of pesticides started. During the first 10–15 years this increase had little impact at the population level because of the slow initial increase in frequency of the resistance alleles. Then came a phase of fast change during which resistance was broadly established in most species. This pattern, which is consistent with the models for gene substitution considered above, suggests that genetic resistance to pesticides is often based on single alleles. In fact, resistance is often determined largely by a single gene, although interaction between several genes also occurs. Our population genetic model predicts that even if pesticide application is continued, the non-resistant alleles will remain in populations at low frequencies for a long time because the final phase of allele substitution is again slow.

Decision makers saw no problem for 10–15 years: that period was slow and long-term for humans, fast and short-term for evolution.

Note that decision makers would not have perceived any problem for 10–15 years, a characteristic of evolutionary processes that makes it difficult for human institutions to deal with them. Most evolutionary processes occur on a time scale that humans perceive as long-term, although in evolutionary perspective they are rapid and short-term. An exception is the evolution of antibiotic resistance in bacteria that cause infections and disease in humans. That happens so rapidly that the medical profession recognizes it as a major problem and has developed countermeasures, with some success.

Genetic change is faster in haploid asexual than in diploid sexual populations

The models also suggest that genetic change is faster in haploid asexuals than in diploid sexuals. Because most micro-organisms are haploid and largely asexual, while most animals and seed plants are diploid and sexual, populations of micro-organisms are expected to respond to the same selection pressure more rapidly than populations of eucaryotes. Because the time scale used in comparing the two genetic systems is in generation-units, the difference is even greater when measured in absolute time, for micro-organisms can have very short generation times.

Pathogenic microbes respond more rapidly to the same selection pressure than do their eucaryotic hosts.

This difference in generation times is important for the coevolution of interacting species. Coevolution of two species occurs when each causes evolutionary changes in the other. Often the partner species in such relationships differ in ploidy and sexual mode, for example, diploid sexual hosts (like humans) and their haploid bacterial pathogens. The simple one-locus model suggests that microbial pathogens have an easier job coevolving with their hosts than do their hosts in responding to their pathogens. This assumes that host and pathogen coevolve by successive gene substitutions and that appropriate mutations occur at the same rate in both. In several plant–fungus relationships virulence alleles in the fungal pathogen are matched by resistance alleles in the plant. In these gene-for-gene systems, the predictions of simple theory are particularly relevant. In contrast, when several loci interact to determine resistance or virulence, coevolution does not necessarily lead to interrelated gene substitutions in the host and the pathogen. Sexual recombination may then play a large role, discussed in Chapter 7.

In sexual diploids the initial spread of a favored allele is much faster when it is dominant than when it is recessive

Dominance has a strong impact on genetic change in diploids.

The most important effect of sexual reproduction on the genetic structure of a population is the creation of new combinations of alleles at different loci by recombination. This aspect of sex is beyond the scope of our model, which describes changes at only one locus. Sex with random mating also affects the population at a single locus, where it distributes alleles over the diploid genotypes in Hardy–Weinberg proportions. This strongly decreases the rate of spread of a favored recessive allele compared to a favored dominant allele. When A_2 is recessive and initially very rare, the selectively favored A_2A_2 genotypes occur at a frequency of q^2, which means they are very rare when q is small. In a sexual population these rare A_2A_2 genotypes will have mostly A_1A_2 offspring, because almost all A_2A_2 individuals will mate with the common A_1A_1 genotype. Thus there is little scope for selection of A_2 because the recessive allele is not expressed and is 'invisible' to natural selection in heterozygotes. This explains the very slow spread of a rare favorable recessive mutation (Fig. 4.6c).

For dominant genes the situation is different. They are expressed in heterozygotes and selection can 'see' the favored allele both in homozygous and in heterozygous genotypes. Therefore a dominant favorable mutation will spread much faster (Fig. 4.6b).

Because recessive favorable mutations spread much more slowly than dominant ones, it is not surprising that many of the mutations conferring pesticide resistance are dominant, for we see only the ones that responded quickly. Recessive resistance mutations may well be present in insect populations, but they have not yet had time to reach high frequencies.

Quantitative genetic change under selection

Quantitative genetics: many genes, none of which has a large enough effect to create an easily recognizable phenotype.

Unlike the examples in the previous section, many traits vary continuously and quantitatively among individuals in a population. They include traits of ecological importance, such as body size, competitive ability, and running speed. Such traits are determined by several to many genes, and no single allele has a large enough effect to create an easily recognizable phenotype. Because these traits do not segregate into clear phenotypic classes in genetic crosses, we cannot use Mendelian genetics to infer genotype–phenotype relations. The frequency diagrams of such traits measured on many individuals are often bell-shaped curves.

Two causes of phenotypic variation: genetic and environmental.

Such phenotypic variation usually has two causes, genetic variation among the individuals and variation in the environments they experience. That genetic variation is at least partially responsible for phenotypic differences follows from experience in plant and animal breeding. There, applied artificial selection has almost always resulted in heritable change (Fig. 4.7a). We know that the environment also affects phenotypic variability because individuals from inbred lines or from the same clone still vary phenotypically despite being genetically similar or identical (Fig. 4.7b).

We cannot use population genetics to analyze genetic change caused by selection on quantitative traits, for we have no idea what genotypes are present and cannot describe genetic change in terms of genotypic frequencies. Instead we take an approach that analyzes phenotypic variation to predict evolutionary change in quantitative traits. It makes several simplifying assumptions.

Basic quantitative genetics

An assumption: quantitative traits are affected by many loci with similar impact.

One important assumption is that a quantitative trait is affected by alleles at many loci and that most of these alleles have a small effect on the trait. For example, suppose that variation at 10 loci, each on a different chromosome, affects differences in height among human males. At each locus the population contains two alleles, denoted + and –. The first has a positive effect and the second a negative effect on height. The allele

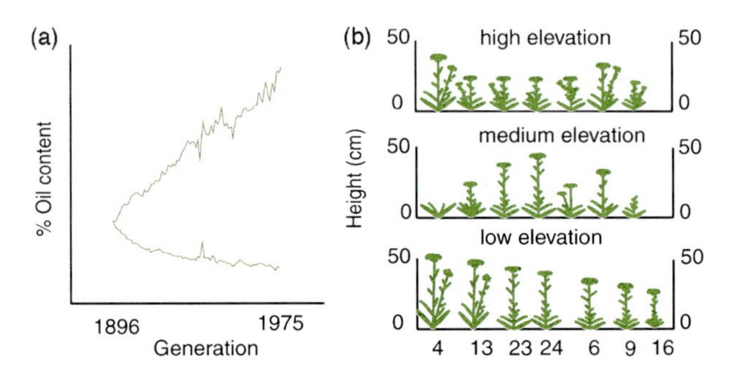

Fig. 4.7 Both genetic and environmental variation can cause phenotypic differences. (a) Since 1896 a population of corn (*Zea mays*) has been selected for high and low oil content. The response to selection, particularly in the high line, points to the continued presence of genetic variation within this population. (From Dudley 1977.) (b) Cuttings of seven different genotypes of *Achillea* were grown at three different elevations. The resulting phenotypes show that variation within a single environment is caused by genetic differences and that variation within a single genotype is caused by environmental differences. The numbers on the *x*-axis are labels for the different genotypes. (From Griffiths *et al.* 1996.)

frequencies may differ between the loci but are assumed to be intermediate, say 0.2 to 0.8. At some loci the + allele may be more frequent in a population, at others the – allele. Under these assumptions most individuals will have a genotype with roughly equal numbers of + and – alleles, and only rarely will genotypes occur that contain mostly + or mostly – alleles. Thus, this model is consistent with the approximately bell-shaped phenotypic distribution of height (Fig. 4.8). Note that we cannot conclude the reverse, that bell-shaped phenotype distributions are caused by variation at many loci, for they also result from variation in the environment.

How much phenotypic variation is caused by genetic differences? The answer determines the response to selection.

How can we find out how much phenotypic variation is caused by genetic differences? This is an important question, for selection on parents will only result in phenotypic change in offspring if some of the phenotypic variation is heritable. Only then will the greater reproductive success of particular phenotypes result in genetic change (Fig. 4.9).

Genetically and environmentally caused variation add to produce total phenotypic variation.

In experiments there is a straightforward procedure to estimate what part of the total phenotypic variance in a population is due to genetic and what part to environmental variation. The underlying assumption is that the phenotype, P, is the sum of genetic effects, G, and environmental effects, E:

$$P = G + E \qquad [4.3]$$

P, G, and E are measured in phenotypic units, for example height measured in centimeters or inches. There is no biological justification for assuming that genetic and environmental factors act additively. That assumption has the advantage of simplicity, and in plant and animal breeding, predictions based on this assumption explain experimental results reasonably well. When different genotypes respond differently to environmental variation, the theoretical predictions are complicated by genotype–environment ($G \times E$) interactions (see Chapter 6), and the additive assumption (Equation 4.3) does not hold. If $G \times E$ interaction

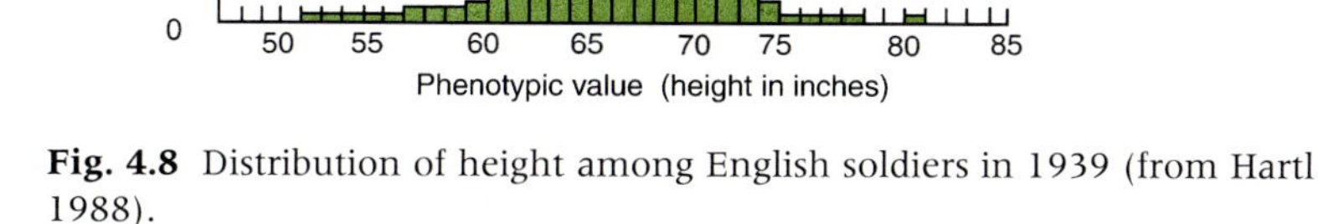

Fig. 4.8 Distribution of height among English soldiers in 1939 (from Hartl 1988).

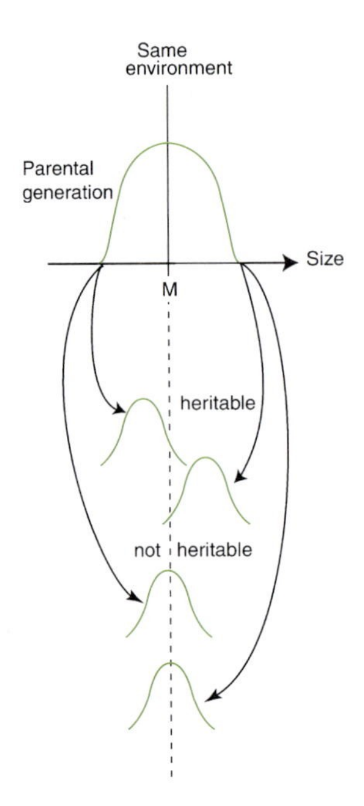

Fig. 4.9 Experimental procedure to test the heritability of a trait. Crosses are performed among extreme individuals (small × small; large × large). If the offspring of the two extreme parent groups are distributed differently and towards the parental means, the variation in the parental generation must have been caused in part by genetic differences (upper two offspring curves). If the two groups of offspring do not differ, the parental variation must have been environmental. (From Griffiths *et al.* 1996.)

is negligible, genes and environment act independently, and for the variances of the variables we may then write:

$$\mathrm{Var}(P) = \mathrm{Var}(G) + \mathrm{Var}(E) \qquad [4.4]$$

This formula suggests how to estimate what part of the total phenotypic variance is due to genetic variation in the population and what part can be ascribed to environmental variation. In controlled experiments, either $\mathrm{Var}(G)$ or $\mathrm{Var}(E)$ can be made practically zero. When genetic variation is held constant, the phenotypic variance, $\mathrm{Var}(P)$, estimates the environmental variance, $\mathrm{Var}(E)$; when environmental variation is held constant, it estimates the genetic variance, $\mathrm{Var}(G)$.

Generally genetic variation is easier to reduce than environmental variation, because one often does not know which environmental variables should be controlled. Genetic variation can be reduced to near zero in plants by using cuttings from a single plant so that all experimental plants have the same genotype. In animals the most practical procedure is to generate inbred lines by continued inbreeding for many generations. Among individuals of an inbred line there is very little genetic variation, and crosses between two highly inbred lines yield virtually the same heterozygotes. This procedure measures how much of the phenotypic variation is caused by genetic variation, the so-called **broad-sense heritability**,

$$H^2 = \mathrm{Var}(G) \ / \ \mathrm{Var}(P) \qquad [4.5]$$

(The squared symbol, which may look a bit strange, derives from the 1920s when H was used as the corresponding ratio of standard deviations.)

The broad-sense heritability is of limited value for estimating the response to selection in sexual populations, for selection on some types of genetic variation will not result in changes in the genetic composition of populations and therefore not in a phenotypic response. This is mainly a consequence of interactions in genotypes, such as dominance, as can be seen in the following example. Suppose that selection favors cold-resistant plants and that a population shows genetic variation for this trait. In particular, suppose that at a locus that contributes to cold resistance the A_1A_2 heterozygotes are more cold resistant than either homozygote. Among the plants favored by selection and producing most of the offspring will be many A_1A_2 genotypes. But only half of the offspring from these heterozygotes will be heterozygous themselves, while the other half will consist of less well-adapted genotypes. Even if only heterozygotes breed to form the next generation, 50 per cent of the offspring will be homozygous, preventing a further response to selection.

Broad-sense heritability is the genetically caused portion of total phenotypic variation; its value for estimating response to selection is limited.

Therefore, we need to estimate the part of the genetic variance that actually results in a response to selection. This part of the genetic variance is called the **additive genetic variance** Var(A), because it is the variance resulting from genotypic differences caused by additive allelic effects. Additive allelic effects result, for example, if the heterozygote A_1A_2 is exactly intermediate in cold resistance between the homozygotes A_1A_1 and A_2A_2. There the alleles A_1 and A_2 are acting additively—neither is dominant, and their combined impact on the phenotype is the numeric average of their independent effects. In this case a response to selection for cold resistance is expected for as long as there is genetic variation at the locus. The additive genetic variance expressed as a fraction of the total phenotypic variance is called the **narrow-sense heritability**,

Additive genetic variance: the part of genetic variation that determines the response to selection.

$$h^2 = \mathrm{Var}(A) \ / \ \mathrm{Var}(P) \qquad [4.6]$$

The narrow-sense heritability is a useful measure of the expected response to selection. How can we estimate it?

Narrow-sense heritability: the additive genetic portion of total phenotypic variation. It determines the rate of response to selection.

Estimation of h^2

Because narrow-sense heritability predicts the response to selection, one way to estimate h^2 is from the response to selection itself. Figure 4.10 depicts a form of selection called **truncation selection**, commonly used in plant and animal breeding to improve traits of economic importance. Only individuals with values of the trait more extreme than a certain value (the truncation point) are selected to breed the next generation. Figure 4.10 explains how h^2 can be estimated from the results of truncation selection.

Traits more relevant to fitness show lower heritabilities.

Table 4.4 shows a summary of 1120 experimental estimates of narrow-sense heritabilities for wild, outbred animal populations, gathered by Mousseau and Roff (1987). They grouped the traits for which heritabilities had been estimated into four categories: life history traits, such as fecundity, viability, survival, and development rate; physiological traits, such as oxygen consumption and resistance to heat stress; behavioral traits, such as alarm reaction and activity level; and morphological traits, such as body size and wing size. The average heritabilities in all pairs of categories except physiology and behavior differed significantly.

In general, traits directly connected to reproductive success have low heritabilities and traits less relevant to fitness have higher heritabilities. Part of the reason is that continued selection on a trait will exhaust additive genetic variation by fixing advantageous alleles, causing the heritability to decrease, so long as beneficial mutations do not occur frequently. Because fitness is under continuous selection in natural populations, we expect low heritabilities for traits that are highly correlated with fitness—and we find them.

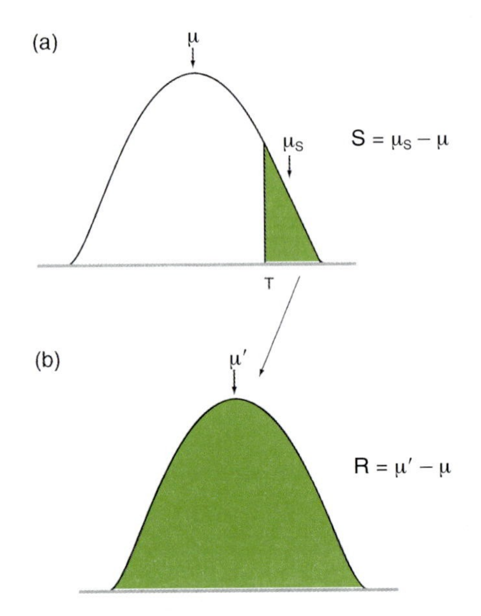

Narrow-sense heritability can be estimated from response to selection as the ratio between the selection response R and the selection differential S

Fig. 4.10 Truncation selection, in which only individuals larger than a certain threshold are allowed to reproduce. (a) The individuals represented by the shaded portion of the frequency curve reproduce, the others do not. The mean of the whole population is μ. T marks the threshold of the trait above which individuals reproduce. The mean of those reproducing is μ_s and the **selection differential** $S = \mu_s - \mu$. (b) As a result of selection, the frequency distribution moves to the right. The mean of the next generation is μ', and the **selection response** $R = \mu' - \mu$. The heritability can be estimated from the response to selection as $h^2 = R/S$.

Table 4.4 Summary of 1120 heritability estimates (from Mousseau and Roff 1987)

Trait category	Life history	Physiology	Behavior	Morphology
Mean hereditability	0.262	0.330	0.302	0.461

The heritability concept: a word of caution

Heritability tells us about the contribution of genetic *variation* to the phenotypic variation in a population. It does *not* tell us to 'what extent a trait is genetic or environmental'. Taken literally, this phrase is meaningless. Try for example to explain (if you can!) the meaning of the following statement: 'intelligence in humans is determined 70 per cent by genes and 30 per cent by environmental conditions such as upbringing and education'. Clearly, both genes and suitable environmental conditions are necessary for a human to exist and to possess any trait. We can only infer the extent to which *differences between individuals* are caused by genetic or environmental factors.

Heritability does *not* tell us to what extent a trait is genetic or environmental;

It is also not correct to apply an estimate of heritability from a particular population to the whole species. When environmental conditions are well controlled in an experimental population, there will be little environmental variance, and consequently heritability estimates will be higher than when environmental conditions are allowed to vary between individuals. Also, when a population is inbred, as is often the case in the laboratory or in domestic plants and animals, there will be relatively little genetic variation, and heritability estimates will be lower than in an outbred population. Estimates of heritabilities are only reliable for the population and environment in which they are measured (see Chapter 6).

and an estimate of heritability applies only to a particular population and a particular environment.

Evolutionary implications of quantitative genetics

The model developed above predicts that under directional selection the rate of phenotypic change depends on the narrow-sense heritability. When much of the total variability in the population is due to additive genetic effects, the response to selection can be rapid and strong. Little additive genetic variance implies a poor response to selection. Artificial selection experiments support the validity of the model. For example, Yoo (1980) got an impressive response to selection on abdominal bristle number in *Drosophila*, a trait with a high heritability (Fig. 4.11). After 90 generations of intense selection the mean increase in bristle number was 16 times the standard deviation in the original population (recall that less than one per cent of a normally distributed population is more than 3 standard deviations from the population mean). On the other hand, selection on rate of egg production in chickens, a trait with low heritability, has been largely unsuccessful (Nordskog 1977).

When heritability is high, the response to strong directional selection can quickly produce very large phenotypic changes.

Many examples from plant and animal breeding testify that strong

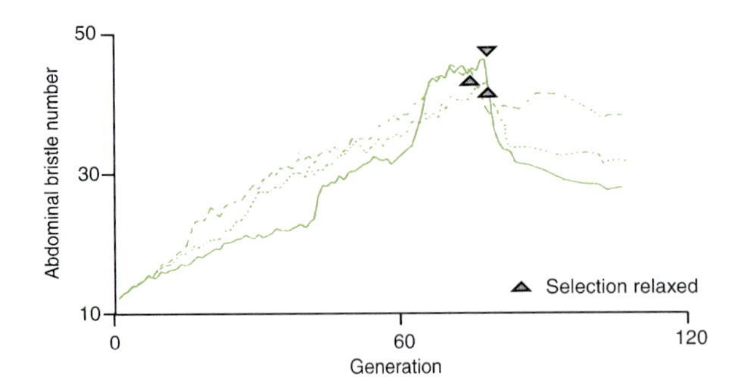

Fig. 4.11 Artificial selection on a trait with a high heritability can produce an impressive response. Here the result is shown from a long-term selection experiment on abdominal bristle number in *Drosophila*. The tendency to return to lower values after relaxation of selection indicates that the continued upward selection produced some negative side-effects. (From Yoo 1980.)

directional selection can produce large phenotypic changes. Just think of the different breeds of dogs and the many crops that are so different from their wild ancestors. For example, in domestic dogs, body mass can range from that of chihuahuas (weighing less than 1 kg) to Newfoundland dogs that weigh 80 kg. This range of body sizes is far greater than can be found among the wild species in the genus *Canus*, greater even than in the family Canidae (0.5–40 kg). These cases of fast change under strong directional artificial selection contrast sharply with what we know of long-term evolutionary change under natural selection, where rates of change estimated from the fossil record over very long periods of time are many orders of magnitudes slower (Gingerich 1983).

Continued strong responses to directional selection may be rare in nature because selection changes direction . . .

How can we explain these different rates of evolutionary change? One possibility is that continued strong directional selection, as applied in artificial selection, is rare in nature. A well-documented example is from the Grants' study of the seed-eating Galapagos finch *Geospiza fortis*. During a severe drought in 1977, large, hard seeds were more frequent, individuals with larger beaks survived better, and the strong selection for large beaks resulted in an increase of some 4%, or about 0.5 standard deviations. Seven years later, weather conditions had changed, small seeds were more frequent, and after this season, beak size decreased by about 2.5%, or about 0.2 standard deviations. This study demonstrated strong natural selection producing rapid phenotypic change, but the direction reversed within 7 years. The occasional occurrence of strong selection of variable direction may look like very slow change when averaged over long time periods.

. . . and because the response to selection in one trait can be braked by trade-offs with other traits.

Another reason for the slower rate of evolution in natural systems may be that the extreme trait values produced by directional selection reduce fitness because of negative side-effects of the alleles selected. A common experience in artificial selection is that some selection lines show a decrease in fertility, sometimes even leading to extinction of the line. Also, after relaxation of selection the selected trait often tends to return towards its original value. This indicates that natural selection is opposing artificial selection. Yoo's experiments (Fig. 4.11) are a good example. This expresses in genetic language what trade-offs express in physiological language. Traits involved in trade-offs cannot be selected very far in any direction without causing fitness to be lost through other traits to which they are connected.

Population and quantitative genetics are being integrated

Population genetic models assume one locus or a few loci with genotypic differences visible in phenotypic differences. Quantitative genetic models assume many loci with small additive effects on the phenotype. So presented, the two approaches have little in common. Population genetic theory is framed in genetic language and focuses on genotypes, quantitative genetic theory uses statistical terms and focuses on phenotypes. The two approaches arose and were developed independently, but it has been known since Fisher's work in 1918 that the principles of quantitative genetics can be derived from, and are consistent with, population genetics. Recent developments in molecular techniques are yielding practical applications of Fisher's theoretical insights.

Although population genetics uses genetic language and focuses on genotypes, and quantitative genetics uses statistical language and focuses on phenotypes,

For two reasons, the two approaches will be better integrated in the future. First, molecular analysis of loci affecting qualitative, Mendelian traits has uncovered many alleles with variability in phenotypic effects. This tends to obscure the precise genotype–phenotype relationships whose recognition is essential for population genetics. Because of this, quantitative genetics becomes relevant for the analysis of qualitative characters. Secondly, DNA sequences mapped to particular positions on chromosomes can be used to detect hitherto unknown loci that affect quantitative traits. When some of the genes affecting quantitative traits are known, the population genetic approach becomes relevant. Thus, population genetics is becoming more quantitative genetic, and quantitative genetics is becoming more population genetic. The best examples are from medical genetics and plant breeding.

the two methods are being integrated because molecular genetics describes many genes affecting traits previously thought to be determined by one gene, and helps to locate the genes affecting quantitative traits.

An example from medical genetics

Phenylketonuria (PKU) is a well-known genetic disease in humans, in which the conversion of the amino acid phenylalanine (Phe) to tyrosine (Tyr) is blocked, causing severe mental retardation in untreated patients. It results from homozygosity for a nonfunctional allele at the locus coding for the enzyme phenylalanine hydroxylase (PAH). Thus it is a recessive genetic disease, usually modeled as one locus with two alleles—a population genetic model.

However, molecular analysis of the PAH locus has revealed that it has

Phenylketonuria: a recessive genetic disease with at least 50 different alleles; variation in the disease can be explained by genetic and nongenetic variation.

at least 50 different alleles (Weiss 1993). Most of these seem to be 'normal'; they do not cause the disease. Among 206 European PKU patients, eight PAH alleles accounted for 64% of all PKU chromosomes. The effects of these eight alleles on PAH activity varied. In the homozygous state, four of them caused less than 1% of the normal PAH activity, but the others caused an enzyme activity of respectively 3%, 10%, 30%, and 50%. Since in randomly mating populations with so many alleles at a locus the heterozygosity is very high, most PKU patients are actually heterozygotes. The severity of the disease in heterozygous patients can be predicted reasonably well from knowledge of the diploid genotype and the independently measured allelic effects of the alleles that reduce PAH activity, by assuming additive gene action.

Thus, when the phenotype is measured precisely, we find variation in the disease that we can try to explain from genetic and nongenetic variation. This is basically a quantitative genetic approach, but unlike most quantitative traits, here we already know many of the alleles responsible for the disease.

Detecting quantitative trait loci (QTLs)

To detect quantitative trait loci, find associations between a marker sequence and extreme trait values.

We can now characterize individuals with respect to many molecular markers. These are DNA sequences (such as restriction sites, i.e. sites where an enzyme that recognizes a specific DNA sequence will cut the sequence into two parts) which have a known chromosomal position, the presence or absence of which can be determined in an individual. If we call these markers A, B, etc. and use a capital letter to denote presence and a lower case letter for absence, individuals can be classified as being A or a, B or b, and so forth. These marker sequences are not genes, just positions on the chromosome where variation in nucleotide sequence occurs, but they segregate as normal Mendelian alleles. The basic principle of QTL (quantitative trait locus) detection is to find associations between the marker genotype and extreme phenotypic trait values, indicating that a marker is located closely to a QTL affecting the trait. If the association between the marker and the QTL is sufficiently strong, it may be possible to find the QTL and to infer from its DNA sequence a possible function of the gene product.

QTL analysis converts quantitative genetics into population genetics. The pattern found: quantitative traits are determined by a few major and many minor genes.

Using such methods, we can take a trait whose genetic determination initially appeared to be quantitative, locate the genes that are affecting it, and then understand their evolution as a problem in population genetics. When this has been done, the pattern found so far has been that quantitative traits are determined by a few genes of large effect, sometimes only one or two, and many genes of small effect, which are often called modifiers.

Summary

This chapter describes how to understand and predict evolutionary change by analyzing the genetic changes that underlie changes in traits; for only genetic changes endure.

- The two approaches to analyzing genetic change are population genetics, which can be applied to genes whose phenotypes are visible and discrete, and quantitative genetics, which can be applied to continuously varying traits whose genetic determination is obscure.
- The progress of a rare, advantageous mutation towards fixation is slow–fast–slow: slow when it is rare or common, and fast at intermediate frequencies. The evolution of resistance of insects to pesticides and of bacteria to antibiotics are good examples.
- Genetic change is faster in haploid asexual than in diploid sexual populations.
- In a diploid sexual population, advantageous mutations will spread much more slowly when they are recessive than when they are dominant.
- Quantitative genetics teaches us that the variation in a population is caused in part by genetic variation and in part by environmental variation. The contribution of each can be estimated.
- Additive genetic variation responds directly to selection; the proportion of the total variation for a trait in a population that is additive genetic variation is defined as the heritability of that trait.
- Traits contributing strongly to reproductive success have lower heritabilities, and traits less relevant to fitness have higher heritabilities.
- Quantitative genetics has successfully predicted the results of artificial selection experiments, the responses to selection in domestic plants and animals, and has helped us to interpret evolutionary change in quantitative traits in wild populations.
- Population and quantitative genetics are becoming integrated. Diseases previously thought to be caused by simple Mendelian loci are now being revealed by molecular genetics to be caused by many alleles that combine in ways reminiscent of quantitative traits. The analysis of quantitative traits with molecular markers is revealing quantitative trait loci (QTLs) that behave like Mendelian genes and allow us to use population genetics to understand evolutionary change in continuous traits.

Genetic change can only occur if there is genetic variation. The origin and maintenance of genetic variation are discussed in the next chapter.

Recommended reading

Fisher, R. A. (1930). *The genetical theory of natural selection.* Oxford University Press, Oxford.

Griffiths, A. J. F., Miller, J. H., Suzuki, D. T., Lewontin, R. C., and Gelbart, W. M. (1996). *An introduction to genetic analysis,* (6th edn). W. H. Freeman, New York.

Haldane, J. B. S. (1990). *The causes of evolution,* (reprint edn). Princeton University Press, Princeton.

Hartl, D. L. and Clark, A. G. (1989). *Principles of population genetics,* (2nd edn). Sinauer, Sunderland, Massachusetts.

Roff, D. A. (1997). *Evolutionary quantitative genetics.* Chapman & Hall, London.

Questions

4.1 Suppose in a diploid random mating population 2 alleles occur at a locus A. A_1 is very common (frequency $p = 0.95$). Due to recent environmental change A_1 has become disadvantageous and is selected against. Consider two cases:

		A_1A_1	A_1A_2	A_2A_2
(1) A_1 is recessive;	fitness	$1 - s$	1	1
(2) A_1 is dominant;	fitness	$1 - s$	$1 - s$	1

In which case will A_1 reach a frequency of 0.5 faster? Why?

4.2 A botanist measured seed weight and estimated its narrow-sense heritability in three widely separated *Phlox* populations.

	mean seed weight	narrow sense heritability
population A	15 mg	0.60
population B	12 mg	0.65
population C	17 mg	0.58

Because the heritabilities were high and approximately equal, he concluded that the differences in seed weight between the populations were largely due to genetic differences. Do you agree? Explain.

4.3 Suppose that in a human population a mutation occurs that increases fitness by 10%. Do you think that this mutation will reach a frequency of 0.95 in less than a 1000 years? If not, what is a more realistic time span needed for this change? What factors influence the length of this period?

Chapter 5
The origin and maintenance of genetic variation

Introduction

Without genetic variation, there can be no evolution.

Chapter 4 discussed the genetic response to selection, but it made an important assumption—that genetic variation was present. If all individuals in a population were genetically identical and produced offspring identical to themselves, evolutionary change would be impossible. Adaptive evolution requires heritable differences correlated with reproductive success, and neutral evolution requires heritable differences uncorrelated with reproductive success, to change the genetic composition of a population. Genetic variation is essential for evolutionary change.

The response to selection is limited by the amount of genetic variation with impact on reproductive success.

Moreover, the amount of genetic variation influences the rate of evolutionary change. If there is very little genetic variation, the rate of evolutionary change is limited by rare favorable variants. Most individuals belong to a standard type (termed '**wild type**' by the classical geneticists). Natural selection removes deleterious variants, and occasionally a favorable variant spreads through the population. If, in contrast, a population contains much genetic variability, then individuals differ genetically in many traits. Classification of most of the population as 'wild type' is not possible, and the rate of evolutionary change is not limited by the presence of favorable variants. Instead, the population might face a different problem. If great genetic variability were associated with great variability in fitness, selection could be too strong, removing so large a fraction of the population each generation, due to low survival or fertility, that the population might go extinct.

The correspondence between genetic variation and variation in fitness is a critical issue.

Clearly, how much genetic variation is present, how it is maintained, and how much of it is correlated with fitness are crucial issues. In this chapter we first consider the origin of genetic variation, then its maintenance, and finally its relevance for adaptive evolution.

Mutation generates genetic variation

Optimal mutation rates

All genetic variation originates from mutations, most of which arise from errors during DNA replication. An intermediate mutation rate is optimal for the evolution of adaptations. This rate differs among species and genes.

The evolution of optimal mutation rates is easier to achieve in asexual lineages than in sexual species.

All genetic variation originates from mutation. A mutation is a hereditary change in the DNA sequence or in chromosome number, form, or structure. While mutations are necessary for evolution, too frequent mutation can prevent evolution, for with a very high mutation rate, not enough of the well-adapted genes would be transmitted unchanged to the next generation. Their loss would prevent the evolution and maintenance of adaptations. It follows that there will be an optimal mutation rate: not too few and not too many mutations. This optimal rate need not be the same for all species and all genes.

Sexual and asexual species differ in this respect. In asexual organisms, where the whole genome is transmitted intact to the offspring, evolution of the mutation rate is easy in principle, for the genes that affect the mutation rate stay together with the genes whose mutation rate they adjust. If conditions favor a change in mutation rate, a mutation that changes the mutation rate at all loci enjoys a selective advantage and will increase in frequency because it stays associated with a genotype that is benefitting from a better mutation rate. In sexual organisms, however, a gene affecting mutation rate does not remain associated with the rest of the genome on which it has its effect, because, every generation, recombination can separate the gene determining mutation rate from the genes that mutate. Therefore evolution of the mutation rate to a value that maximizes the rate of adaptive evolution occurs easily in asexual species but is more difficult in sexual species.

Rates of mutation

The average mutation rate per base pair per replication is about 10^{-9} —one per billion—in organisms with DNA genomes, but this figure varies 1000-fold or more between different genes in the same genome. Certain sequences, called mutational hotspots, are particularly prone to mutation.

Mutations of large effect in humans, mice, and *Drosophila* occur with a frequency of about 10^{-5} per gamete.

This figure is derived from measurements on DNA sequences and does not tell us about the phenotypic consequences of the mutations for the fitness. A classical way to determine mutation rates is to observe the spontaneous occurrence of aberrant phenotypes known to result from single allele changes. In this way spontaneous mutations to known human diseases have been estimated to occur with a frequency of about 10^{-5} (1 per 100 000) per gamete. Similar figures have been obtained from studies in mice and *Drosophila*.

Another approach has been used by *Drosophila* geneticists (Mukai 1964; Mukai *et al.* 1972; Houle *et al.* 1992). They accumulate recessive (or partially recessive) mutations for many generations on a chromosome that is kept heterozygous and prevented from recombining. At regular intervals the fitness effect of the chromosome is measured in the homozygous condition. The results suggest that the deleterious mutation rate in *Drosophila* is about one mutation with a small deleterious effect per zygote and that mildly deleterious mutations greatly outnumber lethal ones.

Mildly deleterious mutations occur in *Drosophila* at a rate of one or more per zygote . . .

Because humans have much more DNA than *Drosophila*, and mutation rates per locus per generation are similar in humans and *Drosophila*, an average human might carry tens of new mutations, but many of them would be in DNA that did not code for proteins. A recent study analyzed the amino acid changes in 46 proteins in the human ancestral line after its divergence from the chimpanzee (Eyre-Walker and Keightly 1999). The results suggested an average of 4.2 new amino acid altering mutations per person per generation, of which at least 1.6 were deleterious. As further discussed in Chapter 7, this is considered to be close to the upper limit that can be tolerated by a species with limited reproductive capacity, unless there is a stage in the life cycle that eliminates most mutants. The atresia of egg follicles in human females, which reduces the number of potential gametes from millions to hundreds, may have this function.

and in humans at about the same rate. This is close to the upper limit that can be tolerated.

Mutation rates in males and females

Recent molecular data on human genetic diseases suggests higher point-mutation rates in males than in females, in some genes. Extreme examples are achondroplasia and Apert syndrome, two dominantly inherited disorders. In both, all new mutations occurred in the father in more than 50 cases. The higher male point-mutation rate may be related to the much higher number of cell divisions in the male than in the female germ line. That point mutations are associated with cell division makes this explanation plausible.

More mutations occur in males than in females.

The effect of recombination on genetic variability

Recombination during meiosis in a sexual individual creates haploid genotypes (gametes) that differ genetically from the gamete haplotypes that formed the individual. Thus an *AaBb* individual, grown from a zygote that resulted from the fusion of an *AB* and an *ab* gamete, will produce gametes both with the parental haplotypes *AB* and *ab* and with the recombinant haplotypes *Ab* and *aB* (Fig. 5.1).

When two DNA sequences are located on different chromosomes, they segregate independently at meiosis. When they are on the same chromosome, they segregate together unless a crossover occurs between them. In either case recombination between genes may result in new combinations of genes on a chromosome or in an individual. When the DNA sequences are parts of the same gene, recombination will be rare,

Recombination during meiosis creates tremendous genetic diversity among offspring. It is a diversity of combinations of loci.

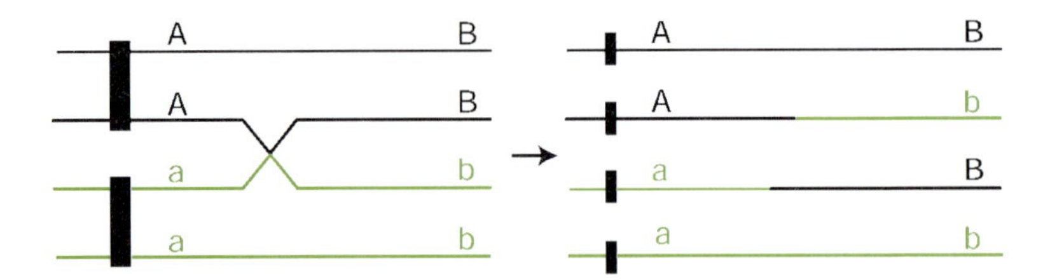

Fig. 5.1 Crossing-over of nonsister chromatids during meiosis, resulting in and exchange of homologous chromosome parts by breakage and reunion, may produce chromosomes with a new allelic composition. Recombination produces new genetic variants

because the probability of crossing over between closely linked sequences is very small. When intragenic recombination does occur, it may produce a new allele, not present before in the population. Thus recombination affects genetic variation among individuals when combinations of many loci are considered. Only rarely does it affect variation at a single locus.

Recombination generates new genetic combinations that produce phenotypes outside the range of the starting population.

The effectiveness of recombination at converting potential into actual variability is spectacular in animal and plant breeding where, starting from a uniform population established by crossing two inbred lines, individuals can be selected in a few generations with phenotypes well outside the range of the original population. The examples mentioned in Chapter 4 of large and rapid phenotypic change under directional selection in *Drosophila* and in dogs illustrate this point. Selection can be so effective because the traits are affected by genes at several loci whose recombination generates great multi-locus variability.

The amount of genetic variation in natural populations

We need to know how much genetic variation there is that affects fitness.

To understand adaptive evolution, we must know how much genetic variation there is in natural populations and how much of it affects individual reproductive success. Since about 1920, this has been one of the major questions of population genetics.

But it is hard to measure because the genotype–phenotype and phenotype–fitness relations are poorly understood.

Attempts to answer this question have been hampered by two related problems. Both stem from our ignorance of the relationships between genotypes and phenotypes. First, except where large phenotypic differences show Mendelian segregation patterns in crosses, we do not know the genetic variation that underlies the observable phenotypic variability. Secondly, we can only measure fitness effects of individual genetic variations when they are fairly large. The first problem—not knowing the genetic variation underlying phenotypic differences—was the main obstacle to estimating the amount of natural genetic variation until molecular methods were introduced in the 1960s. Then the problem of measuring fitness effects took priority.

The use of molecular methods

Molecular methods revealed tremendous genetic variation in natural populations.

In the mid-sixties biologists started to apply the biochemical technique of protein electrophoresis to samples of individuals from natural populations of animals and plants. The technique separates proteins on the basis of their mobility through a gel under influence of an electric current. Similar proteins that differ in their net electrical charge move at different speeds. This can be observed by staining the proteins after they have moved through the gel for some time. The technique greatly improved estimates of genetic variation, for it made visible variation at loci that was invisible in unmanipulated phenotypes. The amount of genetic variability in populations is usually measured by the **genetic diversity**, *h*, defined as the probability that two alleles chosen at random from all alleles at that locus in the population are different. The easiest way to compute this probability is by seeing that it equals one minus the probability that two randomly chosen alleles are identical. Denoting the frequency of allele *i* by x_i, we get

$$h = \sum_{i \neq j} x_i x_j = 1 - \sum_i x_i^2 \qquad [5.1]$$

When the population mates randomly, the two alleles at a locus in an individual form a random pair. Therefore under random mating the genetic diversity, *h*, equals the actual **heterozygosity**, *H*, the proportion of the population that is heterozygous at a locus. If we average over loci, *H* is also interpreted as the average proportion of loci that are heterozygous per individual.

Electrophoretic heterozygosity is about 10% and varies among populations and species.

Protein electrophoresis has been applied to many samples from populations of many species. The results of many protein electrophoretic studies suggest that heterozygosity, *H*, is about 10% and varies between populations and species (Fig. 5.2).

Since about 1980 more refined molecular techniques have yielded measurements of genetic variability at higher resolution. One approach is to isolate DNA and cut it with restriction enzymes that recognize particular short sequences. The resulting DNA fragments can be separated by gel electrophoresis according to molecular weight and visualized as stained bands. Differences between homologous chromosomes in the location of restriction sites (the short nucleotide sequences recognized by the restriction enzymes) can thus be measured. Another approach is to sequence the DNA to get the nucleotide sequence itself (e.g. . . . AATGCTTCGA . . .). This became practical with the development of the polymerase chain reaction (PCR), which amplifies small amounts of DNA, even the DNA from a single cell.

Both restriction analysis and sequencing allow us to estimate genetic variability at the level of nucleotides. The genetic diversity, *h*, is not a

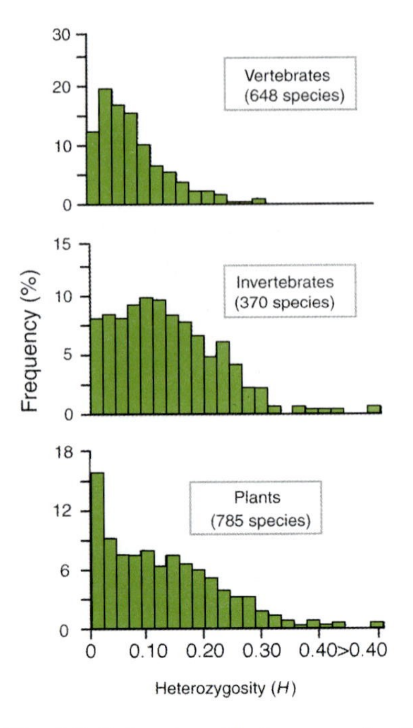

Fig. 5.2 Estimates of heterozygosity, H, based on protein-electrophoretic surveys in many different species (from Avise 1994).

Nucleotide diversities are about 0.0001–0.01 within populations.

good measure of the variability of DNA sequences, for when long homologous sequences are compared, all sequences differ from each other, and h is close to 1. A better measure is the **nucleotide diversity**, π, the average number of nucleotide differences per site between randomly chosen pairs of sequences:

$$\pi = \sum_{ij} x_i x_j \pi_{ij} \qquad [5.2]$$

where x_i and x_j are the frequencies of the pair of DNA sequences i and j, and π_{ij} is the fraction of nucleotide differences within this pair of sequences. Nucleotide diversities are typically in the range 0.0001–0.01.

Molecular methods do not *solve* the problem of deducing the underlying genetic variation from observed phenotypic variability: they *circumvent* it. Molecular methods give direct access to genomic information without using phenotypic variation to draw conclusions about the genotype. They tell us how much genetic variation is present in a particular part of the genome, but they do not tell us how this genetic variation affects phenotypic variation. In a sense, the patterns of stained bands in a gel representing protein variants or pieces of DNA are phenotypes made visible by molecular techniques, but the relationship of those bands to fitness is rarely clear.

How much molecular sequence variation is selectively neutral?

Thus molecular methods have revealed enormous genetic variation. How much of this variation causes fitness variation and serves as a substrate for adaptive evolution? Or, to put the same question the other way round, how much molecular genetic variation is selectively neutral? This is an empirical question. In Chapter 3 we discussed Kimura's neutral theory, which claims that most variation at the molecular level is neutral. His theory caused considerable controversy about the relative importance of genetic drift and adaptive evolution in molecular evolution. We next discuss some attempts to measure fitness consequences of molecular genetic variation. Then we consider some models that aim to understand how mutation, genetic drift, and natural selection affect the level of genetic variation in a population.

Evidence of natural selection from DNA sequence evolution

There is good molecular evidence for natural selection at functionally important sites in DNA molecules.

No one believes that all genetic variation is selectively neutral. The abundant evidence of adaptation through natural selection (Chapter 2) must be reflected in DNA sequences. The question is how much of the variation in DNA sequences can be considered neutral. In a few cases strong indirect evidence of adaptive evolution has been obtained from comparisons of homologous DNA sequences. For example, Hughes and Nei (1989) compared the DNA sequences of the antigen recognition sites of major histocompatibility (MHC) genes of humans and mice (i.e. genes involved in immune responses). Rates of substitution were estimated by counting the number of nucleotide differences between homologous stretches of DNA, and synonymous and nonsynonymous substitution rates were distinguished. Nonsynonymous substitutions change the amino acid coded; synonymous substitutions do not. Synonymous changes are usually more frequent than nonsynonymous ones, because amino acid replacements often reduce protein function and are selected against. Based on 36 protein-coding genes, the mean rate of synonymous substitution had been estimated to be five times higher than for nonsynonymous substitution (Li and Graur 1991). Hughes and Nei, however, found that in the antigen recognition region more *non*synonymous than synonymous substitutions had occurred, indicating natural selection for fast change in the antigen recognition properties of histocompatibility genes. They found variation in DNA sequences that was not neutral.

Genetic drift and adaptive evolution can both explain most molecular evolution data

But the debate on whether molecular evolution is largely neutral or adaptive is unresolved.

A few other cases are known where DNA sequence evolution probably reflects adaptive evolution, but most DNA sequence data allows both an adaptive explanation, invoking the action of natural selection, and an explanation based on nearly neutral mutations that are governed largely by genetic drift. It is remarkable that such different explanations overlap so much in their predictions. Despite much data, whether molecular evolution is largely neutral or largely adaptive is still not resolved.

Equilibrium models of the maintenance of genetic variation

Most explanations of the maintenance of genetic variation assume a balance between the forces that increase and decrease it.

The theories that try to predict the quantitative effects of genetic drift and selection on genetic variability all share the important assumption that populations are in genetic equilibrium. They therefore concentrate on the equilibrium states in the genetic change caused by mutation, selection, migration, and drift. There are two reasons for this. One is mathematical convenience. The other is that the periods during which the system is in a state of change, for example when an advantageous allele is spreading through a population, are short compared to the periods when an allele is fixed in the population. Whether this assumption is valid remains to be seen, and we will return to it. For the moment we assume equilibrium, which implies a balance between the forces that increase and decrease genetic variation.

Genetic diversity at mutation–drift balance

Mutations with no impact on fitness that are not linked to genes under selection only experience drift.

To analyze drift we assume that every mutation is unique.

In 1968 Kimura postulated that most of the evolutionary changes at the molecular level are neutral or nearly neutral. This idea is plausible, for many nucleotide substitutions do not cause phenotypic change because they do not change an amino acid or because amino acid changes do not affect protein function. Mutations that do not change the fitness of their carriers are subjected solely to genetic drift.

We now consider the fate of such mutations as predicted by the unique mutation model, also known as the **infinite-allele model**. This model assumes that every mutation is unique in the sense of not yet existing in the population, a reasonable assumption if we characterize the mutations not by their phenotypic effect but by their DNA sequence. Because the probability of a specific nucleotide change in DNA is extremely small (10^{-9} per replication), the same nucleotide change in a different individual would probably represent a new allele, for some of the 1000 or so surrounding nucleotides forming the complete gene would also differ. Thus the assumption is plausible.

The fate of neutral unique mutations

Every new, unique mutation will eventually either disappear from the population or become fixed. Fixation of a mutation means that all homologous chromosomes in the population carry a copy of the mutation. The process whereby a new mutation eventually becomes fixed is called **gene substitution**. The probability that a new mutant allele will reach fixation is the **fixation probability**, and the time it takes to become fixed is the **fixation time**, measured in generations. In the infinite-allele model several theoretical results have been derived for these quantities.

One important and remarkably simple result is that the substitution rate for neutral mutations equals the mutation rate. This can be shown as follows. The probability that an allele becomes fixed by genetic drift is equal to its current frequency. Thus, in a diploid population of N individuals a new mutation has a probability of $\frac{1}{2N}$ to become fixed and of $1 - \frac{1}{2N}$ to be lost. Thus most new neutral mutations will be lost. If a new mutation occurs with probability u per locus per generation, there will be $2Nu$ new mutations in the population each generation. Since each of these mutations will be fixed with probability $\frac{1}{2N}$, the substitution rate for neutral mutations equals $2Nu \cdot \frac{1}{2N} = u$. Thus one neutral mutation will reach fixation on average per locus every $\frac{1}{u}$ generations, which in many species is a very long time because u is of the order of 10^{-5} or 10^{-6}. The substitution rate is independent of population size, for the higher fixation probability of a mutation in a small population precisely compensates for the smaller total number of mutations in smaller populations.

The substitution rate for unique neutral mutations equals the mutation rate.

That neutral mutations are fixed at a constant rate, equal to the mutation rate, supports the concept of a molecular clock (Chapter 3), which states that the molecular evolution of proteins occurs at a constant rate in all lines of descent. Molecular clocks play an important role in the estimation of divergence times and the construction of phylogenetic trees (Chapter 12).

For the mutations that do eventually become fixed, the expected time to reach fixation is $4N_e$ generations. Here the symbol N_e stands for **effective population size**. In this context this is the size of an ideal population consisting of individuals with equal reproductive success that would experience the same amount of genetic drift as a real population of size N. In real populations individuals differ in reproductive success. Real populations may also be subdivided and vary in size, factors that also influence the rate of genetic drift. In natural populations the effective size is therefore often much smaller than the actual size.

The time to fixation by drift is four times the effective population size.

There is a big difference between the fixation time of the mutations that get fixed and the survival time of the mutations that get lost. In a population averaging 10^6 individuals, the fixation time is about 10^6 generations, whereas mutations that are lost survive on average less than 10 generations. Thus the theory of unique neutral mutations yields a picture of a continual flux of genetic variants through a population. There is a steady input of new mutations, most of which are quickly lost, while a few drift very slowly to fixation, a balance between mutation and drift (Fig. 5.3).

It takes much longer to fix a neutral mutation than to lose it, and many more are lost than fixed.

Genetic diversity at mutation–selection balance

If we characterize mutations not by their DNA sequence but by their phenotypic effect, we can exchange the model of unique mutations for one of recurrent mutations. In some cases, several alleles that differ in their nucleotide sequence all produce the same phenotype, e.g. phenylketonuria (Chapter 4). When these alleles are lumped into one category, a recurrent mutation model is justified. This is the classical population genetic approach to calculate the expected frequency of deleterious mutations.

Classical—as opposed to molecular—population genetics assumes recurrent, not unique, mutations.

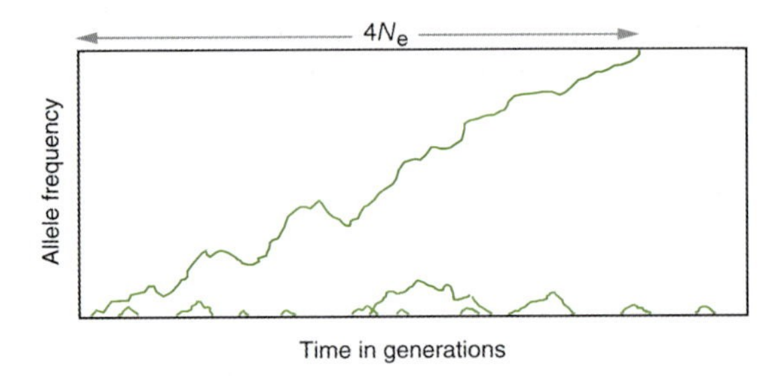

Fig. 5.3 The fate of neutral mutations. Most get lost very quickly, while a few drift very slowly to fixation (expected time to fixation is $4N_e$ generations). (From Nei 1987.)

The fate of recurrent deleterious mutations

There are simple formulae to predict the frequency of genes held in selection–mutation balance,

We now calculate the expected frequency of a recurrent deleterious mutation at mutation–selection balance in a haploid population. If the population consists of N individuals and the mutation rate per locus per generation equals u, each generation Nu new mutations will occur at a given locus. At mutation–selection balance, the same number of mutations must be removed by selection each generation. If the average fitness of mutant individuals is $1 - s$ relative to a fitness 1 of nonmutant individuals, and the relative frequency of mutant individuals is given by q, selection will remove Nsq mutant alleles each generation. Therefore at equilibrium $Nu = Nsq$, from which it follows that the frequency of mutant alleles at mutation–selection balance is

$$q = \frac{u}{s}$$

For a diploid population a similar calculation yields

$$q = \sqrt{\frac{u}{s}}$$

for recessive mutations and

$$q = \frac{u}{s}$$

for dominant mutations.

but difficulties in estimating dominance and selection make it hard to see how well the formulae work.

Because the selective disadvantage s of a mutation is hard to estimate accurately, it is not easy to test this theory in real populations. The most reliable estimates are for mutations where affected individuals do not reproduce ($s = 1$), as in several human hereditary diseases. Unfortunately, a small fitness effect in heterozygotes causing a deviation from strict recessiveness or dominance can change the expected equilibrium frequency considerably. In *Drosophila*, strictly recessive mutations are rare; frequently there is some effect in heterozygotes. Moreover, some mutant phenotypes may be caused by mutations at several loci. For example, the chlorate resistance phenotype in the fungus *Aspergillus nidulans* can be caused by mutations at any of about 10 different loci. Similarly, several human genetic diseases which until recently were thought to be caused by a mutation at one locus, have been shown to be genetically heterogeneous: mutations at several loci can cause the same disease.

The assumption of an equilibrium between mutation and selection may also be questionable for many populations. Migration and genetic bottlenecks can move a population well away from equilibrium, and the return to equilibrium can take a long time. Many human genetic diseases caused by single-gene mutations, such as enzyme deficiencies leading to metabolic disorders, show strikingly different frequencies in different populations. This suggests that many populations are not in mutation–selection equilibrium for these loci, probably due to historic population events. For example, phenylketonuria (PKU) has a higher incidence in eastern than in western Europe. The Scandinavian countries, especially Finland, and Japan show a very low PKU frequency, whereas the disease is found at high frequency in Ireland.

Whether or not it leads to equilibrium, the mutation–selection balance is a powerful mechanism for generating genetic variation. It guarantees a certain amount of genetic variability at all loci. As we will see below, this is particularly important for quantitative traits, which are affected by many loci.

Genetic diversity at a balance of different selection forces

So far we have considered the contribution to genetic diversity of alleles that are either selectively neutral or selected for or against. Neutral alleles occur at intermediate frequencies because they drift; selected alleles because they are on their way to fixation or loss. In both cases the intermediate frequencies do not persist for a long time. However, selection may also favor intermediate frequencies, maintaining variation in the long term. Then different selection forces pull in opposite directions so that their effects balance.

Two ways to achieve a balance of selective forces are **heterosis** and negative **frequency-dependent selection.** Heterosis means that the heterozygote is fitter than either homozygote. Selection against the two homozygotes (and thus against both alleles) balances at a certain frequency. Negative frequency-dependent selection favors rare alleles over common ones, causing a rare allele to increase in frequency and a common one to decrease until both have equal fitness.

Two situations, heterosis and negative frequency-dependent selection, create a balance of forces that maintain alleles at intermediate frequencies.

Heterosis

The possibility of a stable allelic polymorphism due to superior fitness of heterozygotes has been recognized since the 1920s. For the derivation

Heterosis: heterozygotes have higher fitness.

we use Table 4.2, from which are taken the assumptions about genotype fitnesses and the formula for frequency change reproduced below.

Genotypes	A_1A_1	A_1A_2	A_2A_2
Relative fitness	1	$1 + hs$	$1 + s$

Change in frequency of allele A_2 over one generation:

$$q' = q\left(\frac{1 + s(q + hp)}{W}\right), \text{ with } W = p^2 + 2pq(1 + hs) + q^2(1 + s).$$

Heterosis leads to stable intermediate allele frequencies that tell us nothing about the intensity of the selection maintaining them.

Assuming $s > 0$, heterosis implies $h > 1$. At the balance between selection against both alleles we expect a stationary allele frequency: $q' = q$. Substitution into the above formula for allele frequency change yields after some rearrangement (remembering that $p^2 + 2pq + q^2 = 1$) the following formula for the equilibrium frequency $\hat{q}$ of allele A_2:

$$\hat{q} = \frac{h}{2h - 1}. \qquad [5.3]$$

Because Equation 5.3 does not include s, the equilibrium value tells us nothing about the intensity of selection. Strong and weak selection forces balance at the same allele frequency.

A classical example: sickle-cell hemoglobin.

The classical example of heterosis is provided by malaria-associated hemoglobin variants in humans. The components of hemoglobin molecules are coded by the α-globin cluster on chromosome 16 and the β-globin cluster on chromosome 11. Several mutant hemoglobin variants produce more- or less-serious forms of anemia as homozygotes. The best-studied variant is sickle-cell hemoglobin (HbS), which takes its name from the altered shape of the red blood cells that contain it. Where malaria is common, the homozygote for 'normal' hemoglobin is susceptible to malaria, the homozygote for the HbS allele suffers from serious anemia, but the heterozygote has little anemia and is protected from malaria. This heterosis explains the high frequencies of the HbS allele in parts of Africa and India where malaria is common, and the low frequencies of the allele in nonmalarial environments. Several different mutant hemoglobin alleles have been found that give a similar protection from malaria in the heterozygous condition (Fig. 5.4). The main ones are HbS (mainly occurring in Africa and India), HbC (West Africa), HbE (mainly South-East Asia), and HbD (northern parts of India and Pakistan).

But the sickle-cell alleles are not in equilibrium.

The dynamics of competition between the hemoglobin alleles is complex and much depends on the poorly known fitness effects of the various homozygotes and heterozygotes. Moreover, because populations containing the different alleles are not well mixed, the long-term outcome is difficult to predict. One of these mutant alleles may ultimately out-compete the others; a stable polymorphism involving two or more alleles is also possible. As with mutation–selection balance, the equilibrium assumption is not fulfilled.

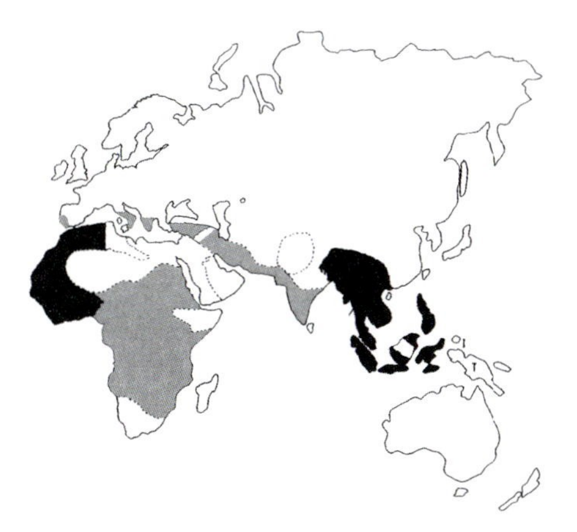

Fig. 5.4 Areas where particular alleles for abnormal hemoglobins are common (from Vogel and Motulski 1979).

Frequency-dependent selection

When selection favors rare types, rare types increase and common types decrease until a balance is reached.

Many traits of organisms are involved in interactions between individuals, both conspecifics and individuals of other species. Such traits include those involved in sexual, social, and antagonistic behavior, and in predator–prey, host–parasite, or mutualistic interactions. Natural selection works on these traits in a frequency-dependent fashion, when the success of a particular behavior or interaction depends on what others are doing. Selection favoring rare types may then be common. Consider, for example, two genotypes that differ in the type of prey they eat. With a constant supply of the two types of prey, whenever one of the genotypes is rare, its fitness is high because plenty of food is available, whereas the fitness of the common genotype is low because of strong competition for food. Thus rare genotypes become more frequent, common genotypes become less frequent, and at some population composition the two balance.

An example: scale-eating fish in Lake Tanganyika.

A beautiful example of negative frequency-dependence is provided by Hori's (1993) study of scale-eating fish in Lake Tanganyika. Several species of the scale-eating cichlid fish genus *Perissodus* have asymmetrical mouth openings (Fig. 5.5), some individuals being 'left-handed' (sinistral) and others 'right-handed' (dextral).

This lateral asymmetry of the mouth is an adaptation for efficiently tearing off the scales of prey. *Perissodus microlepis* approaches its prey from behind to snatch scales from the flank. Dextral individuals always attack the victim's left flank, and sinistral ones the right flank. Handedness appears to be determined genetically by a simple one locus–two allele system, with dextrality dominant over sinistrality. Hori examined the ratio of handedness in the lake *P. microlepis* over an 11-year period. He found that, on average, both types were equally frequent, with small temporal oscillations in the ratio with a period of 5

Fig. 5.5 Asymmetrical mouth opening of a scale-eating cichlid fish, *Perissodus microlepis*. A right-handed individual is attacking its prey from the left side. (From Hori 1993.)

years. During periods when sinistral individuals were more numerous, the prey suffered scale-eating from dextral individuals more often than from sinistral individuals, and vice versa. When the prey were more alert to the common type of predator, they let predators of the rare type be more successful than those of the common type: a clear case of negative frequency-dependent selection.

Negative frequency-dependence in a model.

That negative frequency-dependent fitnesses can maintain genetic variability is easily demonstrated in a simple population genetic model that can be analyzed graphically (Fig. 5.6). Assuming haploid organisms (diploids require a slightly more complicated model), we can let the relative fitnesses of the two genotypes decrease linearly with frequency. This is a simple way of ensuring that fitness is negatively frequency-dependent.

Genotypes	A_1	A_2
Relative fitnesses	$1 - ap$	$1 - bq$

Figure 5.6 shows that when A_1 is rare, it has a higher fitness than A_2 and therefore increases in frequency. Similarly, at high frequency it decreases, so that when both fitnesses are equal a stable frequency equilibrium must exist.

How much evidence is there for negative frequency-dependent selection? The botanical and agricultural literature reports that the performance of a genotype or variety often depends on the neighboring types. If a plant surrounded by plants of its own type performs worse than if surrounded by plants of different type, there is an advantage to rarity, because rare plants mostly have neighbors of another type. The mechanism is probably that different types have different resource requirements, so that resource competition is strongest between plants of the same type.

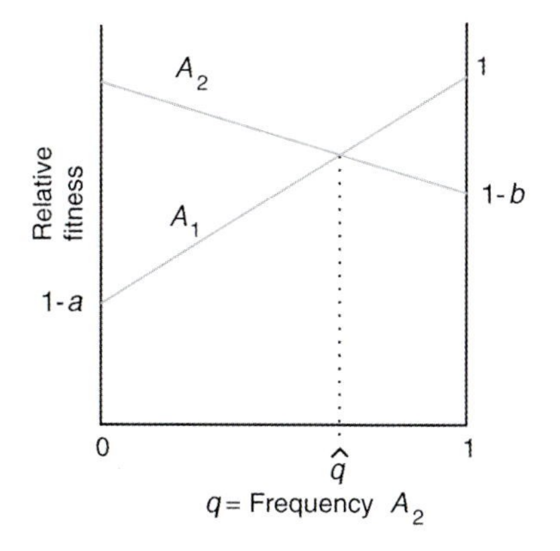

Fig. 5.6 The relative fitnesses of genotypes A_1 and A_2 in a haploid population are shown as linearly decreasing functions of their relative frequency (respectively $1 - ap$ and $1 - bq$). Such negative frequency-dependent fitnesses generate a stable allele frequency equilibrium at $\hat{q} = a/(a + b)$.

One of the clearest experimental demonstrations comes from the work of Antonovics and Ellstrand (1984). They planted clonal tillers of the grass *Anthoxanthum odoratum* into natural sites so that some test plants were surrounded by their own genotype and some by plants of a different genotype. In two different experiments, clones in minority situations had approximately twice the fitness of those in majority genotype conditions. They did not discover the reasons for the minority advantage.

Competition among genotypes in grass: good evidence of selection for rare types.

Genetic diversity of complex quantitative traits

Stabilizing selection is very common

Many quantitative characters have optimum properties: some intermediate value is the best, smaller and larger values reduce function. The basic reason is probably trade-offs between different functions of a trait: a higher metabolic rate provides more potential for growth and activity, but a lower rate requires less food and resources; stronger bones give better support for the body, but lighter bones are cheaper to make and require less energy to carry; higher blood pressure promotes rapid transport of substances via the bloodstream, but lower blood pressure is better for the vascular system; laying more eggs means more potential offspring, but laying fewer eggs costs less and allows better care of each offspring.

Many quantitative characters experience selection for an intermediate optimum.

Artificial selection shows that genetic variability exists to change body size: compare the sizes of the different breeds of dogs. Much of the genetic potential for variation in body size in dogs must have been present in wolves, from which dogs descend. Yet wolves have been of a constant size for millions of years, which must mean that bigger and smaller wolves are selected against.

If artificial selection of wild species can produce big changes, both up and down, then natural selection must have been stabilizing.

Natural selection often favors some intermediate trait value at which the net benefit of the ensemble of different functional aspects of the trait is highest. This means that often selection is stabilizing: deviations from the optimum phenotype are selected against (Fig. 5.8). This is even likely to be true when paleontological evidence indicates a long-term directional change. For example, horses have evolved from ancestors that were about the size of a dog 50 million years ago, but the mean rate of change was so slow that it can be explained by very weak directional selection or even by genetic drift. At all points in that long history, the body size of horses could have been predominantly under stabilizing selection.

There are many well-documented examples of stabilizing selection, such as on birth weight of babies (Fig. 5.7).

Much of the genetic variation for quantitative traits is probably maintained by mutation

Artificial selection on practically any trait will produce a selection response; populations contain much genetic variation for quantitative traits.

There is genetic variation for almost all quantitative traits in natural populations. Plant and animal breeders know that artificial selection on practically any trait will produce a selection response, confirming the presence of genetic variability for the trait. As explained in Chapter 4, the extent of this genetic variation cannot be expressed in terms of the genetic diversity or heterozygosity at the loci involved, because those loci are almost always unknown. Instead, we can estimate what fraction of the phenotypic variation is due to genetic variation. So although we cannot point to the genes that vary, populations do contain much genetic variation for quantitative traits.

How is genetic variation maintained under stabilizing selection? In

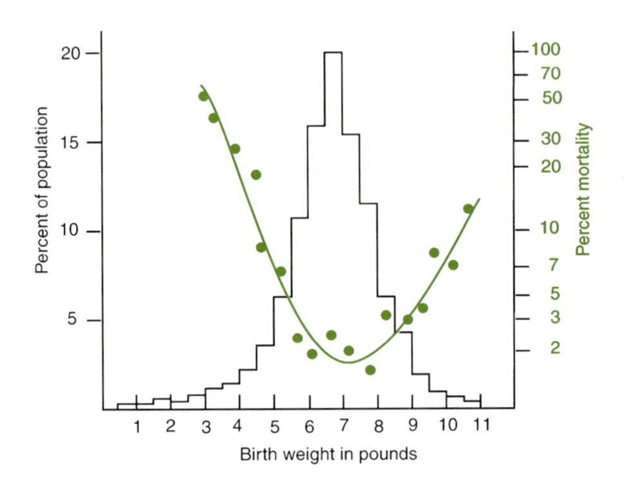

Fig. 5.7 The distributions of birth weight and early mortality rate among 13 730 babies. These data (from Karn and Penrose 1951) indicate stabilizing selection on birth weight, since the optimum birth weight is associated with the lowest mortality. (From Cavalli-Sforza and Bodmer 1971.)

theory, heterosis or negative frequency-dependent selection may be responsible. It is hard to tell as long as the genes involved are not known, but on a priori grounds it does not seem likely that these mechanisms predominate for the large number of loci involved in the many quantitative traits under stabilizing selection. Because firm evidence for single locus heterosis is scarce (the sickle-cell polymorphism in malarial areas is one of the few cases), assuming that it occurs at many loci affecting quantitative traits is unjustified. It is also hard to see why the fitness of alleles affecting a quantitative trait under stabilizing selection would be frequency dependent.

Heterosis and frequency-dependent selection are not likely to explain this variation.

A more likely explanation is mutation–selection balance. A mutation that slightly increases or decreases the optimal phenotype will experience very weak selection, because with stabilizing selection fitness declines away from the optimal value (Fig. 5.8). A mutation that slightly increases the phenotypic value will be advantageous when it is present in an individual with a phenotype lower that the optimum, just as often as it is disadvantageous in an individual with a higher than optimal value. Similar reasoning applies to a mutation that slightly decreases the phenotypic value. Thus the average effect of a mutation on fitness will be very small, and such mutations may reach relatively high equilibrium values at mutation–selection balance. Although the mutation rate per locus is small, there may be quite a bit of mutational input of genetic variability, because most quantitative traits are influenced by many loci.

Mutation–selection balance is more plausible. Under stabilizing selection the average effect of a mutation on fitness is small and can reach high equilibrium frequencies.

The mutation–selection balance explanation does not exclude other explanations for the maintenance of genetic variability. There may be heterosis at some loci, or frequency-dependent selection. Changes in the direction of selection may also play a role. The optimal phenotypic value may also shift with environmental conditions (remember selection on beak size in Darwin's finches, Chapter 4), contributing to the maintenance of alleles by increasing the frequency at which they shift between being advantageous and disadvantageous. In conclusion, much of the genetic variation for quantitative traits under stabilizing selection is probably maintained by mutation.

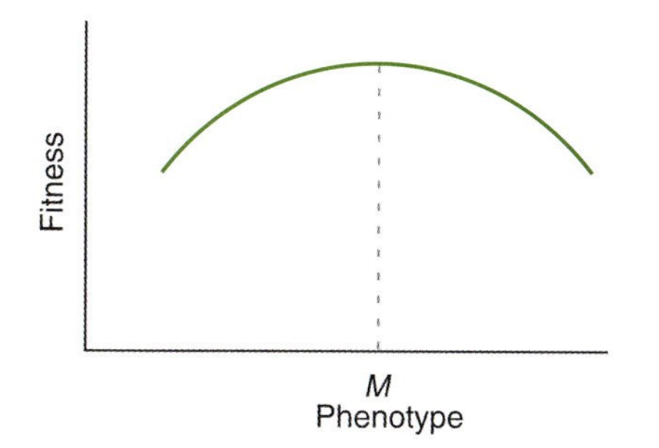

Fig. 5.8 A model of stabilizing selection. Fitness decreases in proportion to the square of the distance from the optimum phenotypic value, *M*. (From Crow 1986.)

Summary

This chapter considers the origin and maintenance of genetic variation, which is necessary for evolution to occur.

- Ultimately, all genetic variation derives from mutation. The mutation rate appears to be adjusted to an optimal level. Both too little and too much mutation would limit the rate of evolution; the former by a shortage of genetic variability and the latter by destruction of well-adapted genotypes.
- Adjustment of the mutation rate to a level that maximizes the rate of adaptive evolution should occur more easily in asexual than in sexual organisms.
- Mutation rates may differ between DNA nucleotide sequences, between genes, and between different types of organisms. In *Drosophila* the rate of slightly deleterious mutations is about one new mutation per zygote, in humans it is even higher.
- Meiotic recombination increases the genetic variability between individuals in a population by creating a large variety of multi-locus genotypes. The variation of alleles at single loci is little affected by recombination.
- The amount of genetic variation in natural populations has been long debated by population geneticists. With the advent of molecular techniques precise measurements uncovered great variation in DNA sequences, most of which is consistent with both natural selection and neutrality. Because these two wholly different explanations overlap remarkably in their predictions, the debate on the amount of genetic variation available for adaptive evolution is still unresolved.
- The most important processes affecting the maintenance of genetic variation are genetic drift and selection. Theories that explain the maintenance of genetic variation all assume that populations are in genetic equilibrium: that the forces increasing and decreasing variation balance each other. If this assumption does not hold, the theories are at best only approximately valid.
- The theory of neutral mutations gives a picture of a continual flux of genetic variants through a population. Most of these are quickly lost; a few drift very slowly to fixation.
- Selection can maintain genetic variation if the effects of different selection forces balance. The two main possibilities are heterosis (selection favoring heterozygotes) and negative frequency-dependent selection (selection favoring rare alleles). The evidence suggests that the latter is more widespread.
- Many quantitative characters are under stabilizing selection, favoring intermediate trait values. Artificial selection provides ample evidence of the presence of genetic variation for almost any quantitative trait. Much of this variation is probably maintained by mutation.

It is not enough that genetic variation be present and be maintained in populations for evolution to occur. Genetic variation must also be expressed in the phenotype; this is the subject of the next chapter.

Recommended reading

Hartl, D. L. and Clark, A. G. (1989). *Principles of population genetics*, (2nd edn). Sinauer Associates, Sunderland, Massachusetts.

Lewontin, R. C. (1974). *The genetic basis of evolutionary change.* Columbia University Press, New York.

Li, W.-H. and Graur, D. (1991). *Fundamentals of molecular evolution.* Sinauer, Sunderland, Massachusetts.

Nei, M. (1987). *Molecular evolutionary genetics.* Columbia University Press, New York.

Questions

5.1 Meiotic recombination can result in the formation of new haplotypes (haploid multi-locus genotypes). Consider two populations: one with a high and one with a low genetic diversity. In which will recombination result in more novel haplotypes?

5.2 Many rare human genetic diseases are caused by an abnormal recessive allele. Usually between 1 in 10 000 and 1 in 100 000 persons have such a disease, while typical mutation rates are in the range 10^{-4}–10^{-6}. Do you think these diseases can be explained as an equilibrium between mutation and selection?

5.3 Cystic fibrosis is a serious, recessive, genetic disease caused by a single gene, characterized by malfunction of the pancreas and lung, and found in about 1 in 4000 newborn babies. Do you think mutation–selection equilibrium is a likely explanation of its frequency? What are some alternative explanations?

5.4 If cystic fibrosis patients do not reproduce, due to early death or infertility, how much heterosis (superior fitness of heterozygotes) would be needed to explain the observed disease frequency?

Chapter 6
The expression of variation

Introduction

Development structures the variation in traits that is presented to the action of selection. Related organisms share developmental mechanisms and patterns of expression of genetic variation.

Chapters 4 and 5 discussed how genetic variation responds to selection and how genetic variation originates and is maintained. They discussed important differences in systems of genetic transmission and focused on what causes gene frequencies to change. However, evolution occurs simultaneously in both information (genes) and matter (organisms). The genes replicate information, and organisms interact with each other and the environment to transmit the genes to the next generation. In Chapters 4 and 5 organisms were implicitly viewed as the genes' method of making more genes. Organisms do not, however, simply transmit genetic information. Through their development they modify the expression of genetic variation and structure the variation in traits that leads to selection. This complicates the connection between genotype and phenotype in interesting ways. Because developmental mechanisms are broadly shared among related organisms, they constrain genetic responses to selection in similar ways in the organisms that share them. This explains why large groups of organisms resemble their common ancestors in their basic body plans and why ancestry at that macroevolutionary level can also influence microevolutionary responses to selection.

The developmental connection of genotype to phenotype profoundly affects the evolution of traits.

Development gives to organisms a role in evolution missed by a strictly gene-centered approach. Not just the genes' method of making more genes, organisms determine, through their development and through their interactions with the environment, how genes will be expressed and which genes will survive. Genes must build organisms out of materials and complex systems of chemical and biological interactions. In so doing they can only work through the properties of those materials and those systems; they cannot directly control every detail of the phenotype. This chapter describes how trait evolution is affected by the developmental connection between genotype and phenotype.

We begin with two observations:

(1) the genetic variation of a trait depends on the environment in which that trait is measured;
(2) if homologous traits are measured in two different lineages in the same environment, the genetic variation detected depends on the lineage.

Phylogenetic effects on the expression of genetic variation are caused by developmental mechanisms, shared within a lineage, that permit the expression of only certain phenotypes.

Thus patterns of expression of genetic variation can have at least two causes: developmental mechanisms specific to a given lineage, and the interactions of developing organisms with their environments. Phylogenetic effects on the expression of genetic variation are caused by developmental mechanisms specific to a lineage that only permit the expression of certain phenotypes. Those lineage-specific constraints, together with the laws of physics and chemistry, influence the interactions of developing organisms with their environments. We introduce this chapter with two examples that illustrate **lineage-specific developmental mechanisms** and show how they interact with variation in the environment to structure phenotypic variation.

Patterns in butterfly wings

Butterfly wings are among the most beautiful products of evolution (Fig. 6.1). They are also the product of a well-understood, phylogenetically conserved, developmental mechanism (Nijhout 1991). Butterfly wings

Butterfly wings are produced by a phylogenetically conserved, developmental mechanism controlling a few simple patterns in wing cells.

Fig. 6.1 The beauty and diversity of butterfly wings are produced by broadly shared developmental mechanisms (photograph by Ingrid Singh; butterflies courtesy of the Basler Naturhistorisches Museum).

develop during pupation from wing disks that are easily accessible for manipulation through the walls of the cocoon. They are also nearly two-dimensional sheets, with an upper and a lower side subdivided into compartments or cells by wing veins and the wing margin, giving them a standard geometry, allowing us to specify patterns at precise positions, and making them an ideal model system for the study of pattern formation.

Mutations affecting wing patterns only produce changes within this framework, not something totally new.

Many butterfly wing patterns are variations on a single basic plan. The units of this plan are the wing cells, and the formation of a pattern within each wing cell, on the upper and lower surfaces of that cell, and in the fore and hind wings can be controlled independently. The pattern of the whole wing consists of repetition with variation of the patterns in the cells. Working with the basic plan of the butterfly wing (Fig. 6.2), evolution has been able to produce tremendous diversity by varying the expression of each element.

This developmental pattern has been conserved in the radiation of thousands of species of butterflies, functioning as a filter that determines what phenotypes are possible. Mutations affecting wing patterns only produce changes within this framework, not something totally new. Thus the developmental mechanisms that control this pattern shape the expression of genetic variation.

Discontinuous growth in arthropods

Discontinuous growth combines with size thresholds for maturation and environmental variation to produce surprising consequences . . .

Arthropods must grow discontinuously by molting because they have hard exoskeletons. In the waterflea, *Daphnia*, discontinuous growth combines with a threshold size at which maturation is initiated. Individuals vary in their size at birth for genetic, environmental, and maternal reasons, and they vary in their growth rates because of variation in food and temperature. The lineage-specific growth pattern

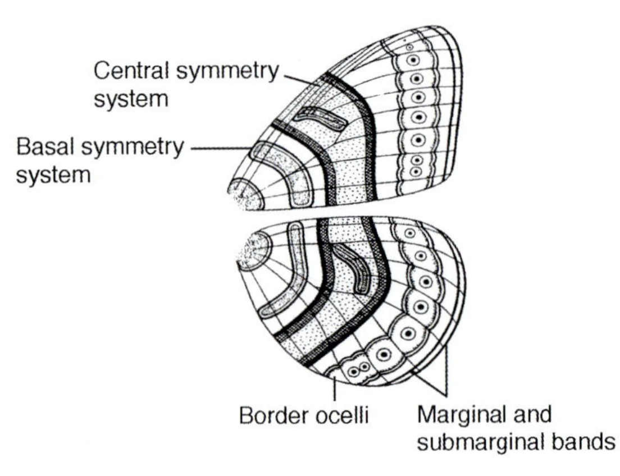

Fig. 6.2 The nymphalid ground plan. Variations on these basic elements have generated the wing designs of thousands of butterfly species. (From Nijhout 1991.)

interacts with variation in size at birth and growth rate to produce surprising consequences (Ebert 1994).

Consider a series of newly hatched *Daphnia* of increasing size (Fig. 6.3). The growth line for the smallest animal crosses the size threshold for maturation in the third molt; it matures after two more molts, having had five juvenile instars. The next four *Daphnia* also cross the threshold in their third molt, have five juvenile **instars**, and define, with the first individual, an instar group. Within this group, size at maturity increases with size at birth. The sixth individual, however, is enough larger at birth to cross the threshold in the second molt and matures after only four juvenile instars. With the next four individuals, it forms another instar group. The sixth individual is larger at birth than the first, but smaller at maturity. Thus discontinuous growth coupled with a size threshold for maturation can—between instar groups—reverse the normal relationship between size at birth and size at maturity.

reversing the normal relationship between size at birth and size at maturity,

This pattern of growth, found in many arthropods, generates phenotypic variation that interacts with size-specific selection in surprising ways. Consider Figure 6.4. In its upper portion are plotted the sizes at birth and sizes at maturity of seven individuals in a six-instar group and seven individuals in a five-instar group. Within each group, size at maturity increases with size at birth, but the relation is offset between the two groups. Mean size at maturity of each group is about 3.3 mm. Now, what happens when there is size-specific mortality on adults and a range of sizes at birth, as indicated by the shaded bars in the lower portion of Figure 6.4? If a predator were preferentially to eat *Daphnia* larger than 3.3 mm, that would select for small size at birth if there was only one instar group generated by a narrow range of birth sizes; for within each instar group there is a consistent relationship between size at birth and size at maturity above 3.3 mm. However, if the range of birth sizes was large enough to

and allowing directional selection on size at birth to translate into stabilizing, disruptive, or directional selection on size at maturity, depending on the circumstances.

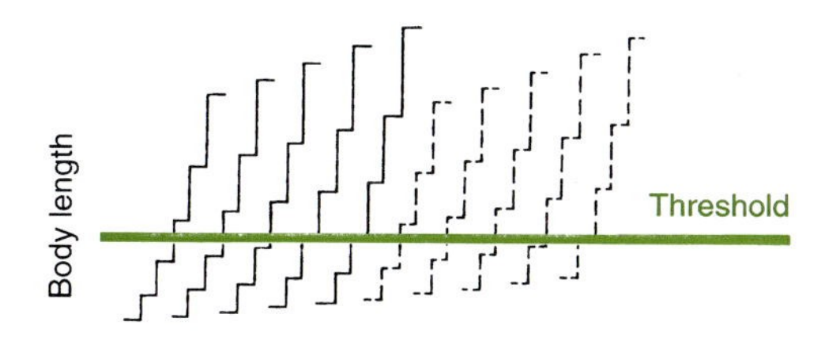

Fig. 6.3 The growth curves for 10 hypothetical *Daphnia* individuals from the first instar until the first egg-laying instar. Each step represents one instar, and the length of the next instar is 1.3 times the length of the previous instar. Maturation is initiated when the threshold length is reached. Because egg production takes two instars, the first eggs are laid at the beginning of the third instar after the threshold is passed. Individuals that were smaller at birth need more instars to mature than individuals that were larger at birth. The group of five individuals on the left needed six instars to reach maturation; the group of five individuals on the right needed only five, because they were larger at birth. The lines have been shifted to the right so that they do not lie on top of each other. (From Ebert 1994.)

generate two instar groups, as is usually the case, then selection against large adults could produce stabilizing selection on birth sizes for one range of birth sizes (0.9–1.2 mm), and disruptive selection on birth sizes for another range (0.7–1.1 mm). Thus the kind of selection operating on size at birth in *Daphnia* depends in a complex way on the range of offspring sizes at birth and on growth rates.

Size at birth in *Daphnia* varies genetically among females, increases with age within each female, and is affected by the food supply. Because offspring of various sizes are always entering *Daphnia* populations, there can never be a single, stable type of selection on offspring size, driven by selection on adult size. Here a pattern of growth and maturation shared by many arthropods acts again as a filter, introducing into the distributions of phenotypes patterns that can only be partially and indirectly affected by changes in genes.

Lineage-specific developmental mechanisms determine the kinds of variation that will be presented to natural selection.

The butterfly wing plan and discontinuous growth in arthropods exemplify lineage-specific developmental mechanisms that determine the kinds of variation presented to natural selection. Other mechanisms

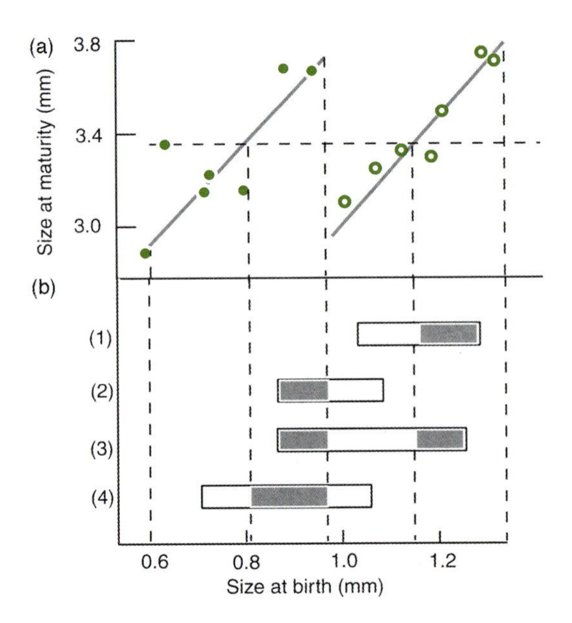

Fig. 6.4 (a) The actual size at maturity of seven individual *Daphnia magna* in a six-instar group and seven individuals in a five-instar group. Within an instar group, size at maturity increases with size at birth. Between groups, there is no clear relation. (b) The type of selection that would be experienced on size at birth given threshold selection against adult individuals larger than 3.3 mm at maturity, for four different ranges of size at birth. (1) Directional selection against larger adults selects against larger newborn (individuals in the dark size-range die, those in the white size-range survive); (2) selects for larger newborn; (3) selects for intermediate-sized newborn (stabilizing selection); (4) selects for very small and very large newborn (disruptive selection). (From Jones *et al.* 1992.)

have evolved primarily to produce phenotypes appropriate for the environment in which development occurs, mechanisms that rely on receiving information from the environment to determine the kind of phenotype that will be produced.

Induced responses: one genotype can produce several phenotypes

Induced responses are adaptations that illustrate the capacity of one genotype to produce quite different phenotypes, depending on the kind of environment encountered either by its mother or during its own development. According to Williams (1966) and Curio (1973), who sought clear definitions of adaptations, an adaptation is the ability of a genotype to produce a change in a phenotype that occurs in response to a specific environmental signal and improves reproductive success. Otherwise, the change does not take place. Many induced responses are adaptations by this definition.

Induced responses are changes in a phenotype that occur in response to a specific environmental signal and have a clear functional relationship to that signal that results in an improvement in growth, survival, or reproduction.

Waterfleas threatened by midges and backswimmers

Waterfleas of the genus *Daphnia* respond to dissolved molecules that indicate the presence of invertebrate predators (midge larvae of the genus *Chaoborus* and backswimmers of the genus *Notonecta*) by developing tail spines, helmets, and neck teeth. Midge larvae and backswimmers prey less effectively on spiny, helmeted *Daphnia*. Helmets and spines are costly, individuals that do not produce them have higher reproductive rates, and therefore when predators are not present, the spines and helmets are not produced (Dodson 1989).

Some *Daphnia* develop helmets and spines that protect them against predators, but only when they detect predators.

Barnacles threatened by predatory snails

Barnacles of the genus *Chthamalus* react to the presence of a predatory snail, *Acanthina*, by altering their development. If the snail is present, the barnacles grow into a bent-over form that suffers less from predation but pays for it with a lower reproductive rate. If the snail is not present, the barnacles develop into a typical form with normal reproduction (Lively 1986).

Some barnacles develop an alternate morphology to resist predators and pay a price in reproductive performance.

White birch defoliated by caterpillars

Birch trees subjected to natural defoliation by caterpillars and artificial defoliation by scientists develop resistant leaves. Resistance is measured by bioassay: caterpillars reared on trees previously subjected to defoliation pupate at smaller sizes than caterpillars reared on trees with no defoliation. The reaction persists in trees treated with fertilizer, suggesting that the leaves are not simply of lower quality because the trees have been damaged, but contain chemicals harmful to caterpillars (Haukioja and Neuvonen 1985).

Defoliated birch trees develop herbivore-resistant leaves.

Thus induced responses demonstrate that one genotype can produce several phenotypes in reaction to environmental signals.

Methods for analyzing patterns of gene expression

Variation in genes does not map linearly into variation in traits.

A naïve view of the relation of genes and traits is that variation in genes has a direct relation to variation in traits, and that a given amount of genetic change would result in an equivalent amount of phenotypic change. However, the examples just given show that this view is incorrect. We need conceptual tools that enable us to think clearly about the roles of genetic variation, developmental mechanisms, and environmental variation in producing patterns of phenotypic variation. This section discusses some helpful concepts.

A reaction norm, a property of a genotype, describes how development maps genotype into the phenotype as a function of the environment.

The first tool is the concept of the **reaction norm**. The reaction norm of a trait is a property of a single genotype and, in the simplest case, a single environmental factor. It is measured by raising individuals from one clone at different levels of the environmental factor, measuring the trait at each level, and plotting it as a function of the environmental factor. The resulting line (Fig. 6.5a) describes how development maps the genotype into the phenotype as a function of the environment. A population of genotypes can be described as a bundle of reaction norms (Fig. 6.5b), and the average reaction of the population to the environmental factor can be described as the population mean reaction norm. When Woltereck (1909) first described reaction norms, he conceived of them as the potential reactions of all traits to all possible environmental variation. We define a reaction norm for one trait, one environmental factor, and one genotype to be precise and clear.

The second tool is the concept of **canalization**. Canalization describes the containment or limitation of phenotypic variation by developmental mechanisms. A trait can be canalized against environmental disturbance, genetic perturbations such as mutations, or both.

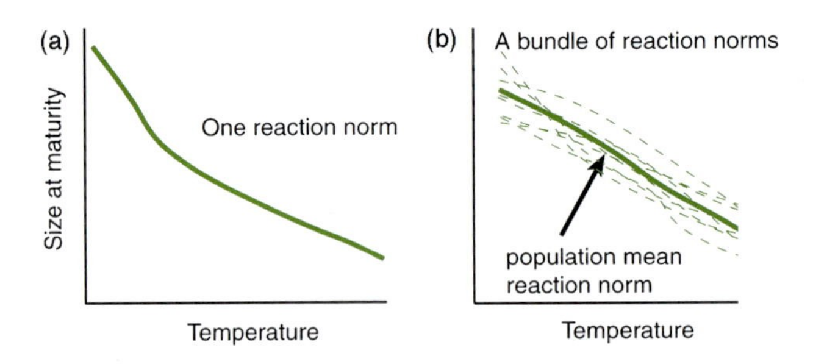

Fig. 6.5 (a) An example of a reaction norm, which is a property of a single genotype: individuals that all belong to a single clone mature at smaller sizes when reared at higher temperatures. (b) An example of a bundle of reaction norms. Each dashed line represents the sensitivity of a single genotype to temperature; the solid line represents the population mean reaction norm.

Canalization was discovered by Waddington (1942) when he found that he could reveal hidden variation in some traits in *Drosophila* by disturbing the mechanism that had been repressing—canalizing—that variation. Two genes have recently been shown to have roles in canalization. One codes for a protein that is produced when organisms are stressed, a heat-shock protein (Hsp90: Rutherford and Lindquist 1998); the other is a developmental gene involved in the control of segmentation (*Ultrabithorax*: Gibson and Hodgness 1996). Mutating these genes releases hidden genetic variation in the traits whose development they help to control. We thus appear to be on the verge of understanding the molecular mechanisms that control canalization.

Canalization is the reduction by developmental mechanisms of phenotypic variation caused by environmental or genetic disturbance.

Here we extend Waddington's concept to include the repression of variation in traits at the population level by developmental mechanisms. At this level, canalization and reaction norms are not opposites: they combine in canalized bundles of reaction norms (Fig. 6.6).

A bundle of reaction norms can be canalized (Fig. 6.6a) or uncanalized (Fig. 6.6b); we can tell the difference only by comparison. A bundle of reaction norms can be canalized in just a narrow range of the environmental factor (Fig. 6.6c). A single reaction norm can be sensitive to the environment (with positive or negative slope), or insensitive to the environment (with zero slope).

A bundle of reaction norms can be canalized or uncanalized.

The analysis of nature and nurture

A trait that is developmentally fixed and expressed in the same state in all individuals of a species, no matter what their genetic background and no matter what the environmental conditions, is a special, limiting case: a perfectly canalized, completely insensitive, perfectly flat reaction norm (Fig. 6.7a). Most traits, however, express some genetic variation and respond somewhat to environmental variation (Fig. 6.7b). Their state is determined by a combination of genetic and environmental influences.

Traits are determined by a combination of genetic and environmental influences.

Depicting trait variation as a bundle of reaction norms is an effective way to see at a glance how genes and environments interact to determine the trait. Consider an artificial example of three genotypes (G1, G2, and G3) sampled from a population of parthenogenetic lizards, reared as clones, and raised at three population densities, low, medium, and high

Bundles of reaction norms show us how genes and environments interact to determine a trait.

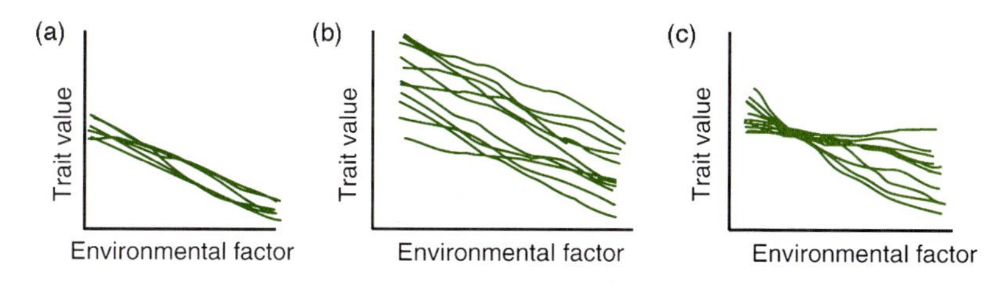

Fig. 6.6 (a) A canalized bundle of reaction norms. (b) An uncanalized bundle of reaction norms. Which bundle is canalized is decided by comparing them. 'Canalized' and 'uncanalized' are relative concepts. (c) A bundle of reaction norms that is canalized only over a narrow range of the environmental factor.

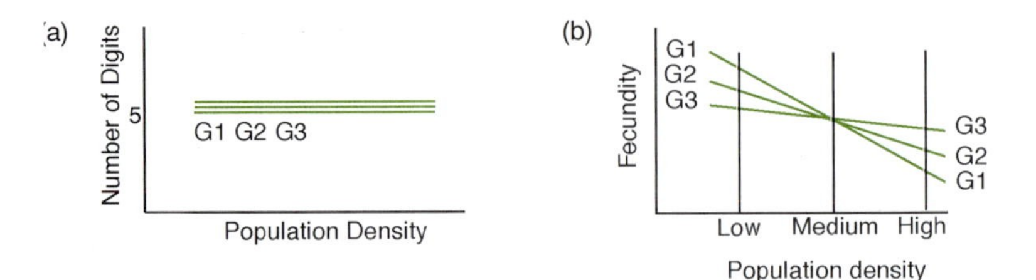

Fig. 6.7 (a) The number of digits in the hand of a lizard is not sensitive to population density and does not vary genetically; it has a flat, canalized reaction norm. (b) The fecundity of the same three genotypes would be sensitive to population density; here sensitivity to density varies with genotype.

(Fig. 6.7). We measure two traits, number of digits per foot and fecundity. Figure 6.7a depicts the reaction norms of the three genotypes for number of digits per foot. In fact, they would lie on top of one another, for every individual in the entire population has five digits per foot at all population densities, but in the figure they are separated to show that three genotypes were measured. These are perfectly canalized, perfectly flat reaction norms. The trait is insensitive to environmental variation, expresses no genetic variation, and cannot respond to selection.

For fecundity, the situation is quite different (Fig. 6.7b). All three genotypes reduce their fecundity at higher population densities, but the sensitivity of the three genotypes to changes in density differs. G1 is quite sensitive. It has the highest fecundity at low population density and the lowest fecundity at high population density. G3 is not very sensitive. It has the lowest fecundity at low population densities and the highest fecundity at high population densities. G2 is intermediate.

Plotting reaction norms shows how measurements of genetic variation depend on the environment in which they are made.

If we were to measure the genotypes only at low or only at high population densities, we would say that the heritability of the trait for fecundity was high. If we were to measure them only at medium population densities, we would say that the heritability of the trait was near zero and that there was little genetic variation for fecundity in the population. Thus plotting reaction norms is an easy way to see how measurements of genetic variation depend on the environment in which they are made. A trait like fecundity in Fig. 6.7b can respond to selection at low and high population density, where it expresses genetic variation, but not at medium density.

The plasticity of a trait can itself respond to selection.

There is also genetic variation for the sensitivity of fecundity to changes in density. Selection to increase the sensitivity of fecundity to density would increase the frequency of G1. Selection to decrease the sensitivity of fecundity to density would increase the frequency of G3. The sensitivity of a trait to change in an environmental factor measures its **phenotypic plasticity**. Thus natural selection can shape phenotypic plasticity if there is genetic variation for the slopes of reaction norms.

The analysis of reaction norms sheds light on the contributions of genes and environment in determining the phenotype. Sensitivity to environmental variation raises the question of the adaptive value of such sensitivity.

How much of the plastic response is adaptive?

Experiments can estimate what part of a plastic response is recently adapted and what part is ancestral.

Deciding whether a plastic response is an adaptation is not straightforward, for some responses may be natural consequences of physical, chemical, and ancestral biological processes that would have occurred whether or not evolution had adapted the population to its current circumstances. We know that some plastic responses are not adapted (they might be fortuitously adaptive) because many organisms react to environmental variation never encountered in evolution. It is therefore important to be able to decide how much of a plastic response, a reaction norm with a non-zero slope, is the product of adaptive evolution and how much is just a by-product of other processes.

One way to determine how much of the plastic response has evolved and is adaptive is to perform a reciprocal transplant. For example, an experiment looked at two isolated populations of frogs, one adapted to low and the other to high temperature (Berven *et al.* 1979). The frogs from the low-temperature population matured later and at a larger size than those from the high-temperature population. How much of the difference between the two populations represents an inevitable plastic response that would have been elicited in ancestors that had not yet adapted to local conditions, and how much is caused by local adaptation? To answer this question, frogs from each population were reared in both high- and low-temperature environments. Some of the results are presented in Figure 6.8.

Individuals from both populations matured earlier at higher temperature. When frogs from the high-temperature population were raised at low temperature, they matured later than frogs from the low-temperature population raised at low temperature. When frogs from the low-temperature population were raised at high temperature, they matured earlier than frogs from the high-temperature population raised at high temperature. Those differences are indicated by the thick vertical arrows. If the larval period of local frogs results in higher fitness than the larval period of nonlocal frogs, then the vertical arrows represent the direction of local adaptation at each temperature. Our estimate of the amount of local adaptation depends on our assumption about the ancestral state. If the ancestor lived at high temperature, then the low-temperature population is in the derived state, and the degree of adaptation is measured along the arrow on the left. If the ancestor lived at low temperature, then the high-temperature population is in the derived state, and the degree of adaptation is measured along the arrow on the right.

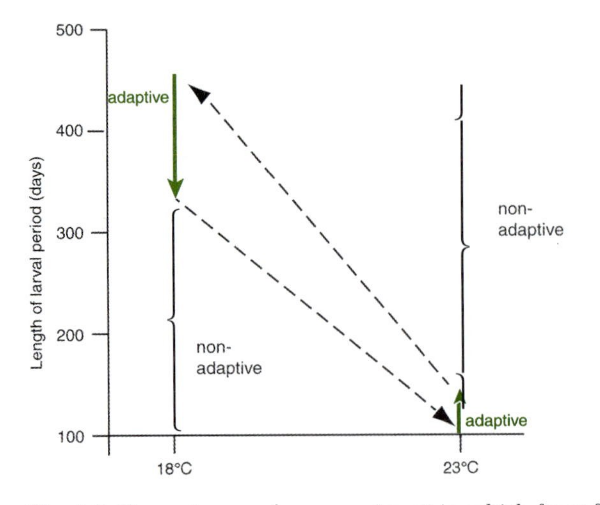

Fig. 6.8 The outcome of an experiment in which frogs from populations that historically experienced either low or high temperatures were reared at both temperatures. The frogs with an evolutionary experience of low temperatures mature earlier at the higher temperature, and the frogs with an evolutionary experience of higher temperatures mature later at the lower temperature. If developmental time at low temperature is the derived state, then the degree of adaptation is measured by the vertical arrow at the left; if developmental time at high temperature is the derived state, then the degree of adaptation is measured by the vertical arrow on the right. (From data in Berven *et al.* 1979.)

Patterns in bundles of reaction norms can change heritabilities and genetic correlations

The response of a trait to selection depends both on its heritability and on its genetic correlations with other traits.

The response to selection of a single quantitative trait depends on its heritability, the amount of phenotypic variation that can be ascribed to additive genetic effects (Chapter 4). When one trait is selected, but is genetically correlated to another trait because some genes influence both traits, the second trait will respond to selection on the first trait whether it is directly under selection or not. Thus the response of a trait to selection depends both on its heritability and on its genetic correlations to other traits also under selection.

Structured bundles of reaction norms change genetic correlations across environments with consequences for responses to selection.

Furthermore, patterns in bundles of reaction norms change heritabilities and genetic correlations in response to changes in environmental factors. Consider Figure 6.9. In panel (a), reaction norms for three genotypes are plotted from 20 °C to 28 °C for an arbitrary trait, Trait 1. Now assume that the same three genotypes also affect another trait, Trait 2; their reaction norms for Trait 2 are given in panel (b). At 20 °C and 28 °C, the heritability of the trait is high; at about 26 °C it is low. When we plot Trait 1 against Trait 2, in panel (c), we see that their negative genetic correlation at 20 °C changes to a positive genetic correlation at 28 °C. Selecting Trait 1 upwards at 28 °C will produce a correlated upward response in Trait 2, but selecting Trait 1 upwards at 20 °C will produce a correlated downward response in Trait 2.

Example: spadefoot toads in Texas.

Up to this point we have dealt with reaction norms as properties of genotypes in organisms that can be cloned. Reaction norms can also be estimated in sexually reproducing organisms as the reaction of the mean value of siblings to changes in an environmental factor. Newman (1988) made such estimates by taking eggs from five families of spadefoot toad tadpoles, rearing half of each family in ponds of short duration and half of each family in ponds of long duration (two levels of an environmental factor), and measuring body length and age at metamorphosis (two traits). In ponds of short duration, the means of the two traits were negatively correlated among families, whereas in ponds of long duration, they were positively correlated (Fig. 6.10).

Selection for more rapid development in ponds of short duration would select for larger body size at metamorphosis. Selection for more rapid development in ponds of long duration would select for smaller body size at metamorphosis. This illustrates the points made in Figure 6.9: because both the heritabilities of single traits, and the genetic correlations of pairs of traits, can change with the environment in which they are expressed, the potential response to selection of a set of correlated traits depends on the environment.

How important is the influence of the environment on the expression of genetic variation? It depends on the trait examined. Weigensberg and Roff (1996) reviewed the heritabilities of morphological, life history, and behavioral traits measured in the field with heritabilities measured in the laboratory for organisms of many kinds. The environment in the laboratory is usually assumed to differ from the environment in the field, but

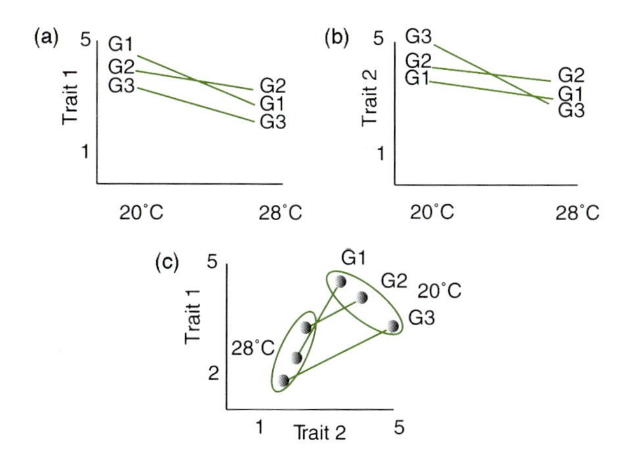

Fig. 6.9 We can combine the reaction norms for two traits to visualize changes in genetic correlations geometrically. (a) The reaction norms of three genotypes for Trait 1 as a function of temperature. (b) The same for Trait 2. (c) Here we plot Trait 1 against Trait 2 for each of the three genotypes and each of the two temperatures. The genetic correlation shifts from positive at 28 °C to negative at 20 °C.

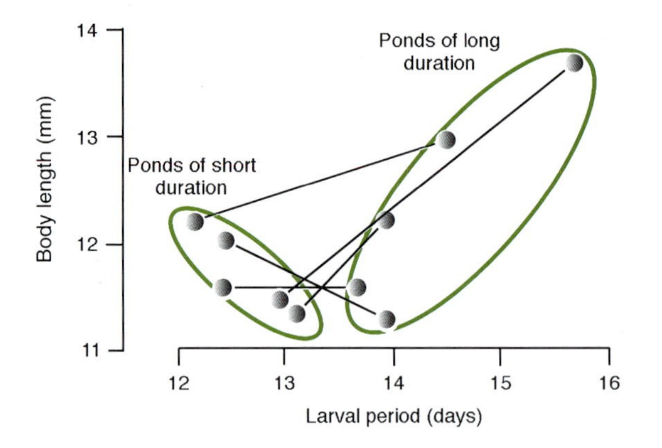

Fig. 6.10 The reaction norms of spadefoot toads reared in ponds of short and long duration behave like the hypothetical reaction norms depicted in Fig. 6.9: a negative genetic correlation in one environment shifts to a positive genetic correlation in the other. Lines join the means of sibs from split broods reared separately.

in this study heritabilities in the two environments were quite similar, and they were more similar for morphological traits than for life-history traits. Thus although the effects depicted in Figure 6.10 do happen, they do not always happen.

Genotype and phenotype are sometimes only loosely coupled

Changes in the genes do not always cause changes in the phenotype, even when the genes are expressed.

Organisms that are similar phenotypically can differ genetically

Some species look identical but have large genetic differences. A striking example: *Tetrahymena*.

The ciliated protozoans of the *Tetrahymena pyriformis* complex are indistinguishable morphologically but differ both in their genes and even in the proteins that build the indistinguishable morphological structures. The genetic differences are large and indicate that the many members of the complex have been evolving independently from one another for over 100 million years (Nanney 1982), despite which they now look exactly the same. This exemplifies phenotypic parallelism overriding genetic divergence.

Organisms that are similar genetically can differ phenotypically

The reverse of this also occurs. In cases of induced defense, the genome contains the information necessary to produce several phenotypes, depending on the environment. Precisely the same genotype could produce, for example, a barnacle that was bent over and resistant to snail predation or a barnacle that grew straight and had better reproductive performance. To that we can add metamorphosis: from tadpoles to frogs in amphibians, from caterpillars to pupae to butterflies in insects, from underwater to aerial leaves in aquatic buttercups, and all other such developmental transformations in the life of a single individual. In all such cases the genome contains the information needed to produce several strikingly different phenotypes.

Many genomes contain the information needed to produce several discrete phenotypes, or a continuous range of phenotypes, depending on the environment.

An extreme example of such genetic potential can be found in cases of environmental sex determination associated with marked sexual dimorphism. If a larva of the marine worm *Bonellia* settles on normal substrate, it metamorphoses into a female and grows into an adult that resembles a 10–20 cm long sausage with a long feeding tube ending in a pair of palps. However, if a larva settles on a female, it metamorphoses into a male that becomes a parasite embedded in the body of the female, reduced to little more than testes making sperm. Thus the same genotype can make either a large, complex, female worm or a tiny, simple, parasitic male.

An extreme example: environmental sex determination of dwarf males.

Reasons for loose coupling of genotype and phenotype

Genotype and phenotype are loosely coupled in eucaryotes for two basic reasons. First, there is a great deal of DNA in the genome that is either never transcribed, or its transcription products have little or no effect on the phenotype. Some of it is parasitic or symbiotic, including transposons and selfish genetic elements; some of it appears to be accumulated junk with no clear role; and some may play an adaptive role that has not yet been determined. Whatever the reasons for its existence, the fact that much of the eucaryotic genome is not transcribed means that whatever genetic changes occur in that portion of the genome will have little effect on the phenotype.

Changes in the nontranscribed portion of the eucaryotic genome have little effect on the phenotype.

The second reason for loose coupling was suggested above by examples of induced defenses and environmentally determined sexual dimorphism: there must be **regulatory genes** that turn entire developmental pathways on or off. If the same sets of developmental pathways exist in all individuals of a species, and all that is needed to produce different phenotypes is to flip switches at a few branch points, then a single gene on a sex chromosome could switch on male rather than female development, or a signal from a predator could switch on one gene that controlled development of a defensive phenotype. Developmental geneticists now know that minor genetic changes in control genes can produce dramatic morphological effects. For example, a single base substitution in the maternal *bicoid* gene of *Drosophila* can change the entire body plan by reversing the axes and symmetry of the embryo (Frohnhöher and Nüsslein-Vollhard 1986).

When regulatory genes turn developmental pathways on or off as a function of the environment, almost any relation between genotype and phenotype is possible.

In recent years, the location, nature, and action of regulatory genes have been clarified at an impressively rapid rate. Many of them turn out to be strongly conserved—shared by a very broad range of organisms—and appear to function as general mechanisms controlling pattern formation in development. Some function as inducible switches—genes that can be turned on or off depending on the receipt of a signal, known as a transcription factor.

Genes controlling developmental patterns are broadly shared

The genes controlling basic developmental mechanisms are shared by a surprisingly broad range of organisms.

The molecules and genes that regulate development are much more broadly shared than was thought just 10 years ago. It is now commonplace to study the expression of genes homologous to insect genes in vertebrates, and of genes homologous to vertebrate genes in insects. Some genes play the same basic roles in both types of organisms, as demonstrated by genetic transplant experiments in which genes from distantly related organisms functioned appropriately in the transplant context. Such genes have been conserved since insects and vertebrates shared common ancestors, marine worms that lived at least 600 million years ago. One view of the evolution of developmental mechanisms is that 'the history of life since the Cambrian has been dominated by the elaboration of regulatory mechanisms that exploit a common set of genes' (Akam *et al.* 1994*a*).

Hox genes determine the anteroposterior axis in nematodes, arthropods, and chordates.

Some signaling molecules are used analogously in vertebrates and invertebrates; the molecules that initiate development of the heart and the eyes in vertebrates appear to be homologous to molecules that play the same role in invertebrates. Some related genes are clearly involved in laying down the basic features of body plans in many groups. For example, the *Hox* genes are responsible for laying down the anteroposterior axis in nematodes, arthropods, and chordates (Fig. 6.11); their linear order along the chromosome corresponds to the region of the body in which their expression has impact, from head to tail.

The similarity of these developmental control genes suggests a possible solution to one of the great early problems of evolutionary biology. We now accept, practically without thinking about it, that our ability to group organisms together into taxonomic groups—species within genera, genera within families, families within orders, orders within classes, and classes within phyla—is made possible by that fact that organisms are descended from common ancestors. The common ancestors of the species within a genera existed more recently than the common ancestors of the classes within a phylum. Such relationships are established by the **homology** of the traits within a group. For example, the hand of a primate, the wing of a bat, and the fin of a whale are homologous, and establishing that fact helps to place the primates, bats, and whales into a taxonomic group, the Eutheria, all of whose species shared a common

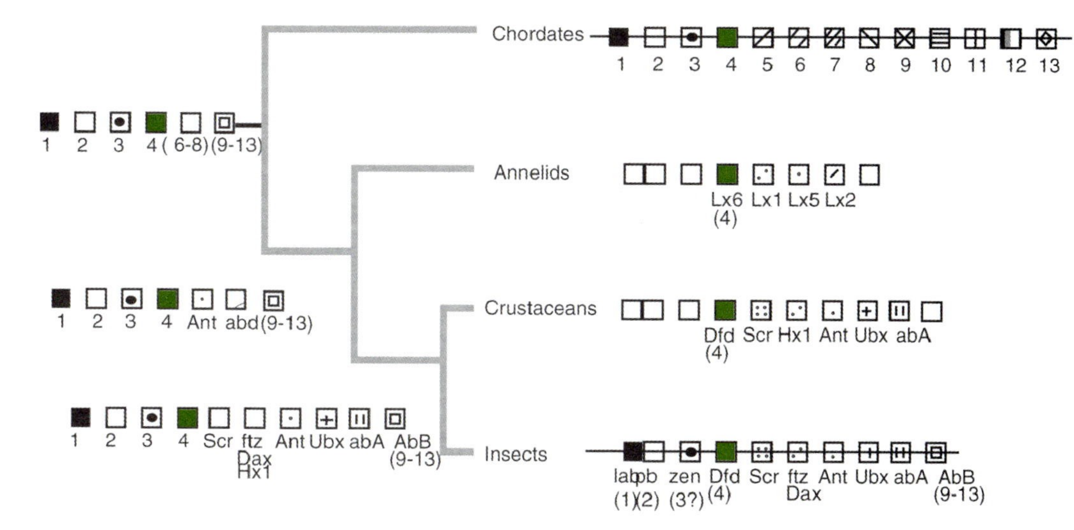

Fig. 6.11 The *Hox* genes of chordates, annelids, crustaceans, and insects share deep homologies. The observed states in living organisms are listed on the right, the inferred ancestral states on the left. Orthologues on the right are oriented in vertical columns; genes whose orthology has been established are named; colinearity is indicated for chordates and insects by the central line through the boxes representing the genes. (From Akam *et al.* 1994*a*.)

ancestor about 100 million years ago. Forging the relationship between shared descent, homology, and taxonomic groups was a triumph of nineteenth-century biology.

Developmental control genes are a new tool for establishing the homology of body parts.

Genes that regulate development provide a new tool for suggesting the homology of body parts. If a gene that initiates eye formation in vertebrates is similar in its DNA sequence to a gene that initiates eye formation in flies, and if expressing the vertebrate gene in a fly causes extra (fly) eyes to form in the fly, then vertebrate eyes and insect eyes, the structures under the control of the similar genes, may also be homologous in some sense, despite their strikingly different morphology. We call such genes **orthologous** and the traits they code for homologous. The recent ability of molecular biology to identify these genes, take them out of one organism, clone them, insert them into another organism, express them, and study the pattern of their expression, allows us to study the evolution of developmental mechanisms directly. Note that having a common regulatory gene does not mean that eyes or other organs necessarily existed in the common ancestor. The ancestral gene, whose sequence has been inherited in both flies and vertebrates, may simply have controlled the production of a light-sensitive cell or its connection to the nervous system. DNA sequence similarity does not imply continuous morphological homology, but it does establish shared descent from common ancestors in which the orthologous genes probably had functions related, if not identical, to their modern functions.

The evolution of insect wings is a good example. It has been suggested

Example: the evolution of insect wings.

that wings originally formed on all segments in the ancestor of winged insects, and that the expression of wings on a particular segment was progressively repressed from posterior to anterior by expression of a highly conserved gene, *Ultrabithorax,* resulting first in four-winged and then in two-winged insects (Carroll *et al.* 1995). If so, many of the genes needed to regulate development were present in the ancestor. The major steps in the evolution of the insect wing were:

(1) changing the pattern of their expression in segments; and
(2) modifying the structure of the wing itself by changes in other genes under the control of the wing-inducing genes.

The spatial pattern of the *Drosophila* wing is determined by the expression of genes orthologous to those that determine the spatial pattern of the butterfly wing and affect the nymphalid butterfly family ground plan.

In the course of evolution, the four-winged condition of butterflies evolved before the two-winged condition of flies. The genes controlling wing development in butterflies and flies still have similar-enough DNA sequences that one can take well-studied genes from *Drosophila,* use their sequences to locate the orthologous genes in butterflies, then study the patterns of expression of the butterfly genes. Such studies have revealed that the spatial pattern of the *Drosophila* wing is determined by the expression of genes orthologous to those that determine the spatial pattern of the wing of the buckeye butterfly, *Precis coenia* (Carroll *et al.* 1994). Moreover, some of the genes found in butterflies by 'fishing' with fruit fly genes known to be involved in wing patterns turned out to be old genes with new roles, that involve them in the construction of elements—midline stripes, chevrons, and venous stripes—in the wing cells of the nymphalid butterfly family ground plan (Nijhout 1991).

The *distalless* gene controls the formation of structures in distal appendages, both butterfly eyespots and fruit fly antennae: a homology of positional information.

Particularly interesting was the discovery of the role played in butterfly wings by a gene called *distalless* in *Drosophila*. *Distalless* was discovered as a mutant that transforms distal antennal structures into distal leg structures, probably by producing a protein that acts as a **transcription factor**, activating the genes needed for distal structures (Wilkins 1993). Thus in fruit flies the gene helps control the morphology of appendages. In butterflies, it has a different role: it helps to control the formation of eyespots on wings. Since wings are appendages and eyespots are formed distally, the gene continues to control the formation of structures in distal appendages, but eyespots are nothing like antennae, so the homology in function between fruit flies and butterflies concerns not what will be produced, but where it will be produced. It is a homology of positional information (Carroll 1994).

The study of the developmental mechanisms that control gene expression is becoming much easier. Although thousands of genes may be involved in the production of a trait, it now seems likely that only a few regulatory genes that produce transcription factors are involved in determining when and where that trait is produced within the body plan, and how the pattern of the trait is determined. Such regulatory genes are surprisingly conserved, which means that one can take a gene out of a fruit fly where its function is well known, clone it, use it to find orthologous genes in some other organism, study their expression, and have a good chance of uncovering a role for that orthologous gene that sheds as much light on the development of that organism as it did in the fruit fly.

The molecular genetics of developmental mechanisms has two important messages for evolutionary biology. First, the mechanisms that build the framework within which genes are expressed can be studied directly; knowledge is accumulating very rapidly; tools are available that allow us to analyze the causes and consequences of changing gene expression. Secondly, developmental control genes have properties that suggest a role in some major transitions in the history of life. This is an area of active research. We return to this second message in Chapter 15.

Molecular genetics reveals developmental mechanisms that build the framework within which genes are expressed. Some of those mechanisms were involved in major transitions in the history of life.

The next section presents an example where the developmental control of gene expression can be followed from genetic and developmental studies in the laboratory to the fitness consequences of changes in expression in organisms released in the field.

Seasonal polyphenism in butterflies

We began this chapter by discussing the plan that determines the expression of patterns in many butterfly wings. There it served as an example of a conserved developmental pattern that influenced how genetic variation could be expressed in a large clade with many species. We then discussed reaction norms and canalization as important patterns in the expression of phenotypic variation. After noting that organisms that are similar phenotypically can differ genetically and those similar genetically can differ phenotypically, we discussed reasons for that loose coupling: the structure of the eucaryotic genome and the genetics of development, particularly the role of regulatory genes, small changes in which can cause big changes in phenotypes. Many regulatory genes are highly conserved, playing similar roles in organisms as distantly related as mammals and insects, and allowing us to use genes whose role is well understood in fruit flies to search for genes with similar DNA sequences and potentially similar roles in other organisms. Such a search of butterfly genes revealed that the gene *distalless,* which controls structures expressed on the distal end of fruit fly appendages, has a homologue in butterflies that controls the expression of eyespots on the distal portion of wings. That observation brought us full circle, back to the nymphalid ground plan, optimistic that new research would soon reveal the conserved molecular mechanisms responsible for the conserved plan of butterfly wings.

At the same time that some of the regulatory genes responsible for laying down the insect body plan were being discovered, a project on the ecological and evolutionary significance of seasonal variation in the patterns of butterfly wings was being carried out in the field (in Malawi) and in the laboratory (in The Netherlands and Scotland), by Brakefield, French, and their colleagues. They study tropical butterflies in the genus *Bicyclus,* which display striking seasonal variation, or **polyphenism**. *Bicyclus anynana,* for example, has a wet-season form with large, striking

Tropical butterflies in the genus *Bicyclus* display striking seasonal polyphenism: a wet-season form with prominent and a dry-season form with reduced eyespots.

yellow–black eye spots, and a dry-season form with greatly reduced eyespots (Fig. 6.12).

This system has been analyzed with methods ranging from molecular genetics to field ecology.

This is a remarkably useful system in which to analyze the causation of wing patterns with molecular genetics, quantitative genetics, transplant experiments in developing pupae, and field trials of the survival of manipulated phenotypes, to test hypotheses about adaptation. Briefly, here is what is known so far.

First, natural selection favors cryptic forms with small eyespots in the dry season and striking forms with large eyespots to deflect predator attacks in the wet season (Brakefield and Larsen 1984).

Seasonal variation in eyespots is adaptive and behaves as a continuous reaction norm.

Secondly, the wet- and dry-season forms are not discrete phenotypes; they represent the end points of continuous reaction norms that react to the temperature of the environment during development (Windig 1993). The discrete appearance of the wet- and dry-season forms is thus caused by the difference of temperatures in the field, not by a developmental switch.

The eyespot pattern responds to selection as an integrated unit.

Thirdly, the size of the eyespots responds to artificial selection; when one eyespot is selected, others respond as well; and selection of eyespots on the ventral surface (the more striking surface) does not produce very strongly correlated responses on the dorsal surface (Holloway *et al.* 1993). This suggests that the developmental regulation of eyespot production extends to all the eyespots on a wing surface, resulting in an eyespot pattern that behaves as an integrated unit; that there is partial independent control over the development of the ventral and dorsal wing surfaces; and that there are many genes that modify the intensity of the basic pattern.

The genes for larger or smaller eyespots affect both the eyespot focus and the response of the surrounding cells to the signal sent by the focus.

Fourthly, the developmental foci that produce eyespots can be grafted into new positions on the wings of the same or different individuals. Such grafts have been done both from lines selected for large eyespots into lines selected for small eyespots, and in the other direction. When

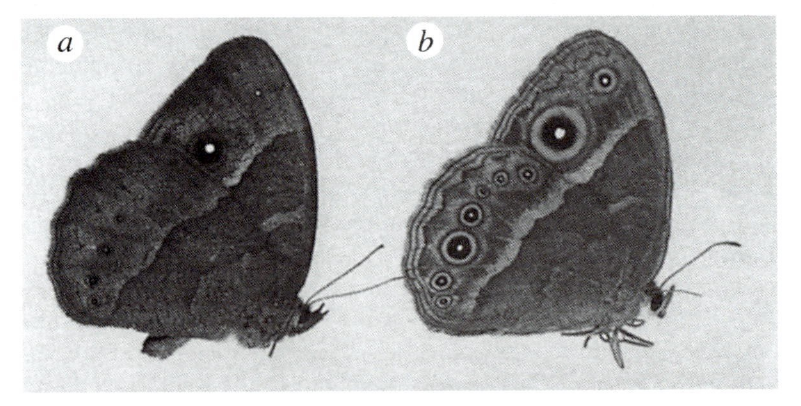

Fig. 6.12 (a) Dry- and (b) wet-season forms of the African butterfly *Bicyclus anynana,* a polyphenism in the field that is a reaction norm when tested across a set of intermediate conditions in the laboratory. (Courtesy of Paul Brakefield.)

an eyespot focus from a line selected for large eyespots is grafted into a line selected for either large or small eyespots, it produces an eyespot with more than twice the area than does a similarly grafted focus from a line selected for small eyespots. The host environment also influences the size of the eyespots, for either type of focus produces an eyespot in a line selected for large eyespots that is at least 50% larger than it does in a line selected for small eyespots. This demonstrates that the genetic changes involved in a response to selection for larger or smaller eyespots influence both the characteristics of the eyespot focus and the response of the surrounding cells to the signal sent by the focus (Monteiro *et al.* 1994).

Much more will be learned from this system, but it has already delivered important messages. First, it shows how a phylogenetically fixed developmental mechanism that produces the nymphalid ground plan is controlled by identified genes of known effect. More of these genes will be found and their interactions will be studied. Secondly, it shows how organisms with a conserved developmental mechanism interact with the environment to produce continuous reaction norms of phenotypic variation constrained by the framework of the ground plan. What sets the limits on plastic reactions needs to be worked out. Thirdly, it shows how the expression of the elements of the ground plan can be modified by many genes with local effects, such as effects only on the dorsal or the ventral surface of the wing. What determines the size of a local element and how control can be established over gene expression within a subfield need to be better understood. Fourthly, and perhaps most importantly, it shows how one can study the flow of causation in a single system in both directions, from the genes outward through development to the phenotype, and from natural selection acting on variable phenotypes in variable environments in the field inward to changes in genes. In such a system, making artificial distinctions between molecular biology, developmental biology, ecology, and evolution makes little sense, for they all contribute important and complementary elements of the explanation.

Causation flows from the genes outward through development to the phenotype, and from natural selection acting on phenotypes in the field inward to changes in the genes.

Adaptive plasticity regulated by plant phytochromes

The phytochromes, light-detecting molecules found in most plants, are an important link in the chain of mechanisms that produce adaptive responses to light conditions, including seasonal timing, germination, and shade avoidance (Smith 1995). They can be used to monitor the intensity, quality, direction, and duration of light and to modify development appropriately.

There are several phytochromes, labeled A, B, and so on, each of which can exist in two states. For example, phytochrome B is tuned to respond to red and to far-red wavelengths. Its two forms are called Pr (for phytochrome-red) and Pfr (for phytochrome-far red). When the Pr

The phytochromes are the 'eyes' of plants. The information they receive influences germination and growth.

form is exposed to red light (600–700 nm), it switches to the Pfr form, and when the Pfr form is exposed to far-red light (700–800 nm), it switches to the Pr form. Because there are many copies of these molecules per cell, the ratio of the number of molecules in each state measures the red:far-red ratio (R:FR) in the environment. That ratio is informative, for chlorophyll preferentially absorbs red light. Beneath a leaf, or beneath a forest canopy, where red light is scarce and far-red light is relatively abundant, the R:FR ratio is low. It is also low in light reflected off a leaf. In the open—for an isolated plant or a plant growing in a gap in the forest—the R:FR ratio is high. Thus the R:FR ratio provides information on shading and on the nearby presence of other plants competing for light. In this sense, the phytochromes are the 'eyes' of plants.

Phytochrome B helps to mediate the germination response. For a seed of a forest tree lying in the soil, waiting to germinate, it would be useful to be able to sense whether it is in a light gap in which it might survive to adulthood if it germinated. By monitoring the R:FR ratio in its tissues it could start germinating when the ratio passes a certain threshold.

The sensory capacities that phytochromes give plants allow them to respond adaptively to encroachment by neighboring competitors and alter morphology from sun to shade.

Phytochrome B also helps plants to detect light reflected from neighbors (Schmitt and Wulff 1993). A plant, growing in full sunlight, that starts to detect light reflected off a rapidly growing neighbor can use that information to change the direction of growth before being directly shaded. In this way it can respond much more effectively, for it takes time to grow and it is harder to grow out from under the direct shade of a neighbor than it is to grow away from an encroaching neighbor while still in full sunlight.

Plants can also use the phytochrome system to change their morphology from sun-adapted to shade-adapted forms. For example, the ribwort plantain (*Plantago lanceolata*) forms heavier rosettes with more, larger, broader leaves in the shade than it does in the sun. The response appears to be mediated by several hormones, each of which mimics some elements of the phytochrome response (van Hinsberg 1997).

The phytochrome system reveals surprising sensory capacities in plants. The different phytochromes have discrete properties, are differentially expressed, and mediate the perception of different light signals. In *Arabidopsis*, for example, mutations deficient in four of the five phytochrome genes have been isolated; their physiological functions overlap considerably, but each appears to trigger a different downstream signaling pathway (Whitelam *et al.* 1998). Such insights into the molecular and physiological genetics of phytochromes show how the phytochrome system can, like the butterfly wing, integrate genetics, biochemistry, and physiology with the ecology and evolution of induced responses. Here the mechanisms underpinning adaptive plasticity can also be laid bare.

Summary

This chapter summarizes how to think about the developmental connections between genotype and phenotype in an evolutionary context.

The critical questions are, how does development structure the phenotypic variation that causes natural selection, and how does it filter the genetic variation that determines the response to selection?

- Butterfly wing patterns and discontinuous growth in arthropods exemplify structured phenotypic variation and filtered genetic variation. They are controlled by lineage-specific developmental mechanisms.
- Induced responses, many of which are adaptive, illustrate the capacity of a single genotype to produce several phenotypes in response to a specific environmental signal and thereby improve reproductive success.
- Reaction norms and canalization help us to think clearly about connections between genotype and phenotype. A reaction norm, a property of a genotype, describes how development maps the genotype into the phenotype as a function of the environment. Canalization describes the containment or limitation of phenotypic variation by developmental mechanisms. A population of genotypes is a bundle of reaction norms that can be canalized, uncanalized, or canalized in some environments but not others.
- The sensitivity of a trait to change in an environmental factor measures its phenotypic plasticity. Natural selection can shape phenotypic plasticity when there is genetic variation for the slopes of reaction norms. Not all plasticity is adaptive. By comparing an ancestral with a derived population, one can decide how much of the plasticity is adaptive by rearing both types in two or more environments.
- Because both the heritabilities of single traits and the genetic correlations of pairs of traits can change with the environment in which they are expressed, the potential response to selection of a set of correlated traits depends on the environment and can be visualized by plotting reaction norms.
- Organisms that are similar genetically can differ phenotypically, and organisms that are similar phenotypically can differ genetically. Genotype and phenotype are partially uncoupled, or only loosely connected. Because some genes are not always expressed, and because a few developmental control genes have large effects, there is huge variation from gene to gene in the impact of genetic changes on phenotypes.
- The genes that control development are shared among large groups of organisms. Their DNA homology allows rapid identification of developmental control functions inherited from common ancestors by distantly related organisms, such as butterflies and flies, even arthropods and vertebrates.
- Seasonal polyphenism in butterfly wings and the phytochromes in plants both exemplify systems in which the molecular genetic

mechanisms underlying adaptive plasticity can be studied in detail. Here specialties are integrated—evolution, development, genetics, and physiology—that traditionally have been separated. The result is added insight.

The next chapter discusses the evolution of sex, which has just as much impact on the generation and maintenance of genetic variation as development has on its expression. We conclude our discussion of microevolution with chapters on life-history evolution and sex allocation and on sexual selection. A chapter on speciation then forms the transition to macroevolution and major patterns in the history of life.

Recommended reading

Akam, M., Holland, P. Ingham, P., and Wray, G. (ed.) (1994). *The evolution of developmental mechanisms.* Development Supplement, The Company of Biologists Limited, Cambridge.

Raff, R. A. (1996). *The shape of life. Genes, development, and the evolution of animal form.* University of Chicago Press, Chicago.

Schlichting, C. D. and Pigliucci, M. (1998). *Phenotypic evolution. A reaction norm perspective.* Sinauer Associates, Sunderland, Massachusetts.

Questions

6.1 In Fig. 6.7b replac[illegible] pulation density by per capita i[illegible] thenogenetic lizards by three hu[illegible] triplets separated at birth and a[illegible] th one of the triplets raised at eac[illegible] estion, 'Is intelligence determin[illegible] e any meaning? Is it clearly [illegible] Does it make appropriate assumptions?

6.2 If organisms with different body plans—such as arthropods and vertebrates—share homologous developmental control genes, then how might developmental mechanisms cause body plans to be shared within each of those large groups but not between them?

6.3 Discuss the roles of external and internal factors in evolution. Are both necessary for adaptive or neutral change? Is either sufficient for either type of change?

Chapter 7
The evolution of sex

Introduction

In Chapters 4 and 5 we considered how selection changes allele frequencies and what maintains genetic variation in populations. Chapter 6 described how development controls the expression of genetic variation, thus playing an important role in determining the response to selection. In this chapter we look at the major feature of genetic systems: whether organisms are sexual or asexual. Our first important point is that sex did not evolve for reproduction. Sex is a method to produce genetically diverse offspring.

Whether an organism is sexual or asexual is the major characteristic of its genetic system. Sex produces genetically diverse offspring.

Genes that increase reproductive success will spread through a population, whereas genes that lower reproductive success will disappear. Thus most, if not all, traits have proven their usefulness for reproductive

Reproduction is the center of biology.

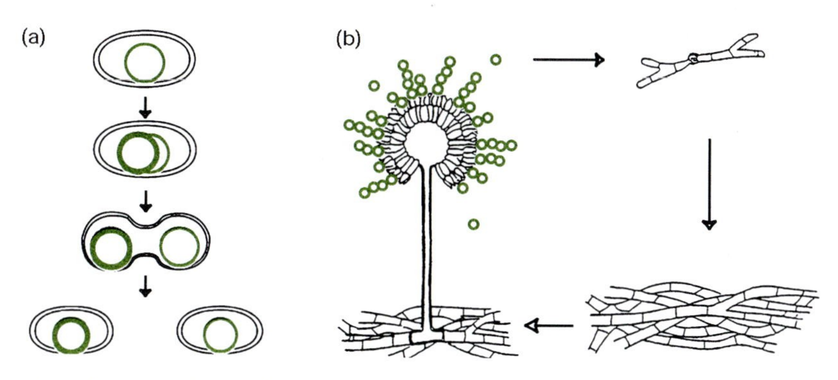

Fig. 7.1 Asexual reproduction. (a) In a unicellular organism, such as a bacterium, first the genetic material is copied and then the cell divides into two daughter cells, each receiving one copy of the genetic information. (b) In a multicellular organism, such as the fungus *Aspergillus niger*, a specialized structure walls off spores by successive mitotic divisions.

success and have been shaped by natural selection to contribute directly or indirectly to reproduction. Viewed in this way, reproduction is the center of biology, everything else—including development, physiology, behavior, and genetics—being at its service.

There are many types of reproductive systems. Single-celled organisms such as bacteria or yeasts reproduce by division into two daughter cells (Fig. 7.1a); many multicellular algae and fungi bud off single cells (spores) that develop into new multicellular individuals (Fig. 7.1b).

Both examples are cases of asexual reproduction, a type of reproduction that an engineer might design. It is basically a copying process: copy the genetic instructions necessary for the organism to develop and function, and package them in a suitable carrier from which the new organism can start to grow.

In sharp contrast to this 'logical', asexual process is sexual reproduction. Here the genetic instructions from two lines of descent fuse to create a new individual (Fig. 7.2).

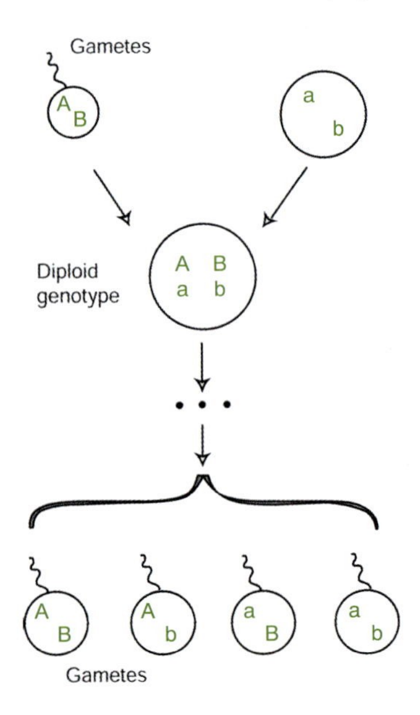

Fig. 7.2 Scheme of sexual reproduction. Two haploid gametes, mostly derived from different individuals, fuse to form a diploid zygote. In multicellular organisms the zygote develops into an adult individual which then produces gametes by meiosis. In many unicellular organisms the zygote directly undergoes meiosis. In the latter case, the meiotic products are vegetative offspring, which at a later stage may form gametes by mitotic divisions. At meiosis two important processes occur, recombination and segregation. By recombination, gametes are formed containing allelic combinations which are different from those in the original parental gametes that formed the individual. Segregation refers to the distribution of the genetic material over the haploid meiotic products in such a way that each cell receives a complete haploid set of the chromosomes and that, normally, the paternally and maternally derived alleles are equally represented among the haploid cells.

Sex occurs in two types of life cycles: diplontic and haplontic.

There are two basic types of sexual life cycles, as discussed in Chapter 4. For clarity, part of Fig. 4.1 is repeated in Fig. 7.3. In a **diplontic life cycle** the adults are diploid and the haploid gametes are produced by meiosis. In a **haplontic life cycle** haploid adults produce gametes by mitosis and the newly formed diploid zygotes immediately undergo meiosis to produce haploid individuals.

Sex is complex, inefficient, and costly.

No sensible engineer would ever propose such a process when asked to design a reproduction machine. In fact, sex probably did not evolve for reproduction. The intimate linkage of reproduction and sex in many organisms should not distract us. It is a **derived** state. In its essence—two genomes merge, recombine and segregate—sex has nothing to do with reproduction, for it does not increase the number of individuals.

Sex is less efficient and more complicated than asex, for it takes longer than simple mitotic division and involves more complicated mechanisms. Several difficulties with sex are discussed below; we mention an obvious one here—logistics. An asexual organism can reproduce by itself, but a sexual organism must find a partner to mate. This may be a big problem, especially for sessile organisms and for small, short-lived organisms that live at low population densities. Many elaborate adaptations have evolved to help mating organisms find each other, including the swarming of palolo worms (Chapter 1) and the chemical substances (sex pheromones) produced by female butterflies and detected with extraordinary sensitivity by males. Other organisms can even be used to bring together the gametes of sessile organisms, as in the many plants that depend on insects for pollination.

Why is there sex? A puzzle not yet solved.

This incomplete list of complex, inefficient, and costly aspects of sex suggests why evolutionary biologists are fascinated by the widespread occurrence of sexual reproduction. 'Sex is the queen of problems in evolutionary biology' (Bell 1982). Sex is very common, especially among

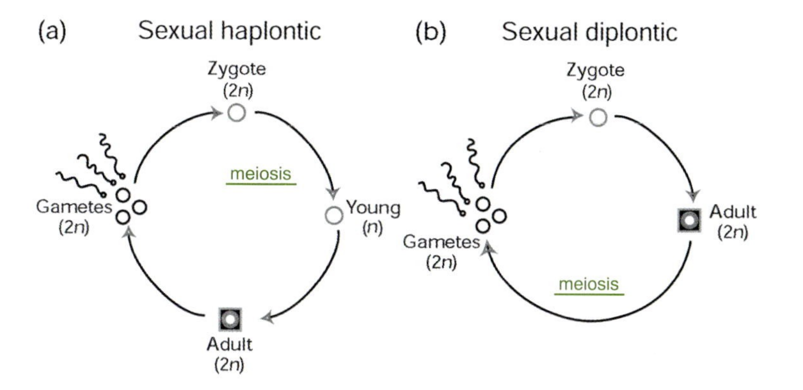

Fig. 7.3 Diplontic and haplontic sexual life cycles. The essential difference between these two types of life cycle is the position in the cycle where meiosis occurs. If zygotes directly undergo meiosis, the resulting cells function as vegetative (and obviously haploid) offspring. If zygotes do not undergo meiosis, they develop into multicellular (and diploid) offspring. Then meiosis occurs in the adult offspring.

large, multicellular organisms, and affects their biology deeply. In a world without sex there would be no males and females with their characteristic differences, no flowers and no insects specialized in pollinating or feeding on them, no extravagant color and form like the peacock's tail, and no behaviors aiming at finding and selecting mates. It is, however, far from obvious why sex should be better than asex. This is the great riddle of the evolution of sex. Something so striking and prevalent, with such large effects on almost every aspect of biology, must surely have an obvious adaptive explanation, but so far we have had a hard time finding this 'obvious' reason.

Variation in sexual life cycles

Although the basic principles of sex in **eucaryotic** species are the same, organisms differ greatly in the extent to which reproduction depends on it and in its frequency of occurrence. In **procaryotes**, processes similar to eucaryotic sex also result in genetic recombination. To illustrate these differences we describe sex in several organisms.

In bacterial conjugation genetic material is transferred from a donor into a recipient cell which gets a recombinant genome. Conjugation is rare under natural conditions.

The bacterium *Escherichia coli* reproduces by asexual division (Fig. 7.1a). However, in a process called **conjugation** (Fig. 7.4) genetic material from a donor cell is transferred into a recipient cell that ends up with a recombinant genome. To that extent bacterial conjugation is analogous to eucaryotic sex, but in bacteria the homologous DNA that recombines almost never consists of two complete single genomes. Instead, an intact genome makes contact with small 'foreign' DNA fragments that recombine at homologous places by crossing-over, breakage, and reunion (see Fig. 7.5). One product of recombination is a partial genome that is usually lost during subsequent cell growth. Thus in bacterial sex, unlike meiosis, there is no segregation. Donor cells that can transfer part of the bacterial chromosome into a recipient cell are rare under natural conditions.

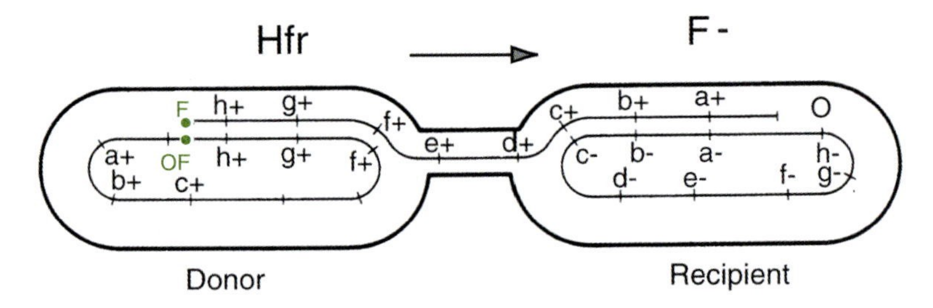

Fig. 7.4 Bacterial conjugation. A plasmid (in *E. coli* the so-called F factor) integrates into a copy of the bacterial chromosome; during cell-to-cell contact with a bacterium lacking the F factor, the plasmid can transfer part or all of this bacterial chromosome into the recipient cell; the transferred segment may then recombine with homologous parts of the recipient cell's chromosome. The letters label identifiable genes; + genes come from the donor, – genes are in the recipient. (From Griffiths *et al.* 1996.)

The unicellular green alga *Chlamydomonas eugametos* lives in small freshwater ponds and in wet soils. Asexual reproduction occurs by binary fission under conditions favoring growth. Under other conditions, including nitrogen limitation, vegetative cells may turn into gametes. There is no morphological differentiation of gametes into eggs and sperm, but there are two genetically determined mating types, called + and –. All cells of a clone have the same mating type, and sexual fusion to yield zygotes is only possible between gametes of different mating type. Thus no sex is possible in a local population that derives clonally from a single cell. Zygotes develop a thick protective wall, and germination of zygotes requires special conditions and is not easily achieved in the lab.

The green alga, *Chlamydomonas*, has two genetically determined mating types, + and –. Sexual fusion resulting in zygote formation is only possible between gametes of different mating type.

The ascomycete fungus, *Aspergillus nidulans*, lives on organic substrates in dry soils and reproduces by forming both asexual conidiospores and sexual ascospores. Ascospores differ from conidiospores in having a thick, protective wall, which enables them to survive temporarily unfavorable conditions. The fungal soma consists of hyphae—long, multinucleate, cylindrical cells that grow at one end. By branching and intra-individual fusion, the hyphae form a network called a mycelium. A mycelium contains millions of haploid nuclei, which are—barring mutations—all identical. The species is self-fertile: a single individual can form sexual spores. From a genetic point of view selfing in a haploid organism differs essentially from selfing in diploids, for the fusion of identical haploid nuclei and subsequent meiotic segregation produces ascospores that are genetically identical to the parental nuclei and to the asexually produced conidiospores. Thus sex in this self-fertile haploid organism does not involve genetic recombination. Selfing in diploids, as in many plants, may involve recombination, for fusion can occur between genetically different gametes. When different *A. nidulans* individuals are grown together, some outcrossing occurs and results in recombinant ascospores, but how often such outcrossing occurs under natural conditions is not known.

The ascomycete fungus, *Aspergillus*, is self-fertile: a single individual can form sexual spores. Sex does not lead to genetic recombination.

Taraxacum officinale, the common dandelion, is geographically widespread in Europe, occurring from the Mediterranean to the Arctic. Near the Mediterranean most plants are diploid and sexual, whereas asexual triploid forms predominate to the north. Only asexual forms occur in North America. Like the sexuals, the asexual plants produce bright yellow flowers, but reproduction, by a process called **apomixis**, produces seeds with genes identical to those in the parent. Asexual plants are thought to arise repeatedly but infrequently in a sexual population, thus producing asexual lineages (clones).

Asexual dandelions arise in a sexual population repeatedly but infrequently.

Arabidopsis thaliana, thale cress, is a tiny, annual flowering plant much used in genetic studies. Unlike many other flowering plants that have the option of both sexual reproduction and asexual propagation by vegetative runners, it reproduces exclusively via sexual seeds. Most seeds are produced by self-fertilization.

Thale cress reproduces mostly via self-fertilized sexual seed formation.

Cryptomyzus is a genus of aphids, plant-sucking insects that cause much

Aphids are viviparous and cyclically parthenogenetic.

damage in agriculture. Aphids, in contrast to most other insects, are viviparous: instead of producing eggs, females give birth to small larval aphids. In the ovarioles of these larvae development of new offspring is already taking place; the daughter starts reproduction before her mother is finished producing her. Their life cycle is also marked by **cyclical parthenogenesis**. From spring to autumn, females, without being fertilized, produce daughters that are genetically identical to their mother. In the autumn males and sexual females are produced, mating and sexual reproduction take place, and females lay eggs that can survive the winter.

Mammals are exclusively sexual and cannot become parthenogenetic because of genetic imprinting.

Homo sapiens is a mammal with a world-wide distribution and, like all mammals, exclusively sexual reproduction. Indeed, asexual reproduction does not seem to be possible because of genomic imprinting (Chapter 1). The expression of some genes important during early embryonic development depends on whether they spent the previous generation in a male or in a female germ line. At critical stages of development only the allele that came from the mother is expressed and the paternal allele is inactivated, while at other stages the paternal allele is active and the maternal allele is silent. A parthenogenetic mutation, causing the formation in a female of a diploid egg that could start development on its own, would not survive, for development would cease as soon as an essential maternal allele is inactivated by imprinting and a paternal homologue is required to carry out an essential function. Evolutionary aspects of genomic imprinting are further discussed in Chapter 10.

Patterns of sexual distribution

Comparative studies have yielded several empirical generalizations about the phylogenetic and ecological distributions of sex and asex (e.g. Glesener and Tilman 1978; Bell 1982; Bierzychudek 1987).

Phylogenetic distribution of sex

Exclusively asexual species often have a relatively recent evolutionary origin.

Many organisms can reproduce both asexually and sexually: only rarely is sex wholly absent. Purely asexual species often originated relatively recently; they appear to be the short-lived offshoots of sexual ancestors (Bell 1982). Exclusively sexual reproduction also occurs in a few taxa, most notably mammals. Among the majority that possess both options, asexual reproduction predominates among small organisms with very large populations, whereas sex predominates in large organisms with smaller populations.

Asexuals tend to be found at higher latitudes, in more disturbed habitats, and with a broader distribution than sexuals.

Ecological distribution of sex

In the many species that have both options, asexual lineages tend to be found at higher latitudes and in more disturbed habitats than sexuals. In general, asexuals have a wider distribution than the sexuals, and sexuals are found in habitats where selection may be mainly biotic rather than physical.

Consequences of sex

Direct consequences: segregation and recombination

Genetic recombination is the prime characteristic of sex.

In eucaryotes, sex is essentially a repetition, each generation, of combining two different (normally haploid) single genomes into a double (normally diploid) genome, followed by subsequent reduction to single genomes (Fig. 7.3). Meiosis regulates the reduction from diploid to haploid cells. It differs from mitosis in allowing crossing-over and in halving the number of chromosomes per cell. In the diploid nucleus of a cell undergoing meiosis the chromosomes are arranged in homologous pairs, from each of which one of the two chromosomes is transported to each of the two daughter cells. Because of this meiotic segregation, the haploid cell products of meiosis contain a complete set of genetic instructions. In most eucaryotes, if meiotic segregation is to be accurate, chromosomal crossing-over cannot be avoided. During chromosomal crossing-over nonsister chromatids of homologous chromosomes come into physical contact at a homologous site and exchange chromosome parts by breakage and reunion (Fig. 7.5).

Since the two homologous chromosomes involved in crossing-over often have different alleles at many loci, crossing-over produces chromosomes containing alleles from both parents of the individual in which the meiosis occurs. Crossing-over is not the only reason for recombination. The independent segregation of chromosomes causes the recombination of genes on different chromosomes. Recombination is thought to explain the evolutionary success of sexual reproduction. Because recombination is the key feature of sex, processes causing recombination in procaryotes, such as bacterial conjugation or transformation, are also called 'sexual' (Fig. 7.4).

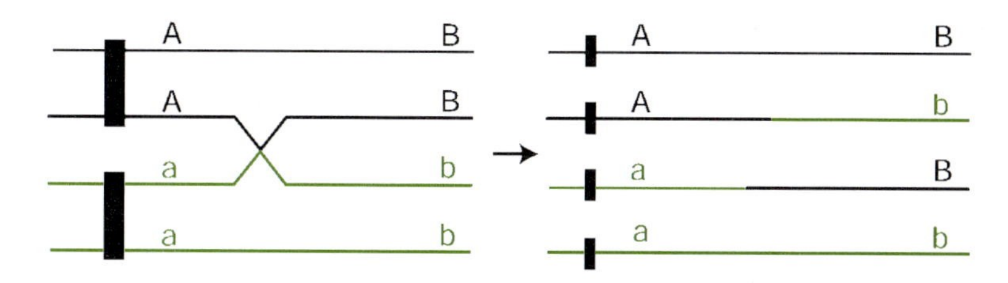

Fig. 7.5 Crossing-over of nonsister chromatids during meiosis, resulting in an exchange of homologous chromosome parts by breakage and reunion, may produce chromosomes with a new allelic composition. Recombination produces new genetic variants.

Indirect consequences: males and females

The male–female dimorphism has major evolutionary consequences.

The main indirect consequence of sex is the evolution of differences between males and females. This male–female dimorphism itself has far-reaching evolutionary effects, for sexual selection is the consequence of the difference in size of male and female gametes, and sexual dimorphism of adult phenotypes is the consequence of sexual selection (Chapter 9).

Sex cells are usually anisogamous, less often isogamous, but isogamy is the ancestral condition.

Recall that eucaryotic sex is essentially the fusion of two haploid nuclei to form one diploid nucleus which, either directly or after several mitotic divisions, produces haploid nuclei by meiosis. There is nothing inherently asymmetric about this process. It is quite conceivable that sexual fusion could occur between similar haploid cells that differ only in their genotype. However, with few exceptions, sexual fusions are asymmetric. The two partners that produce the sex cells often differ in morphology and behavior (male and female), and a more fundamental asymmetry exists between the sex cells themselves, where two cases can be distinguished. The most common situation is **anisogamy**, where the fusing gametes have different morphology and size. The big, scarce gametes are female (egg cells or ovules), and the small, numerous gametes are male (sperm cells or pollen). The less common situation is **isogamy**, where the fusing gametes do not differ morphologically and there is no male–female distinction. It occurs, among others, in ciliates, unicellular algae, and many fungi.

In many isogamous species, the two gametes that form a zygote, although of equal size, do differ in behavioral or in physiological details, each playing a different role in the fusion process. Frequently, isogamous species have two or more different gamete types. Because the words 'male' and 'female' are not applicable here, these are called **mating types** and are designated as '+' and '–', or as 'A' and 'a'. Organisms with more than two mating types include the many mushroom species with large numbers of mating types; mating is only possible between different mating types.

Because eggs are 'expensive' and sperm is 'cheap', females are choosy.

In anisogamy the large female and small male gamete differ in 'value'. A female gamete represents a large investment of resources, the male gamete a small investment. This difference in investment has led to sexual selection, which involves competition for and choice of mates. Since a female gamete is 'expensive' and a male gamete is 'cheap', one expects females to be choosy, selecting sperm of high quality for fertilization, and males to compete with each other for access to females. Thus anisogamy is a necessary prior condition for sexual selection, which has then caused the evolution of many differences between males and females (Chapter 9).

Because all isogamous gametes have the same 'value', no further differences have evolved between isogamous individuals of different mating type, for sexual selection is nonexistent or very weak.

Isogamy appears to be the **ancestral** condition: anisogamy appears to have evolved from isogamy. The models that have been studied to understand what might select for anisogamy assume that individuals may either produce a few big gametes or many small ones: that the product of gamete size times gamete number is the same for every individual. Anisogamy appears to be favored if larger zygotes are much fitter than smaller ones. Then gamete specialization becomes advantageous: one type (the egg) specializes in provisioning the offspring during its early development, and the other type (the sperm) specializes in finding the egg.

Anisogamy evolves when one type of gamete specializes in providing nutrients for the offspring, and the other specializes in finding the egg.

An explanation of the evolution of mating types from symmetric gamete fusion is more difficult. It is discussed further in Chapter 10.

To summarize, a plausible scenario for the evolution of male–female differences starts with selection for better-provisioned zygotes, leading to anisogamy in which two types of gametes are produced: big, expensive female gametes and small, cheap male gametes. This difference in cost of the two types of gametes led to sexual selection that shaped the often striking differences between male and female organisms.

Other indirect consequences of sex follow from the male–female difference. They are caused by natural selection acting on the distribution of the male and female functions among individuals, on the allocation of reproductive effort to male versus female offspring, and on the relative frequency of males and females. These and related issues, collectively termed **sex allocation**, are dealt with in Chapter 8.

The evolutionary maintenance of sex: theoretical ideas

We now turn to one of the most intriguing questions in evolutionary biology: when and why is sexual reproduction favored over asexual reproduction? This question is obvious, and from Darwin onwards biologists have repeatedly tried to answer it. Since about 1970 attempts to reach a conclusive understanding of the significance of sex have intensified, and many books and articles have been devoted to a problem that seems to be very difficult to answer. A clear answer is still not within reach. It is surprising and annoying that a functional understanding of such a conspicuous and dominant trait has been so hard to obtain.

When and why is sex favored over asex?

How should one, in principle, answer the question? Understanding the function of sex requires a form of bookkeeping: listing and quantifying the selective advantages and disadvantages of sex and asex, and then judging the balance between the two. The list of advantages and disadvantages of sex has become so long that a simplifying classification of the hypotheses has become necessary (Kondrashov 1993). Here we describe some of the main ideas.

The twofold cost of sex

Consider a mutant female in a sexual population that reproduces asexually instead of sexually, for example a mutation that suppresses meiosis

Asexual mutations have a twofold fitness advantage: the compensating advantages of sex must be equally strong.

and allows eggs to develop by mitotic division into offspring genetically identical to the mother. If the asexual female produces the same number of offspring as sexual females, and if all offspring have the same average fitness, the asexual mutation would initially have a twofold fitness advantage, would spread rapidly into the sexual population, and would replace the sexual by asexual individuals (Fig. 7.6).

This argument assumes that males contribute only genes to their offspring. Then sexual females 'waste' half their reproductive potential on sons. If males enhanced the fitness of the offspring by providing parental care, the advantage of an asexual mutation would be less, for asexual females would have to provide all parental care themselves.

Recombination creates and destroys favorable gene combinations

Sexual recombination creates new and destroys existing combinations.

Sexual recombination both creates new combinations of alleles and destroys existing combinations. Suppose that in a haploid population the allele combinations AB and ab have positive effects on fitness, while Ab and aB have negative effects. For example, the loci A and B might code for enzymes that function in the same metabolic pathway, and the enzymes specified by the different alleles might differ slightly in chemical characteristics. Since the two enzymes have to cooperate, it may happen that some combinations (AB and ab) function better than other combinations (Ab and aB). The fitness consequences of recombination then depend on the population composition. If AB and ab are common and Ab and aB are rare, then when the alleles are combined independently and mating is at random, recombination destroys more AB and ab geno-

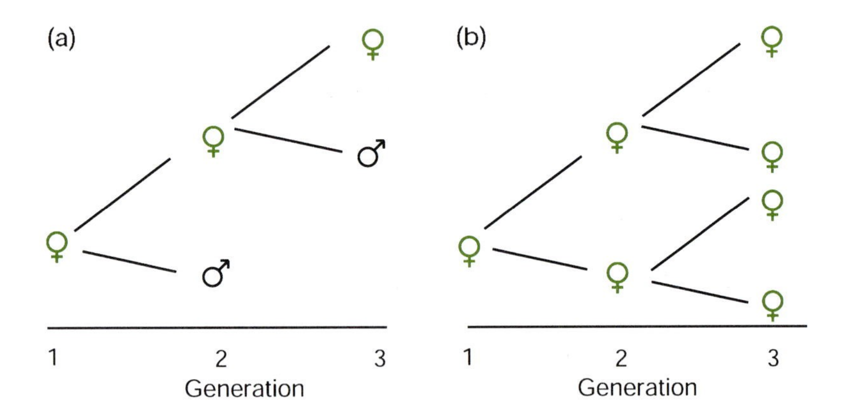

Fig. 7.6 The twofold cost of sex. A mutation that suppressed meiosis and allowed eggs to develop parthenogenetically would enjoy a twofold increase relative to its standard sexual allele, if sexual and asexual females produced on average the same number of offspring with the same average fitness. Here it is assumed for convenience that each female produces two offspring: a sexual female (a) one son and one daughter; and an asexual female (b) two daughters.

types than it creates, while more Ab and aB genotypes are created than destroyed. This is easiest to see in the extreme case of a population that initially consists only of AB and ab genotypes. If the two genes are on different chromosomes, then any mating between AB and ab will produce up to 50 per cent maladapted Ab and aB offspring by recombination. Here recombination has an unfavorable effect. If Ab and aB are initially more common than expected from independent combination, recombination will have a positive effect on average. Thus natural selection may favor recombination or not, depending on the distribution of the two-locus genotypes in the population.

Recombination affects the speed of evolution

In a sexual population, recombination can bring together in the same genome two (or more) beneficial mutations that happened in different organisms, whereas in an asexual population this genotype can only be created when the second mutation occurs in a genotype already carrying the first mutation (Fig. 1.6). In large populations, in which different beneficial mutations occur in different organisms within a short period of time, evolution will often be faster in sexual than in asexual populations.

Recombination can bring together in the same genome beneficial mutations that arose independently.

A related idea is that sexual recombination can liberate an advantageous mutation which occurs in a genome with several deleterious mutations. Under asexual reproduction the advantageous mutation might disappear if deleterious mutations reduced the fitness of its carrier.

Recombination affects the elimination of mutations

Sexual recombination enables selective elimination of deleterious mutations and thus repairs genetic lineages with mutational damage. One way in which this can happen was pointed out by Muller (1964).

Recombination can cleanse genetic lineages of mutational damage.

Muller's ratchet: mutation accumulation in small populations

An asexual individual can only produce offspring with the same or a greater number of mutations than itself. This must be true if reversions (mutations fully restoring the original function) are so rare that we can safely ignore them. Consequently, the number of deleterious mutations in the genome of an asexual lineage will tend to increase over time. At best it could stay constant if the population size is so large that selection maintains genomes in which no new mutations have occurred. In small populations this is difficult because the optimal genotype can be lost by random drift. Therefore, small, asexual populations accumulate mutations; this has been called Muller's ratchet. In a sexual population, in contrast, an optimal genotype lost by drift can be reconstituted by recombination.

Muller's ratchet: In a small population, the number of deleterious mutations in an asexual lineage can only increase over time.

Sexual recombination also contributes to the elimination of deleterious mutations by promoting the elimination of several mutations simultaneously. This principle was also suggested by Muller and has been worked out by several population geneticists (Crow and Kimura 1979;

Recombination can eliminate several mutations at once.

Kondrashov 1982). The basic idea is that every deleterious mutation can only be removed from the population by failure of the individual carrying the mutation to reproduce (a so-called 'genetic death'). This implies that a high rate of deleterious mutations will impose a heavy burden on a population because many genetic deaths are required to prevent accumulation of the mutations in the population.

Calculations suggest that if the mutations reduced fitness independently (and were therefore eliminated independently, as in asexual reproduction), severe problems would arise if the deleterious mutation rate per zygote were about 1.0 or higher. In fact, as mentioned in Chapter 5, the rate of deleterious mutation in humans has recently been estimated to be at least 1.6 per zygote. So how do we cope with deleterious mutations? Sexual reproduction may help by recreating variation in the number of mutations per individual every generation. Each pair of parents produces some zygotes with fewer and some zygotes with more mutations than either of them carries. The calculations suggest that if different mutations tend to enhance each other's negative effects, then sexual recombination is particularly effective in the elimination of deleterious mutations, for individuals carrying many mutations have very low fitness. The 'genetic deaths' will occur more often among individuals carrying several mutations each, so that the elimination of relatively few individuals removes many mutations from the population.

Recombination affects host–parasite coevolution

Almost every living thing is host to parasites and pathogens, organisms that enhance their fitness by exploiting the host and lowering its fitness. Infectious diseases are one example. Host–parasite interactions are often frequency-dependent, for parasites adapt to the most common host type, giving rare host types an advantage. In fact, all interactions between species where the partners affect each other's fitness are likely to result in frequency-dependent selection on the traits that mediate the interaction.

Natural selection makes life difficult for common resistance and virulence types.

Consider the level of virulence of a pathogen and the level of resistance of its host. There is some evidence of genetic variation for both. Natural selection in the pathogen population will favor virulence mutants that have the most reproductive success in interaction with the most common resistance type in the host population. Similarly, natural selection in the host population will favor resistance types that minimize damage in interaction with the most common virulence type among the pathogens. This is simply because a pathogen most often encounters a common host genotype, and a host most often encounters a common pathogen genotype. The response to selection makes life difficult for the common resistance and virulence types. Rare resistance types are favored because selection on pathogens for specific virulence against them is weak, and rare virulence variants are favored because selection on hosts for specific resistance against them is weak—until their success makes them common and natural selection against them intensifies.

Thus negative frequency-dependent selection favors rare genotypes in both sexual and asexual species of hosts and pathogens. Such biotic interactions lead to relatively frequent changes of direction of selection. Because a rare genotype is favored selectively, it increases in frequency. As it becomes common, it is selected against. This again causes it to become rarer and thus selectively favored, and so on. In such a system it is important that a genotype that is temporarily selected against does not completely disappear before it is favored again—selection should not change genotype frequencies too fast at low frequencies.

Here sex enters the story. When selection regularly changes direction, sexual populations, which recreate genotypes by segregation and recombination, retain temporarily bad genotypes for longer than asexual populations under most conditions. A simple model of an extreme situation illustrates this for one locus with two alleles. Suppose that selection often changes direction. In one period the heterozygotes are inviable, but in the next both homozygotes are inviable. An asexual population would go extinct in the second period, but a sexual population could persist because it recreates the genotypes that were previously lethal. Similar models can be devised for multi-locus genotypes.

A sexual population can persist by recreating genotypes whose fitness fluctuates in the face of adapting parasites.

Sexuals are more vulnerable to genetic parasitism

Strict asexual reproduction implies cotransmission of all genes, nuclear and cytoplasmic. A mutation that lowers fitness will not spread, for its fate is completely linked to that of the organism and its other genes. Sexual reproduction implies that chromosomes and alleles segregate and recombine in every generation; not all genes are transmitted together to the offspring. This opens the door to the possible spread of mutants that cause unfair transmission at the expense of their nonmutant colleagues. Such mutations are called selfish because they promote their own spread at the expense of alternative alleles or the host organism. They include nuclear **meiotic drivers** (genes that distort meiosis to their own advantage) and cytoplasmic parasitic genetic elements. Several examples and mechanisms are dealt with in Chapter 10.

Sex opens the door to nuclear meiotic drivers and to cytoplasmic parasitic genetic elements.

The evolutionary maintenance of sex: empirical evidence

In contrast to the abundant theories on the maintenance of sex, only a few empirical tests have been done, for they are difficult. Several types of tests are conceivable: examining whether existing distribution patterns of sex and asex obey theoretical predictions; performing experiments in species that have both a sexual and an asexual life cycle, to see which conditions promote which mode of reproduction; and testing whether the assumptions of a theory are met.

Using the observed distribution patterns is problematic, for they cannot discriminate between alternative explanations. It is true that asexual

Distribution patterns can be explained by mechanisms consistent with several hypotheses.

reproduction is roughly correlated with relatively simple, abiotically extreme and disturbed habitats and that sex is correlated with biotically complex and undisturbed environments. However, this can be explained both by the relative absence of parasites from simple and extreme habitats and by arguing that some correlate of asexual reproduction is responsible for the better performance of asexuals in extreme habitats. For example, the asexual dandelions that predominate at higher latitudes are triploid, sexual dandelions are diploid, and triploid plants can often endure lower temperatures than diploids. Is the distribution caused by sex or ploidy? Who knows?

In New Zealand snails, sex is more frequent in populations with more parasites.

Lively tested the idea that host–parasite interactions favor sex. They sampled many populations of *Potamopyrgus antipodarum*, a diecious freshwater snail native to New Zealand. In this species both diploid sexual and triploid asexual females occur, reminding us of the dandelions discussed earlier. The snails host several parasitic trematode species. Lively (1992) found a positive correlation between the number of males per sample (indicative of the frequency of sexual reproduction) and the level of trematode infection, as expected if a high level of parasitism favors sexual over asexual reproduction.

Of course, the observed correlation may have other causes. Sexual individuals may be more easily infected than asexual females, for sex may help to transmit the trematodes. The authors argue that these alternative explanations are unlikely, that the simplest explanation of their findings is that asexual clones have displaced sexuals from habitats with low parasite risk, and that sexual individuals can persist in habitats where parasites are common.

The distribution of mutational effects in *Chlamydomonas* suggests that sex would be helpful in removing mutations in this species.

Another approach is to test experimentally whether the conditions necessary for a theoretical advantage of sex are satisfied. One of the conditions that is necessary for sex to eliminate mutations more effectively than asex is that deleterious mutations enhance each other's effects on fitness. De Visser reasoned that if this were the case, then the offspring at the low end of the fitness distribution should have lower fitness than would be expected if mutational effects were independent. To test this idea, de Visser *et al.* (1997) measured the fitness distribution of large numbers of offspring from crosses between two strains of the green alga *Chlamydomonas moewusii.* The parental strains probably differed in many mutations that had accumulated independently in their genome, for they had been cultured separately by asexual reproduction for hundreds of generations. If the parental mutations segregated independently, they should be binomially distributed over the offspring. They did find evidence for the required synergism between mutations, but the results did not rule out all other alternatives, and further work is necessary.

Discussion

The main theoretical ideas about the maintenance of sex do not exclude each other. They should all enter the bookkeeping scheme which quantifies the advantages and disadvantages of sex and asex. But how should we quantify them? Are they all valid for all species? If not, can there be one evolutionary explanation for sex? Should we take the pluralistic view that in every species sex is maintained by a different combination of factors? Or is some intermediate between these two extremes the best we can hope for: identifying a few factors that dominate in the maintenance of sex in most organisms, with many special cases of species where some particular factor plays a dominant role? These questions are not easily answered, and evolutionary biologists are still struggling with them. A pluralistic explanation of sex is not attractive, for it would imply that one of the most conspicuous and influential processes in biology does not exist because it has one or two important general functions, but because it affects many processes, with history largely determining which functions are relevant in which species. Nevertheless, the pluralistic view cannot be excluded, and if it is correct, then the search for a general explanation of sex may have taken so long because it is an illusion.

A pluralistic explanation of sex is not attractive, cannot be excluded, and may be correct.

There are certainly good arguments for a form of weak pluralism. For example, in mammals genomic imprinting appears to preclude any success of parthenogenetic mutants. Because this seems to be an absolute constraint, it is meaningless to discuss the potential optimality of sexual or asexual reproduction in mammals. Perhaps natural selection would favor asexual reproduction in humans if it could occur. We are forced to conclude that sex occurs among mammals for historic reasons that fixed genomic imprinting in the lineage, making the asexual alternative impossible, not because natural selection favors sex for some reason.

Sexual reproduction often occurs in life cycles with confounding elements, such as genomic imprinting . . .

Another example is the fungus *Aspergillus nidulans*. Because it has both a sexual and an asexual life cycle, the two modes of reproduction compete: both have to be actively maintained by natural selection. Four of the six theoretical advantages and disadvantages of sex identify genetic recombination as the key characteristic of sex. However, arguments based on recombination lose their force for *Aspergillus nidulans*, which only rarely produces sexual offspring by outcrossing. Most sexual spores are formed by selfing, which in a haploid fungus implies that diploid nuclei formed by fusions between two identical haploid nuclei undergo meiosis, only to produce haploid nuclei that are completely identical (except for mutation) to the nuclei at the start of the process. Here recombination is unlikely to play a significant role.

Another feature of the sexual cycle in *Aspergillus nidulans* is a better candidate for the maintenance of sex. Only sexual spores can resist harsh conditions for a long time. This ecological factor must enter the bookkeeping for this species, for it may be among the most important factors maintaining sex, but it is hard to see that producing resistant survival structures is an essential general function of sex. It may be another example of a historical contingency: for whatever reason, once established the association between sex and resistance of spores may be difficult to break and prevents the loss of sex.

and resistant stages.

We do not yet know why sex evolved.

At this point you are likely to be confused. No clear evolutionary explanation of sexual reproduction, one of the most striking features of organisms, seems to emerge. You have read about many different sexual life cycles and many different explanations, but—as so often in biology—the diversity obscures general principles and explanations. This is the present state of this field, whether we like it or not. The confusion you experience reflects the confusion among evolutionary biologists about the evolution of sex.

Summary

This chapter discusses why evolutionary biologists are puzzled by the existence of sexual reproduction and the solutions they have proposed for that problem.

- Asexual copying is an efficient and simple way to reproduce; sex is more complicated, takes more time and energy, and requires finding and selecting a good partner. Nevertheless most organisms are sexual.
- Evolutionary biologists have failed to find an obvious general explanation of why natural selection has produced and maintains sex. Many theories have been suggested. Most concentrate on genetic recombination at meiosis.
- Recombination may speed up adaptive evolution, it may be help in purging the genome of deleterious mutations, and it may be important in the coevolutionary struggle between hosts and pathogens. None of these seems to be a general explanation.
- Empirical tests of the hypotheses are badly needed but hard to do. We may have to accept that the evolutionary explanation of sexual reproduction is pluralistic, sex being maintained in different groups of organisms by different sets of selective factors.

Sex is one of the most important features of the life cycles of eucaryotic organisms. The next chapter discusses the evolution of the other major features of eucaryotic, multicellular life cycles: life history traits.

Recommended reading

Hurst, L. D. and Peck, J. R. (1996). Recent advances in understanding of the evolution of sex and maintenance of sex. *Trends in Ecology and Evolution*, **11**, 46–52.

Maynard Smith, J. (1978). *The evolution of sex.* Cambridge University Press, Cambridge.

Michod, R. E. and Levin, B. R. (ed.) (1988). *The evolution of sex.* Sinauer, Sunderland, Massachusetts.

Stearns, S. C. (ed.) (1987). *The evolution of sex and its consequences.* Birkhäuser, Basel.

AGA Symposium (1993). Evolution of sex. *Journal of Heredity*, **84**, (5).

Questions

7.1 Many organisms that can reproduce both sexually and asexually share a reproductive pattern. When the environment is favorable for growth, reproduction is asexual. When the environment deteriorates, sex occurs, for example at the end of the growing season. Sexually produced offspring are a survival stage, dormant until better conditions arrive. Which of the theoretical ideas about the function of sex is supported by this observation? And which of the disadvantages of sex are less of a problem?

7.2 In many old biology texts and in some recent popular articles and school books the biological function of sex is not presented as problematic. Sex is 'just' necessary for providing sufficient genetic variability, otherwise species would go extinct as soon as the environment changes significantly. How do you respond to this explanation?

7.3 Williams argued that in species with both sexual and asexual reproduction a long-term advantage of sex to the species would not explain its maintenance, and sexually derived offspring must have an immediate advantage over asexually produced offspring. Do you agree with his argument? Give reasons.

7.4 Recently sheep and mice have been cloned from adult somatic cells. Does this invalidate the idea that there can be no parthenogenetic mammals because genes essential for early development are imprinted differently in the male and female germ lines? What evidence would you need to decide?

Chapter 8
The evolution of life histories and sex ratios

Introduction

Imagine a very flexible zygote. How should it live the life it is about to begin? At what age and size should it start reproducing? How many times in its life should it breed? When it reproduces, how much energy and time should it allocate to reproduction, to growth, and to maintenance? Should its offspring be few in number but high in quality and large in size, or should they be small and numerous but less likely to survive? Should it concentrate its reproduction early in life and have a short life as a consequence, or should it put less into reproduction and live longer? How many of its offspring should be male and how many female? Should that decision depend upon ecological or social circumstances, or be fixed at birth? If it is a sequential hermaphrodite, should it be born as a male and turn into a female later, or the other way around? Those are some of the fascinating questions answered by life-history evolution and sex-allocation theory.

Variation in reproductive success consists of variation in life-history traits: how big organisms are, how many offspring they have, how long they live.

Life-history evolution also has important implications for evolution in general. Reproductive success is achieved through life-history traits, mainly survival, age and size at maturity, and fecundity; and variation in reproductive success is necessary for natural selection. Thus understanding variation in life-history traits is one key to understanding natural selection. The evolution of life-history traits results in basic characteristics of all species—how big they are, how long they live, how many offspring they have, how fast their populations can grow. And life-history traits participate strongly in the growth and fluctuation of populations, which determines how many individuals of a species are present. The population dynamics of interacting species—competitors, predators and prey, parasites and hosts, symbionts and mutualists—in turn contribute strongly to the structure of biological communities. Thus life-history traits participate in key ecological interactions—one reason why separating evolution from ecology would be artificial.

The ideal life history is simply stated: a 'Darwinian demon' matures at birth, immediately gives birth to an infinite number of offspring with the same characteristics, and lives for ever. Such organisms would rapidly fill the planet, but they do not exist. Why not?

The ideal organism matures at birth, immediately has an infinite number of similar offspring, and lives for ever.

Because mass is conserved, an organism cannot produce a mass of offspring greater than its own mass in one reproductive attempt. In practice, the upper limit is about half its own mass, with a few interesting exceptions, such as spiders that cannibalize their mother. How total offspring mass is divided into individual offspring determines the number of offspring and contributes to their 'quality', their ability to survive to reproduce. Being large is advantageous, for larger organisms can produce a greater total mass of offspring and thus more offspring of a given quality. But it takes time to grow large, and the longer one waits to mature while growing, the more likely it is that one will die without reproducing at all. Once reproduction begins, resources must be divided among immediate reproduction, growth or storage and thus future reproduction, and maintenance. These allocations have consequences, for if maintenance is neglected to increase reproduction, mortality rates will rise and life span will decrease, and if growth is neglected to increase reproduction, future reproduction will suffer.

The ideal organism does not exist because of trade-offs and constraints.

Those are the problems of life-history evolution explained in terms of benefits (changes in traits that increase fitness) and costs (changes in traits that reduce fitness). Organisms are expected to evolve to the point where the net benefit, the positive difference between benefits and costs, is greatest. That point—not always attainable—is determined by the connections among life-history traits and by how the environment affects mortality and fecundity.

Organisms should evolve to where the net benefit of trait states is greatest.

The cost–benefit explanation assumes that when the optimal solution is found, it is optimal for the whole population. While that assumption has been successful up to a point, it does not hold in general, not even for core life-history traits such as age and size at maturity. Nowhere is its violation clearer than in problems of sex ratio and sex allocation. If an organism can control how many female and male offspring to produce, then the best offspring sex ratio (sons:daughters) depends on the sex ratios produced by the other organisms in the population. If the others produce only males, it is best to produce only females. If the others produce only females, it is best to produce only males. The evolution of the sex ratio is the classical example of frequency dependence: the population sex ratio determines the mating opportunities for the offspring.

Optimality does not work when selection is frequency dependent, as it is for sex ratios and sex allocation.

The distinction between optimality and frequency dependence deeply affects our view of evolution. The optimality view is of a world at equilibrium in which solutions are reached by individuals acting independently.

The frequency-dependent view is of a world dynamically changing, in which the best strategy depends on what others do. This chapter offers an opportunity to see how well each approach performs in attempting to answer the main questions about life histories and sex allocation:

Four questions about life-history evolution and sex allocation.

1. Why not mature earlier or later, larger or smaller?
2. Why not have more or fewer, larger or smaller offspring?
3. What determines life span? Is aging inevitable?
4. Why not allocate more to male or to female function?

The evolutionary explanation of how organisms are designed

To explain the evolution of life history traits, we need information on:

In Chapters 3–5, we analyzed evolution from the genetic point of view. That is a good way to understand changes in allele frequencies, and it is a necessary part of the evolutionary explanation of the adaptation of life-history traits, but it is not by itself sufficient to explain them. To explain the evolution of life histories, we have to combine information from four sources.

how selection pressures on life-history traits vary with the age and size of organisms—this comes from demography;

1. Selection pressures on traits vary with the age and size of the organisms in which they occur. The field that explains this is *demography*. Demography connects age- and size-specific variation in survival and fecundity to variation in fitness, and thereby tells us the strength of natural selection on life-history traits. For example, natural selection on reproductive performance is stronger in younger than in older adults (what is 'young' and what is 'old' depends on the species). Consider a population at evolutionary equilibrium in which the average values for age and size at maturity, and survival and fecundity rates, are close to the optimum for all age classes. Now let extrinsic mortality rates increase in one age class, for example because of a new age-specific disease or size-specific predator. The change in mortality rates 'devalues' that and all subsequent ages because organisms are less likely to survive to those ages to reproduce. The evolutionary response will be decreased investment in older age or size classes and increased investment in younger ones.

the inheritance and phenotypic plasticity of life-history traits—this comes from quantitative genetics;

2. Life-history traits are influenced by many genes; they are polygenic or quantitative traits (Chapter 4). The insights of **quantitative genetics** are important for life-history evolution. Recall that only a certain part of the genetic variation of a trait determines its reaction to selection; this part is the additive genetic variation. The proportion of the total phenotypic variation of a trait that is contributed by additive genetic variation is its heritability. When heritability = 1.0, the trait has exactly the same value in the offspring as it does in the average of the two parents; when heritability = 0.0, none of the phenotypic variation can be attributed to additive genetic variation, and the trait will not respond to selection. In many species, the heritabilities of life-history traits are in the range 0.05–0.40. Thus most life-history traits that have been investigated could respond to selection.

In its original form, quantitative genetics was applied to laboratory organisms or domestic plants and animals, where one could assume constant environmental conditions. It has been extended to natural populations by including the effects of reaction norms. Structure in bundles of reaction norms alters the relationships of traits to fitness and the expression of genetic variation, changing both the selection pressures on traits and the capacity for a genetic response to selection across environments (Chapter 6).

3. Life-history traits are connected by trade-offs, which exist when a change in one trait that increases fitness is linked to a change in another trait that decreases fitness. The response of life-history traits to a novel selection pressure depends on the strength of the trade-offs present. An improvement in one trait that is linked to high costs in connected traits cannot proceed very far. Important trade-offs include those between the number of offspring and their survival as juveniles, and between reproductive investment and adult survival.

the trade-offs in which the traits are involved—this determines the costs and benefits of evolutionary changes in the traits; and

Trade-offs have both a genetic and a physiological component. The genetic component can be expressed as a genetic correlation, which, like heritability, depends on the additive genetic variance of the traits. If there is a genetic correlation between two traits, then some genes affect both traits. If those effects are on average in the same direction for both traits, the genetic correlation will be positive. If the effects are, on average, positive on one trait and negative on the other, the genetic correlation will be negative.

The physiological component depends on how the organism is constructed and is a mixture of types of connections among traits. Some of those connections are the same for all individuals in a species, have been inherited from ancestors, reflect the phylogenetic history of the species, and differ among taxonomic groups. Other connections among traits vary among the individuals of a species for two reasons: developmental interactions with the environment, which are different for every individual, and variation in the genes that affect the traits involved in the trade-off.

Both causes of variation in traits—genetic and physiological—are constrained by fixed effects expressed in development, e.g. how the nymphalid ground plan, controlled by conserved genes, constrains genetic variation for wingspot patterns in butterflies (Chapter 6).

4. Traits also need to be understood in phylogenetic context. Phylogenetic effects are the contribution to traits shared by all individuals of a species or **clade**. We normally think of them as 'the development' or 'the physiology' or 'the morphology' of a species or higher taxon. To understand how broadly those traits are shared, and where in the history of the lineage they might have originated, we need to compare them with traits in close and distant relatives. The comparative method can tell us how much of a pattern to attribute to history and lineage, and how much to attribute to microevolutionary processes that operated within the local population in the recent past (Chapter 16).

the phylogenetic context in which the traits sit—this represents the effects of history.

Thus to explain the evolution of life-history traits, we need to understand the demographic selection pressures operating on them; the quantitative genetics that determines their response to selection; the trade-offs, both genetic and physiological, that connect traits; and the phylogenetic context in which they sit. This is how we explain the evolution of any trait. The explanation has an intrinsic part (genetics, trade-offs, phylogenetic effects) and an extrinsic part (selection pressures expressed as effects on age- and size-specific mortality and fecundity rates). With that in mind, we discuss next the major life-history traits: age and size at maturity, clutch size and reproductive investment, and life span and aging.

The evolution of age and size at maturation

At evolutionary equilibrium earlier or later maturation would lower fitness.

Age at maturity is a dividing line, for up to maturation natural selection for survival is strong, and after maturation aging begins. Fitness is often more sensitive to changes in age at maturity than to changes in other life-history traits, and at evolutionary equilibrium, either earlier or later maturation would lower fitness. 'Small organisms are usually small not because smallness improves fecundity or lowers mortality. They are small because it takes time to grow large, and with heavy mortality the investment in growth would never be paid back as increased fecundity' (Kozlowski 1992).

Early maturation reduces the chance of dying before reproducing and shortens generation time (benefits),

Both early and delayed maturity have fitness costs and benefits. Early maturity has at least two benefits: it reduces the chance of dying before reproducing, and it reduces generation time. Shorter generations mean that offspring are born earlier and start reproducing sooner. One cost of early maturation, the instantaneous mortality rate of the offspring, rises when females mature earlier (Fig. 8.1). A female that tried to mature very early would be so small and poorly developed herself that she could not produce any offspring at all (aphids are an exception).

but results in fewer offspring, with higher mortality rates (costs).

Two benefits of delayed maturation are often important. If delaying maturity permits further growth and fecundity increases with size, then delaying maturity leads to higher fecundity. If delaying maturity improves the quality of offspring or parental care, it improves offspring survival. Maturation will be delayed for these reasons until the fitness gained through increased fecundity and better offspring survival is balanced by the fitness lost through longer generation time and lower survival to maturity.

Age at maturity appears to be adjusted to an intermediate optimum.

Thus earlier maturation brings shorter generation times and a shorter period of exposure to mortality before maturity; later maturation brings increased fecundity and lower juvenile mortality per time unit. At an intermediate age, an optimum should exist (Fig. 8.2). Models making these assumptions accounted for 80–88% of the variation in natural populations of lizards, salamanders, and fish (Stearns and Koella 1986). Age at maturity appears to be adjusted to an intermediate optimum in many cases.

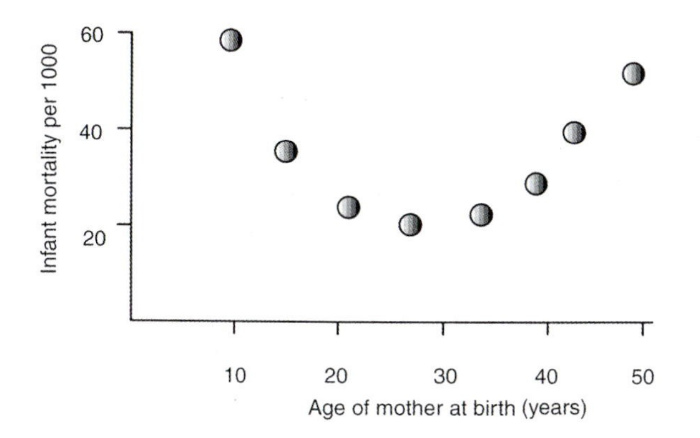

Fig. 8.1 Infant mortality rate as a function of mother's age, humans, United States 1960–61, based on 107 038 infant deaths documented by the National Center for Health Statistics. Each point is the mean of a 5-year age class (after Stafford, unpublished data).

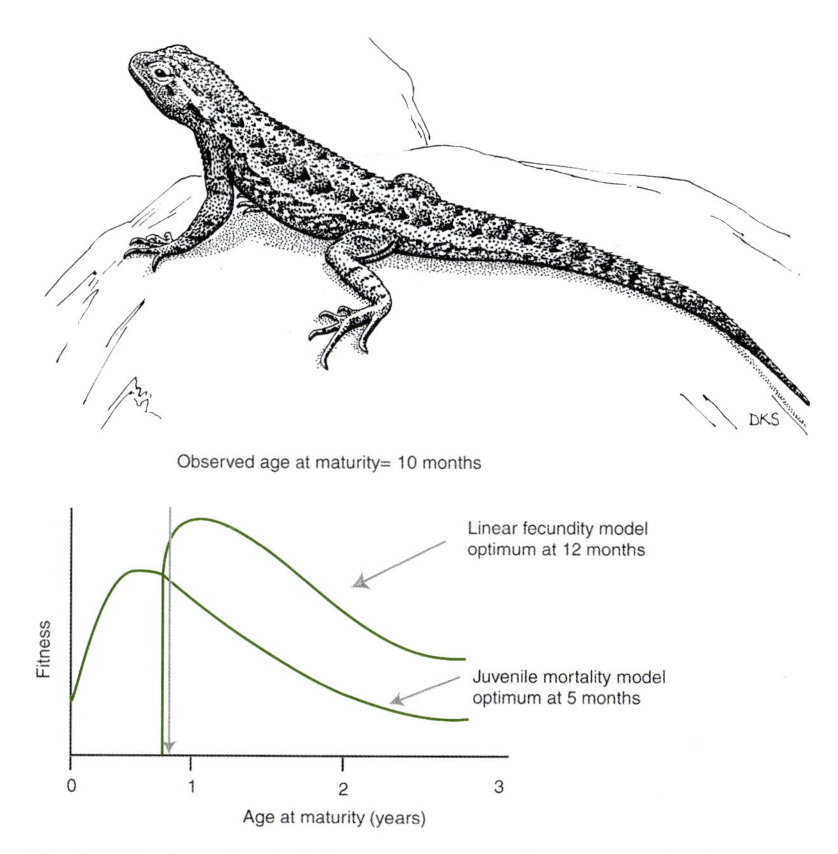

Fig. 8.2 The fence lizard, *Sceloporus*, matures at about 10 months (by Dafila K. Scott). Optimality models for age at maturity are reasonably successful if they assume that the main advantage of delaying maturity is the increased fecundity that comes with larger body size.

Because age and size at maturity vary among closely related species, among populations within species, and among individuals within populations, we know they can respond rapidly to natural selection. Age at maturity responds to artificial selection in flour beetles and fruit flies, and genetic variation for age and size at maturity has been documented in many species. Thus age and size at maturity can be adjusted by natural selection to local conditions within populations. There are also constraints on age and size at maturity imposed by history and design.

The evolution of clutch size and reproductive investment

The second major feature of the life cycle is how reproductive investment varies with age. Reproductive investment is the product of the number of offspring and the investment per offspring. Thus we can ask two questions about its evolution: 'How large should each offspring be (at birth for organisms without parental care, at independence for organisms with parental care)?', and 'How many offspring should be produced in each reproductive attempt?'

Some world records for reproduction.

Before looking at the theory, we describe some variation that exists in reproductive investment. Bats have the largest offspring for their body size in the mammals, the record being held by *Pipestrellus pipestrellus.* It bears twins whose combined weight at birth is 50% of the mother's weight after birth—and she must fly and feed while pregnant. The flightless New Zealand kiwi, a bird the size of a large chicken, lays the largest egg, for its size, of any bird, more than five times larger than the largest eggs of domestic chickens. The caecilian, *Dermophis mexicanus,* a tropical amphibian, gives birth to a clutch that weighs up to 65% of her post-birth weight. If we measure reproductive effort by the weight of offspring when they become independent of their parents, rather than by their weight at birth, then altricial birds (birds that hatch naked, blind, featherless, and dependent) make amazing reproductive efforts. Their combined offspring can weigh at fledging up to eight times the weight of the parents, which must work very hard to feed them. The world record for smallest size and worst juvenile survival is probably held by an orchid seed, weighing less than a microgram and with a chance of about one in a billion of surviving to reproduce. The insects with the best juvenile survival are probably species of dung or carrion beetles that lay only four or five eggs; their juveniles must survive as well as juvenile elephants or whales.

The Lack clutch

The Lack clutch: too many offspring starve.

How many offspring of a given size should an organism produce in a given reproductive attempt? Attempts to answer this question began with David Lack, who suggested that altricial birds should lay the number of eggs that fledge the most offspring. Thus he assumed that only one trade-off was important: the trade-off between the number of eggs laid and the probability that offspring would survive until they left the nest. If the probability of surviving to fledge decreases linearly as the number of offspring increases, and if there are no trade-offs with parental survival or with offspring survival to maturity, then the relation between clutch size and fitness (number of offspring fledged) is a simple parabola with an intermediate optimum (Fig. 8.3).

There have been many attempts to determine whether altricial birds produce clutches that are optimal in the sense just described. Clutches are often smaller than those that would be optimal if the only trade-off were between number of eggs and survival to fledging. What are the reasons for deviations, positive or negative, from the Lack clutch? Answers for deviations in both directions may apply to all organisms, not just birds.

Effects that lead to an optimal clutch smaller than the Lack clutch include additional trade-offs with parental and offspring life-history traits, temporal variation in optimal clutch size, and parent–offspring conflict won by the offspring.

Offspring from and parents of larger clutches sacrifice weight, have worse survival and poorer reproduction.

Much is known about trade-offs between clutch size and other life-history traits. By 1992 there had been 55 studies of the effects on various life-history traits of manipulating clutch size by adding or removing eggs or chicks (Stearns 1992). When clutches were enlarged, weight of fledglings was reduced in 68%, survival of fledglings to the next season was reduced in 53%, the weight of parents was reduced in 41%, the survival of the parents to the next season was reduced in 36%, and the future reproduction of the parents was reduced in 57% of the studies in which they were measured. The reproduction of offspring from larger clutches was not measured often, but every time it was measured, it was reduced in larger clutches. Thus there are often good reasons not to lay more eggs.

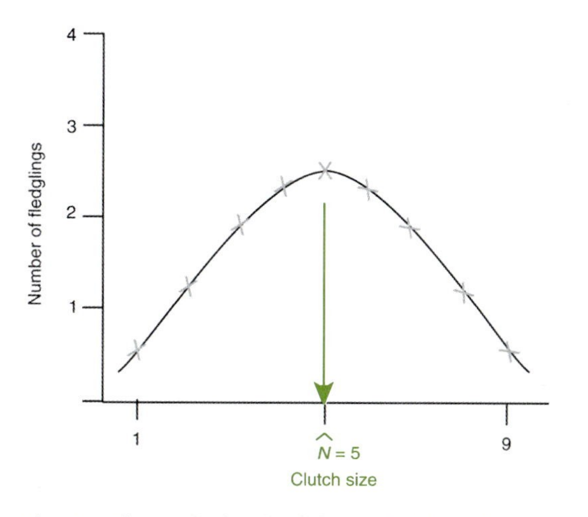

Fig. 8.3 The Lack clutch: if the probability that a young bird will survive to leave the nest declines linearly with the size of the clutch, then it is best to lay an intermediate number of eggs.

Variation in offspring survival rates selects for smaller clutches.

When offspring survival varies from year to year, producing a large clutch in a bad year will yield few survivors. If there is a trade-off between reproductive investment in one year and reproductive investment in the next year, so that a large clutch in a bad year could be followed by a small clutch in a good year, then optimal clutch size will be reduced. Variation in year-to-year juvenile mortality rates appears to have selected for reproductive restraint and long lives in many marine fish and forest trees.

Parent–offspring conflict won by the offspring also reduces clutch size.

Parent–offspring conflict can also reduce clutch size. In species that reproduce several times per lifetime and in which reproduction is costly, the parents will be selected not to invest all their resources in the current clutch but to save something for future reproduction. The offspring, on the other hand, will be selected to extract more investment from the parents than it is in the parents' interests to give. Such conflicts are often especially obvious at weaning or fledging, when parents may force offspring to become independent. If the conflicts are costly, clutch size will be reduced.

Unpredictable resource levels select for increased brood size and flexible brood reduction.

What might lead to an optimal number of offspring larger than that predicted by the Lack clutch? First, if at the beginning of a season the parents cannot predict whether it will be a good or a bad year, and if newborn offspring are relatively cheap, then it may pay to have as many offspring as could survive in the best year possible, then reduce the number of offspring to the actual resource levels encountered, either by neglecting some offspring, by letting the offspring fight with each other, or by directly killing or eating the weakest. Such brood reduction is common in birds of prey and fruit trees.

If parents can selectively abort the less fit offspring, this will also select for increased gamete, zygote, or offspring production.

Another mechanism is more subtle and may be less common. If the fitness of offspring is variable, and if the variations in fitness among the offspring can be identified by the parents when the offspring are small, young, and therefore cheap, then it pays the parents to produce many zygotes, then selectively abort the less fit to concentrate investment on the more fit. This may explain the many zygotes fertilized in pronghorn antelope and nutrias, which greatly exceed, by 10–100 times, the number of offspring that could be born.

Phylogenetic constraints

Clutch size appears to be constrained within some lineages.

Another reason for deviations from the optimal Lack clutch is that reproduction may be constrained within lineages. For example, all birds in the Order Procellariformes lay one egg, in some species not every year. For the largest species, the wandering albatross, which takes up to 33 days and travels up to 15 000 km on a single foraging flight, a clutch size of one may be optimal, for the single chick, which has a special starvation physiology, must wait up to a month for a meal, and two offspring probably could not be fed. However, the smaller, less widely foraging species in the order, including several species of petrel, also have clutches of one egg although they almost certainly could feed two or more chicks. In these smaller species, clutches appear to be phylogenetically constrained.

Lifetime reproductive investment

Lack assumed a single trade-off between offspring number and offspring quality and focused on a single clutch. In contrast, reproductive effort models aim to predict the optimal reproductive effort over the whole life span, taking all reproductive trade-offs into account. They usually assume that if reproductive investment is increased at one age, then the probability that the parents will survive to reproduce again will be lowered, or, that if they do survive, their ability to reproduce in the next season will be reduced, or both. Such models make several predictions:

Reproductive effort models assume that reproduction has costs in terms of future reproduction and survival.

1. If mortality rates increase in one adult age class, then the optimal reproductive effort increases before that age and decreases after it (Michod 1979).
2. If adult mortality rates increase, the optimal age for maturation decreases (Roff 1981).
3. If mortality rates increase in all age classes, then optimal reproductive effort increases early in life and optimal age for maturation decreases (Charlesworth 1980).

The assumptions and predictions of reproductive effort models have been confirmed in experiments on kestrels in The Netherlands, on guppies in Trinidad, and on fruit flies in the laboratory.

The assumptions and predictions of reproductive effort models have been confirmed by experiments . . .

The kestrel study (introduced in Chapter 2) focused on the idea that the fitness of parents consists of the contribution made to fitness by the offspring they are currently producing (the reproductive value of their current clutch) plus all the later contributions to fitness that the parents could make in the rest of their life (their residual reproductive value). The current reproductive value of a female incubating its eggs is its clutch size times the reproductive value of one egg. The reproductive value of one egg is the sum of the probability that the egg becomes an adult that survives to reproduce once, times its first clutch size, plus the probability that it survives to reproduce twice, times its second clutch size, and so forth. The residual reproductive value of a parent is the probability that it will survive to reproduce once more, times the expected number of offspring that it will then have, plus the probability that it will survive to reproduce again a second time, times the expected number of offspring, and so forth.

. . . on kestrels in The Netherlands,

Daan and his colleagues calculated the reproductive value of the clutch and the residual reproductive value of the parents for experimentally reduced and enlarged clutches (Daan *et al.* 1990). If kestrels correctly balance the costs and benefits of reproductive investment, then the total reproductive value of the clutches they actually laid should be larger than that of either reduced or enlarged clutches. The main cost of producing a larger clutch was decreased parental survival, which decreased the residual reproductive value of parents of enlarged clutches. The offspring from control clutches also had better expectations of future reproduction than did those from enlarged clutches, but this effect was not as large as the effect on parental survival. When they accounted for these factors, they found that the clutch size actually laid was the one that yielded the highest reproductive value (Table 8.1).

on guppies in Trinidad,

Reproductive effort models were also confirmed in the field manipulation experiments done by Reznick and his colleagues on guppies living in shallow streams in Trinidad (Reznick *et al.* 1990; this example was introduced in the Prologue). At some sites their main predator is a cichlid fish that can eat large, sexually mature guppies and causes high mortality rates in all size classes. At other sites, their main predator is a killifish that eats mostly small, juvenile guppies and causes lower mortality rates than the cichlid in all size classes. In 1976, before the experiments began, guppies from cichlid sites matured earlier, made a larger reproductive effort, and had more, smaller offspring than did guppies from killifish sites. The differences were heritable, and they fit the predictions of reproductive effort models. To demonstrate that predation caused the pattern, predation pressure was manipulated to decrease the mortality rates on all age classes.

After 11 years, or 30–60 generations, significant evolution was observed, as predicted (Table 8.2). Age and size at maturity, size of the first brood, reproductive effort early in life, and size of offspring in the first two broods had already changed and corresponded qualitatively and quantitatively to the differences found in the unmanipulated populations. Changes in the intensity of mortality rates caused by a manipulation of predation pressure led to rapid evolutionary change in life-history traits in the direction predicted. The speed of evolution was striking. If the unmanipulated populations were at evolutionary equilibrium, then it took just 7–18 generations after the manipulation for most of the traits to reach equilibrium again. The traits that changed most occurred early in life. This was one of several recent studies demonstrating significant, rapid evolutionary change in ecologically important traits and suggesting that separating evolution from ecology would be artificial.

and on fruit flies in the laboratory.

Using fruit flies in a laboratory experiment, Stearns *et al.* (1996) tested the reproductive effort model with two treatments. The treatments dif-

Table 8.1 Direct and indirect costs of reproduction in kestrels

	Reduced	Control	Enlarged
Number of broods	28	54	20
Mean clutch size	5.25	5.19	5.40
Mean number fledged	2.60	3.95	5.84
Reproductive value of clutch	2.52	4.20	5.59
Local parental survival	0.65	0.59	0.43
Residual reproductive value	9.88	8.89	6.49
Total reproductive value	12.40	13.09	12.08

Table 8.2 Divergence of guppy life histories after manipulations

Life history trait	Control (cichlid)	Introduction (killifish)
Male age at maturity (days)	48.5	58.2
Male weight at maturity (mg wet)	67.5	76.1
Female age at first birth (days)	85.7	92.3
Female weight at first birth (mg wet)	161.5	185.6
Size of first litter	4.5	3.3
Offspring weight (mg dry) litter 1	0.87	0.95
Offspring weight (mg dry) litter 2	0.90	1.02

fered in the adult mortality rates, administered by killing a percentage of flies after counting them twice per week: high in the first treatment (the probability of dying within 1 week as an adult was 99%), low in the second treatment (the probability of dying within 1 week as an adult was 36%). Both juvenile and adult densities were the same in all treatments.

Fruit flies are subject to a trade-off that determines much of their response to such differences in selection. They can decrease development time by pupating and eclosing earlier, but to do so they have to pay the price of being smaller, and smaller flies have lower fecundities. Under these conditions wild flies reach peak fecundity when 16–17 days old. In the experiment, they were placed into the population cage when they were 14 days old, after which most of those encountering the high adult mortality treatment only had 1 day in which to lay eggs before they were killed.

Thus the flies evolving under high adult mortality should increase the number of eggs laid on that one day, the fourteenth to fifteenth day of life. They had two options. They could develop faster, eclose earlier, start to reproduce earlier, and thus achieve their peak reproduction by the fifteenth day of life, but pay the price of being smaller and having lower fecundity. Or they could develop more slowly, spend more time eating as larvae, eclose later, have higher fecundity, but pay the price of not yet

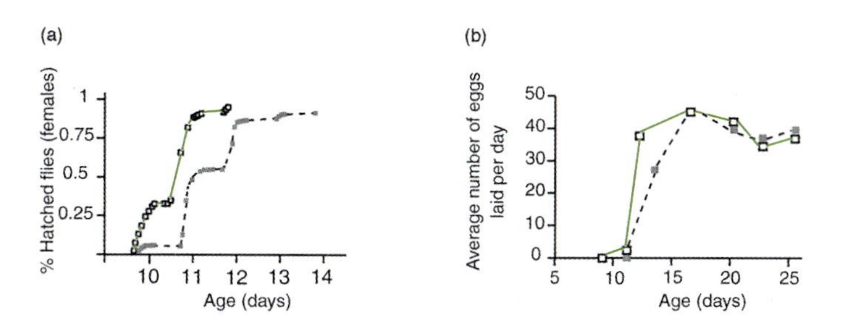

Fig. 8.4 Fruit flies, *Drosophila melanogaster,* that have evolved in response to high adult mortality rates, (a) develop more rapidly and eclose earlier and (b) lay more eggs early in life, than do flies that have experienced lower adult mortality rates. Green lines, flies from the high adult mortality regime; dashed lines, flies from the low-mortality regime.

being at their peak fecundity by the fifteenth day of life. Analysis of the development time–body size–fecundity relationship suggested that they should choose the first option, and they did. Within 2 years of selection they had evolved precisely the differences predicted by a reproductive effort model, and, like the guppies, traits expressed early in life—development time and early fecundity—were the first to change (Fig. 8.4).

The guppy and fruit fly examples confirm a basic assumption of the evolutionary theory of aging, our next topic: selection on reproductive performance earlier in life is stronger than selection on reproductive performance later in life.

The evolution of life span and aging

Life spans evolve.

It is not immediately clear why organisms grow old and die, and why different species have different maximum life spans. The longest-lived invertebrates are sea anemones, lobsters, and bivalves: some clams can live 220 years. The shortest lived are rotifers, insects, and small crustaceans: some live less than a week. The record for mammals in zoos is held by the African elephant at 57 years, with the domestic horse and the spiny echidna tied for second at 50 years. Plants vary in life span from days to many centuries. The variation in life span among species makes it clear that life span has evolved.

The contrast between the mortality of the soma and the immortality of the germ line is striking.

However the most striking puzzle is not why clams live longer than lobsters, or why humans live longer than chimpanzees. The most striking puzzle is why our germ line is so well maintained that it is potentially immortal, connecting us through an unbroken sequence over 3.7 billion years long to the origin of life, while we are so poorly maintained that we age and die, even if we are protected from accidents and given optimal conditions. Our germ line is part of our own body, determined by the same genes, built with the same biochemistry. Why can it survive, apparently for ever, while we must die? The answer is one of the triumphs of evolutionary thought.

Life span as a trade-off problem in life-history evolution

In discussing the evolution of life span, it is helpful to distinguish between extrinsic mortality and intrinsic mortality. Extrinsic mortality is imposed on organisms by their environment, for example, by predators, diseases, or bad weather. Intrinsic mortality is mortality caused by the breakdown of normal physiology and biochemistry as organisms age.

Two ways of thinking about life span both give important insights. On the one hand, we can think of the evolution of life span as resulting from selection acting directly on reproduction and indirectly on adult mortality rates through trade-offs with reproduction. Longer life spans will evolve if extrinsic mortality rates decrease in older organisms, increasing the value of older organisms because of their increased contribution to reproductive success. Longer life-spans will also evolve if extrinsic mortality rates increase in younger age classes, decreasing the value of younger organisms because of their decreased contribution to reproductive success. Thus one can think of average life-span as the result of interactions between age-specific mortality rates and a set of reproductive trade-offs. Such trade-offs are in part fixed effects—effects present in all individuals in the population—caused by the common development and physiology that characterize the species and represent part of its phylogenetic heritage.

Thus life-history theory views the evolution of the reproductive life span as a balance between selection to increase the number of reproductive events per lifetime and trade-offs that increase the intrinsic sources of mortality with age. The first lengthen life, the second shorten it. They combine to adjust the length of life to an intermediate optimum. The selection pressures that lengthen life decrease the reproductive value of juveniles and increase the reproductive value of adults. These include lower adult mortality rates, higher juvenile mortality rates, increased variation in juvenile mortality rates, and decreased variation in adult mortality rates.

Age-specific selection pressures adjust the length of life to an intermediate optimum determined by the interaction of selection with trade-offs intrinsic to the organism whether there is any aging or not.

Aging as a problem in evolutionary genetics

We can also think of the evolution of life span as the by-product of genetic effects. We can divide any population into young and old organisms so that the contribution to reproductive success of younger organisms will be greater than that of older organisms. Even in species with indeterminate growth, that gain fecundity with size, at some age so many will have died that an older age class can no longer contribute as many offspring as a younger age class that contains more survivors. For this reason, selection pressures on survival rates and fecundities must always decline with age. They decline at different rates in organisms with different life histories. In humans in industrialized countries, the selection pressure to improve survival has dropped almost to zero by the time one is 50 years old, a fact that does not bring cheer to the middle-aged (Fig. 8.5).

Life span is also the indirect by-product of an accumulation of genetic effects on reproductive performance.

Now consider a gene that improves the reproductive success of the younger organisms at the expense of the survival of the older organisms. It has positive effects early in life and negative effects late in life. A gene that affects two or more traits is called a **pleiotropic** gene. Effects that increase fitness through one trait at the expense of decreased fitness through another trait are antagonistic. Therefore such genes have **antagonistic pleiotropy**. A mutation that has such effects, positive early in life, negative late in life, will be favored by selection, increase in frequency, and in most cases go to fixation. That is the first sort of gene important for the evolution of aging and life span. There is indirect evidence that such genes exist; the search to locate specific examples is intensifying.

Such genetic effects result either from pleiotropy . . .

Two traits in humans, one in males and one in females, suggest how such genes might act. Prostate cancer occurs at high frequency in males

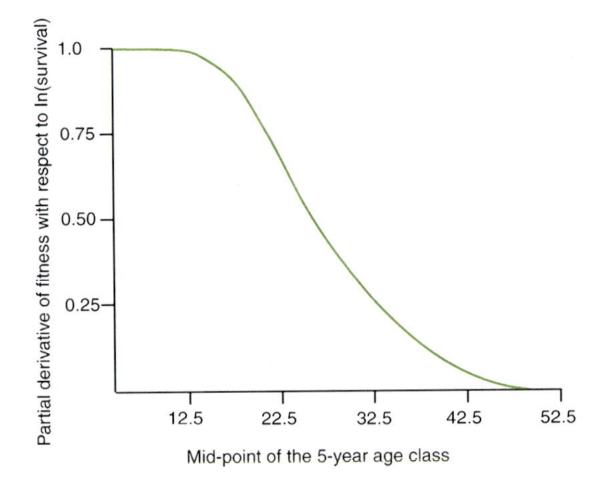

Fig. 8.5 The sensitivity of fitness to changes in survival rates declines with age. The case depicted corresponds to the population of the United States in about 1940. (From Charlesworth and Williamson 1975.)

over 70, but it can prevented by treatment with female hormones or castration. It appears to be a consequence of a long period of exposure to testosterone, a hormone absolutely necessary for male sexual and thus reproductive performance. Osteoporosis, or a loss of bone density, is mediated by estrogens in older women, but estrogens are essential for reproduction in younger women. In both cases the old-age pathologies are associated with age-related changes in the responses of tissues to hormones.

. . . or from age-specific expression of detrimental mutations.

The second kind of genetic effect that may be involved in aging is the accumulation of mutations with age-specific expression. Consider a gene that is only expressed in a certain age group. Selection against a deleterious mutation in such a gene is stronger if it is expressed in younger organisms that contribute more to reproductive success. In a population at evolutionary equilibrium, the number of mutations present for a given trait depends on the per-trait mutation rate and the strength of selection operating on the trait. If we assume that the per-trait mutation rates are the same for traits expressed in young and old organisms, then only the selection pressures differ. They are stronger on traits only expressed in younger organisms, where they reduce genetic variation. Thus we expect to find more mutations with age-specific expression present for older traits. This is the mutation-accumulation effect; it does not depend on antagonistic pleiotropy. The evidence for mutations with age-specific effects is not yet convincing. If they do occur, they contribute to aging.

Thus the evolutionary answer to the question, 'Why age?', has two parts. The force of selection declines with age; after a certain age organisms are irrelevant to evolution. Given this decline, two sorts of genetic effects become possible, the accumulation of genes that benefit younger age classes at the expense of older ones, and the accumulation of mutations with stronger effects on older age classes than on younger ones. Maturation is the point where these effects should start to occur; before then they should not be seen. Aging should follow the onset of reproduction, with diffuse erosion of physiological and biochemical functions caused by many genes that produce aging as a by-product, not as an adaptation. Aging caused by a few genes with large effects is not ruled out but is not expected to be the usual case.

It follows that fitness is maximized at a level of investment in repair that is less than would be required for indefinite survival. That is why we grow old and die. Aging should result from the accumulation of unrepaired somatic damage, and species with different longevities should exhibit corresponding differences in their levels of somatic repair (Kirkwood 1987). Repair is costly. More than 2% of the energy budget of cells is spent on DNA repair and proof-reading, on processes that determine accuracy in protein synthesis, on protein turnover, and on the scavenging of oxygen radicals that damage biological structures. Decreases in external sources of adult mortality will increase the value of investment in repair; coupled to that should be decreased investment in growth and reproduction, leading to longer life spans and lower reproductive efforts. Increases in extrinsic adult mortality will devalue investment in repair, and should lead to less repair and higher reproductive investment.

Fitness is maximized at a level of repair less than that needed for indefinite survival.

These ideas have been tested on fruit flies, nematodes, and mice, with consistent results: when the force of selection on older age classes is increased by only allowing older organisms to reproduce, aging is postponed. In some cases, increased life span is accompanied by reduced fecundity early in life; in others, by changes in larval growth rates, survival rates, and competitive ability. Thus life span responds rapidly to selection in the laboratory in a manner consistent with evolutionary theory, and longer life must be paid for by reductions in performance early in life, either in lower fecundity, smaller body size, or lower juvenile survival and competitive ability.

Life span responds rapidly to selection in the laboratory in a manner consistent with evolutionary theory.

Putting life-history evolution and evolutionary genetics together

The large-scale differences in life span among species are the product of life-history evolution. Aging occurs in many species, but in nature the intrinsic effects of aging have less impact on large-scale differences in life span among species than does life-history optimization. Moreover, in natural populations few organisms survive long enough to encounter the effects of aging. Within a species, however, especially in populations under favorable conditions, such as animals in zoos, fruit trees, domesticated animals, and humans with good nutrition and modern medical care, the effects of aging become noticeable.

One might think that average life span is a problem in life-history evolution and aging is a problem in evolutionary genetics; but we can ask, 'What is the evolutionary origin of the trade-offs that we now detect as

Life span is a problem in life-history evolution, aging is a problem in evolutionary genetics, and the two interact.

fixed, phylogenetic effects shared by all individuals in a population?' In the past many sorts of mutations arose. The ones that had negative effects on both reproduction and survival were eliminated, and those with positive effects on both reproduction and survival were fixed. There were also some mutations with antagonistically pleiotropic effects, positive on one trait and negative on another, where the balance of the mixed effects was positive. These increased in frequency, were fixed, and contributed to the development and physiology that produces the trade-offs important in life-history evolution.

The evolution of sex allocation

Sex allocation concerns the allocation of reproductive effort among male and female offspring or between male and female function.

Many organisms can control the sex of their offspring. Others can mature as one sex, reproduce, then change sex and reproduce as the opposite sex. Sex-allocation theory predicts the allocation of reproductive effort among male and female offspring or between male and female function. It makes strikingly successful predictions and unites previously unrelated patterns as aspects of a single explanation.

Sex-allocation research has its own vocabulary. Species with separate sexes are either **diecious** (plants) or **gonochoristic** (animals). Sequential hermaphrodites may be either **protandric** (born as males and changing later to females) or **protogynous** (born as females and changing later to males).

Diecy (gonochorism) may be advantageous because it is expensive for an organism to maintain both male and female organs, and less costly to specialize in one sexual function. Simultaneous hermaphroditism may, however, be advantageous when mates are hard to find. It is common among sessile or slow-moving animals and plants. Hermaphrodites may fertilize themselves, but this generally happens only if mates are not available, for self-fertilized hermaphrodites often suffer from inbreeding.

The central questions,

Patterns of sex allocation raise two central questions that connect to the evolution of mating systems and social behavior:

1. What is the equilibrium sex ratio for organisms with separate sexes? How many sons and how many daughters should be produced?
2. For sequential hermaphrodites, as what sex should the organism be born and how old and large should it be when it changes sex?

are different aspects of a single insight: male and female function are equivalent paths to fitness.

One fact is the key to answering both questions: every diploid, sexually produced zygote gets half its autosomal genes from its father and half from its mother. This holds for sequential and simultaneous hermaphrodites as well as for gonochores. Now focus on the grandchildren and ask how fitness has been realized when you look two generations ahead. The fitness gained by an individual through male function or male offspring must be compared with fitness gained by other individuals through male function, and the fitness gained by an individual through female function or female offspring must be compared with fitness gained by other individuals through female function. Each path contributes half the genes in every offspring and therefore accounts for half of the contribution to the grandchildren.

Fisher's sex-ratio theory: the classic example of frequency-dependent selection

The evolutionarily stable sex ratio is 50:50 under simple assumptions.

What should be the sex ratio favored by selection under the simplest assumptions? Fisher (1930) answered this question as follows. Consider a large population with two sexes, well mixed, with random mating (no social structure). Suppose there is for some reason an excess of females in the population. Then a mutant that produces more male than female offspring will be favored, for the male offspring will have less competition for mates than would female offspring. In contrast, if there is an excess of males, a mutant that produces more female than male offspring will be favored, for then the female offspring will have less competition for mates than the male offspring. Thus deviations from equal frequencies of the two sexes produce frequency-dependent selection pressures that lead to a stable 50:50 sex ratio. It can be maintained either by having half the females in the population produce only male offspring and half only female offspring, or by having each female produce half male and half female offspring. Note that this is the primary sex ratio, the sex ratio at birth, not the operational sex ratio in the population that is encountered by adults (see questions at the end of the chapter).

The genetic mechanisms that produce that sex ratio are diverse.

This selection pressure probably led to the evolution of sex chromosomes, which in the simplest XX/XY system constrain the offspring sex ratios of individual females to average 50:50. In mammals and fruit flies, females are XX (**homogametic**) and males are XY (**heterogametic**). In birds and butterflies, females are heterogametic and males are homogametic. Genetic sex-determining mechanisms can even vary among populations within a species. In the house fly, for example, females are heterogametic south of the Alps and males are heterogametic north of the Alps (Franco *et al.* 1982). These examples show that selection acting on phenotypes may favor a 50:50 sex ratio in many species, and that the genetic mechanisms that produce that sex ratio can be quite diverse.

A brief overview of mechanisms of sex determination

Sex can be determined by genes , environment, and parasites: if by genes, then by chromosomes, ploidy, or autosomal genes;

Whereas chromosomes determine sex in birds, mammals, and many insects, sex in many other species is determined by quite different mechanisms (Bull 1983). Sex can be determined by genes or the environment, and the sex of some hosts is determined by parasites. Sex chromosomes are just one mechanism of genetic sex determination. Two others are also important. In all of the Hymenoptera—the ants, bees, wasps, and sawflies—and in some beetles and mites, females are diploid and males are haploid. This creates asymmetries in relationships with important consequences for the evolution of social behavior and sex ratios. Other species have sex-determining loci on normal autosomes. In some species these are single loci; in other species sex is determined by many loci.

if by environment, then by temperature, social interactions, or sex first contacted,

Sex can also be determined by the environment. In some reptiles—turtles, crocodiles, and alligators—sex is determined by the temperature at which the eggs are incubated. In many reef fish—primarily wrasses and parrot fish—sex is determined, at least in part, by social interactions. Large, old, high-ranking individuals are male, and younger, smaller, low-ranking individuals are female. The marine worm *Bonellia* and the rhizocephalans, relatives of barnacles that parasitize crabs, have extremely flexible sex determination. Both have planktonic larvae, and if a larva settles on substrate rather than another member of its own species, it metamorphoses into a female. If a larva settles on a female, in the case of *Bonellia*, or on a hermit crab already parasitized by a female, in the case of the rhizocephalans, it metamorphoses into a dwarf, parasitic male consisting of little more than a testis and a duct connecting it to the female reproductive system.

or by cytoplasmic parasites such as *Wolbachia*.

Sex can also be determined by cytoplasmic parasites that are transmitted vertically from parents to offspring through the gametes. A bacterium living in cytoplasm will leave no descendants if it occurs in a male host, for sperm transmit no cytoplasm to the next generation. It is therefore in the interests of cytoplasmic parasites to occur in a female, and some of them have evolved the ability to feminize their hosts, turning males into functional females that reproduce as females. Bacteria in the genus *Wolbachia* are the best understood. They occur widely in arthropods and have been identified as feminizing factors in wood lice, wasps, and ladybird beetles. In experiments with dramatic results, wasps that only produced female offspring were 'cured of their disease' by treatment with antibiotics. Treated females produced offspring with normal 50:50 male:female sex ratios; untreated females produced only female offspring (Stouthamer *et al.* 1990). If *Wolbachia* spread unchecked through a host population, their hosts would eventually produce only female offspring and go extinct. Some populations of one host, a wood louse, appear to have solved this problem by incorporating the sex-determining portion of the bacterial DNA into the nuclear genome, resulting in a new sex-determining wood-louse gene (Juchault *et al.* 1993) and 50:50 sex ratios.

The variety of patterns of sex allocation

The natural history of sex determination reveals a wonderful diversity of mechanisms with many consequences. Now we move from the mechanisms of sex determination to the sex ratios that we expect them to produce. We start with age and size at sex change in sequential hermaphrodites.

Sequential hermaphrodites should be born into the sex that loses less by being small or young, then change into the sex that gains more by being old or large.

Sequential hermaphroditism should evolve, given sufficient flexibility in the mechanisms of sex determination, where the advantage of being a particular sex changes with age or with size. Organisms should be born into the sex that loses less by being small or young, and they should change into the sex that gains more by being old or large (Fig. 8.6). Sex change should occur when the fitness advantage of being one sex when old or large becomes greater than the fitness advantage of remaining the sex that was small and young.

Protandry: females gain more reproductive success as size increases than do males.

Protandric organisms begin their reproductive career as males, then switch sex and function as females. Protandry has evolved where a large female enjoys an advantage in caring for offspring and where females gain more reproductive success as size increases than do males. A small male is able to fertilize a female with a modest amount of sperm and is relatively mobile. A female that cares for developing eggs and embryos may be able to accommodate more offspring as she grows larger. For example, the gastropod *Crepidula,* found in rocky intertidal habitats, is a sequential hermaphrodite that copulates in 'daisy chains' in which the bottom organism is old, large, and fully female, the top organism is young, small and fully male, and the intermediate organisms are in various stages of changing from male to female function, still producing a mixture of both kinds of gametes (Fig. 8.7).

Protogyny: large males have more offspring than large females.

Protogynous organisms begin reproducing as females, then switch sex and function as males. For protogyny, the size-advantage model suggests that small females should have more offspring than small males, large males more offspring than large females. These conditions are realized when big males can win fights to monopolize opportunities to mate with many females. The winning strategy is then to reproduce as a female until large enough to win the fights, then change sex. Such males are polygynous, often holding harems, and when the male dies, the largest female in the group becomes a male. Most species in the marine fish

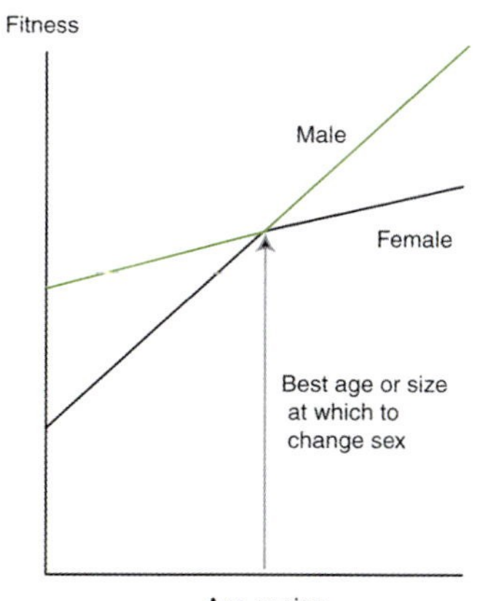

Fig. 8.6 The size-advantage model for sequential hermaphroditism: it is best to be born as the sex that loses less by being young and small, and to change to the sex that gains more by being old and large, at the age and size where one would gain more fitness by being the other sex. The case depicted is protogyny (female first). By switching the labels male–female we get the model for protandry. (After Ghiselin 1969.)

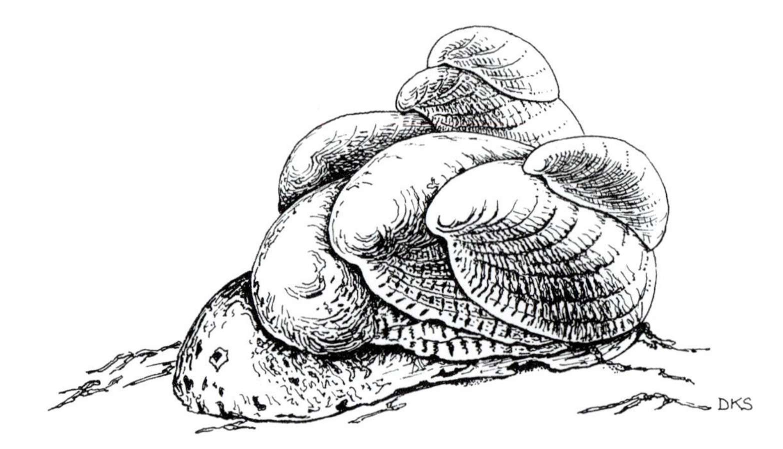

Fig. 8.7 Copulating *Crepidula*. The bottom organism is female, the top organism is male, and some of the intermediates are changing from male to female function. This is a case of protandrous sequential hermaphroditism, where females gain more fitness by being large than do males. (By Dafila K. Scott.)

family of wrasses, which includes the cleaner fish and the anemone fish, are protogynous. The male and female forms that one individual displays in the course of its development are so strikingly different that taxonomists have often classified them as separate species (Fig. 8.8).

High-ranking females or females in good physiological condition should have male-biased litters or clutches. This appears to be true in red deer.

Sex allocation can also vary as a function of social rank (Trivers and Willard 1973). In polygynous species, one male controls access to and mates with several females. Low-ranking females or females in poor condition should have female-biased litters or clutches. High-ranking females or females in good physiological condition should have male-biased litters or clutches. This is because if the social rank of the mother affects the condition of the offspring, offspring will tend to inherit their mother's social rank. Whereas female offspring in poor condition will always bear offspring, male offspring in poor condition will probably not father any offspring because they will not be able to compete successfully for access to females. If female offspring in good condition cannot have many more offspring than female offspring in poor condition, whereas male offspring in good condition will father many offspring by copulating with many females, then the relative reproductive success of male versus female offspring should increase with the social rank of the mother. Thus low-ranking females should invest more in daughters, and high-ranking females should invest more in sons.

Clutton-Brock and Iason (1986) summarized the evidence for adaptive variation in the sex ratio of offspring in more than 30 mammal species. Many claims of sex-ratio variation were based on inadequate evidence, only a few mammal populations had sex-ratio variation strong enough to rule out random causes, and trends that appeared in one population often proved to be inconsistent when checked in other populations. However, at least one example of adaptive variation in the sex

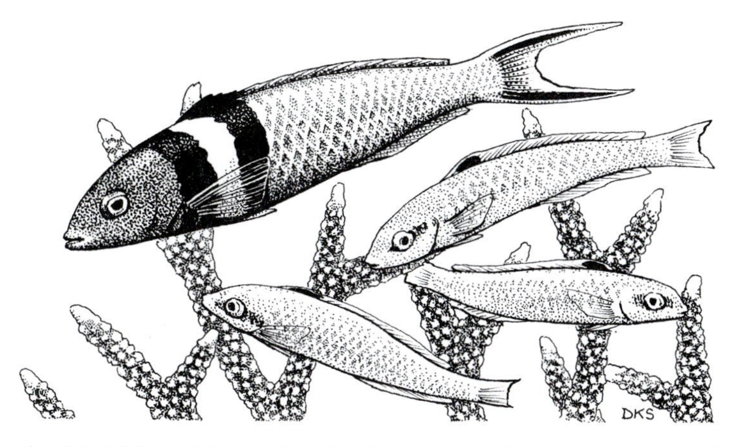

Fig. 8.8 Male and female blueheaded wrasses. The brightly marked fish is a harem-holding male, the smaller fish are female, and if the male were removed, the largest female would change sex and start to function as a male within days, attaining normal male function within weeks. This is a case of protogynous sequential hermaphroditism, where males gain more fitness by being large than do females. (By Dafila K. Scott.)

ratio of offspring is found in red deer, where high-ranked females produce more sons and low-ranked females produce more daughters (Fig. 8.9). Part of the reason is that sons of subordinate females suffer higher juvenile mortality than sons of high-ranked females, and part is that most stags with high reproductive success are the sons of dominant mothers (Clutton-Brock and Iason 1986). In other words, sons inherit some of their mother's social rank. The mechanism that adjusts the sex ratio is not known, but only two are plausible: either sperm are selected before fertilization, or offspring of one sex are selectively aborted after conception.

When all the grandchildren stem from matings between brothers and sisters, there should be one son and as many daughters as possible. Example: a parasitic mite.

Unusual sex ratios result when all grandchildren stem from matings between brothers and sisters, an extreme case of local mate competition among siblings—brothers compete with each other for all possible matings. Because one son can inseminate all the daughters, a second son would be wasted, and the mother can get more grandchildren by producing another daughter than a second son. The optimal sex ratio is therefore one son and as many daughters as possible. An extreme example is *Acarophenax*, a haplo-diploid parasitic mite, in which the one son fertilizes all the daughters inside the mother, who is then eaten by her offspring.

A parasitoid wasp adjusts the sex of its offspring to the size of its host.

Parasitoid wasps display striking short-term flexibility in sex allocation. Sex ratios in some parasitoid wasps depend on prey size in a relative way. If a female encounters a series of prey, such as insect larvae, that are of different sizes, she will lay female eggs in the larger ones and male eggs in the smaller ones. The definition of what is 'large' and what is 'small' depends on the sizes encountered that day, even in the last hour. If she encounters, say, 3 mm and 4 mm larvae, she will lay male

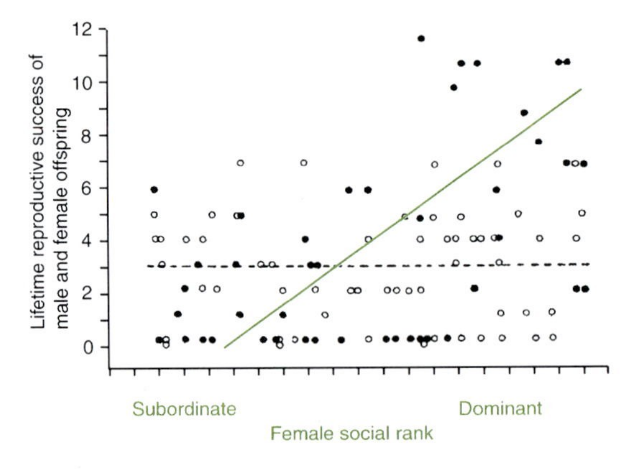

Fig. 8.9 Offspring lifetime reproductive success (LRS), daughters versus sons, for female red deer of different social ranks. Daughters, open circles and dashed line; sons, closed circles and solid line. Mother's social rank has no effect on the reproductive success of daughters, but the sons of high-ranking mothers have significantly higher reproductive success than do the sons of low-ranking mothers. (From Clutton-Brock and Iason 1986.)

eggs into the 3 mm larvae and female eggs into the 4 mm larvae. If she encounters 4 mm and 5 mm larvae, she will lay male eggs into the 4 mm larvae and female eggs into the 5 mm larvae. This resembles the size-advantage model. Male offspring lose less by being small than female offspring, which gain more by being large.

In these examples the connections between sex allocation, mating systems, sexual selection, and life-history evolution are strong. There is no distinction here between ecology, evolution, and behavior. All three fields are associated in how the examples described are explained.

The Shaw–Mohler theorem

Mothers and fathers should gain the same fitness through sons as through daughters. The product of male and female fitness, *mxf*, should be maximized.

Many of these cases of unusual sex ratios can be understood as implications of a single, unifying idea. Recall that in analyzing the standard 50:50 sex ratio, Fisher assumed that males and females are equally good at producing male and female offspring at all ages and sizes, and that mating is at random in a thoroughly mixed large population. When those assumptions do not hold, we need a more general answer to the question, what sex ratio is expected? The answer was given by Shaw and Mohler (1953). They reasoned as follows. Both mothers and fathers should gain the same fitness through sons as through daughters. Consider a population with some average sex ratio, not necessarily the one expected at equilibrium. Will a new mutant that produces a different sex ratio increase in frequency and invade the population? For a mutant with a different sex allocation than the resident population to invade, its total fitness should be greater than that of the residents. It would have to increase the fitness gained through one sex function more than enough to compensate for fitness lost through the other sex function. If a population has long been tested by such mutants, most of the mutants that increase total fitness have already occurred and have been fixed. The population should therefore be close to evolutionary equilibrium. At evolutionary equilibrium, the product of male and female fitness, $m \times f$, should be maximized (Charnov 1982).

Let us analyze a simple case in which producing one less son would allow the production of one more daughter. In this case, fitness gained through male function, m, trades off simply and linearly with fitness gained through female function, f, so that $f = 1 - m$. Then $m \times f = m(1 - m) = m - m^2$, which is maximal when $1 - 2m = 0$, or when $m = f = 0.5$. This is the familiar 50:50 sex ratio. In the *Acarophenax* example of local mate competition, this simple trade-off does not hold, for only one son is needed to inseminate all the daughters. Fitness gained through male function is the same whether there is one son or many, but fitness gained through female function increases in proportion to the number of daughters, so the number of sons should be minimized and the number of daughters maximized.

Summary

This chapter analyzes how natural selection interacts with trade-offs to design life histories and sex allocation for reproductive success.

- At what age and size should an organism mature? The answer depends on how likely it is that it will die as a juvenile, whether that risk differs from the risks it will encounter once it has matured and is adult, and the rate at which it gains potential fecundity as it grows and ages.
- Once it matures, how many offspring should it have? The answer depends on trade-offs between offspring number and the expected lifetime reproductive success of offspring, and offspring number and its own subsequent reproductive success—in short, on fitness gained through current offspring and fitness gained through the rest of its reproductive activities.
- How long can it expect to live if it is not eaten by predators, killed by disease, or subject to accident? The answer depends on the kinds of trade-offs it has inherited. They determine how much it can neglect maintenance to invest more in offspring. The answer also depends on how many genes with positive effects early in life and negative effects late in life have accumulated in its population.
- How much should it invest in male as opposed to female offspring, and if it is a sequential hermaphrodite, as what sex should it start life and at what age and size should it change sex? The answer depends on the sex ratio that it encounters in its local environment and on its capacity to acquire resources and social rank.

The answer to each of these questions should be, whatever maximizes lifetime reproductive success. When these questions have been answered, some important ones remain. How much should the organism invest in structures and behaviors that help it to win the competition for mates? How much should it invest in structures that make it attractive to the opposite sex? What is the opposite sex looking for in a mate? These questions are addressed in the next chapter, on sexual selection.

Recommended reading

Charnov, E. L. (1982). *The theory of sex allocation.* Princeton University Press, Princeton.

Roff, D. A. (1992). *The evolution of life histories.* Chapman & Hall, New York.

Stearns, S. C. (1992). *The evolution of life histories.* Oxford University Press, Oxford.

Questions

8.1 If we compare red deer to protogynous wrasses, it appears that the red deer have made the best of a bad situation. If they were sequential hermaphrodites, all would be born as females, and only those that grew rapidly and acquired good physiological condition and large body size would change sex and reproduce as males. The population would contain fewer males and more females, and the risk that any given individual would not reproduce at all would be lower. Why are red deer not sequential hermaphrodites?

8.2 In the kestrel study, the control birds had an average selective advantage of about 0.07 over the birds with manipulated clutches (13.09/12.08 = 1.08, 13.09/12.40 = 1.06). Assume that the population consists only of individuals producing one egg less than the optimum, that there is a single gene that can affect clutch size, and that a dominant mutant allele arises at this locus that increases clutch size by one egg. Using methods from Chapter 4, and assuming that selection is directional and that the frequency of the mutant at the start is 0.01, calculate and plot the curve showing the increase of the mutant. How long does it take the population to reach the optimum? What difference would it make if clutch size were polygenic and selection were stabilizing?

8.3 Imagine a large population that consists of 25% adult males and 75% adult females. Mating is at random; mating success depends directly on the frequency of appropriate partners. Now contrast two sex-allocation strategies. One type of female gives birth to three male and one female offspring. The other type of female gives birth to two male and two female offspring. Survival of offspring is independent of the sex-allocation strategy of the parent. All females in the population produce four offspring per lifetime. Which type will have more grandchildren?

8.4 Fisher's sex-ratio theory predicts a 50:50 primary sex ratio, the sex ratio in the newborn offspring. Why does it not predict a 50:50 operational sex ratio?

Chapter 9
Sexual selection

Introduction

Sexual selection is a component of natural selection in which mating success trades off with survival.

Sexual selection is selection for traits associated with mating success and partner choice. Mating success trades off with other components of fitness, primarily adult survival, just as do other reproductive traits, such as fecundity. If a change in a trait increases lifetime reproductive success by improving the ability of an individual to attract mates and fertilize them, it will be favored by selection although it lowers survival probability. Sexual selection will change traits influencing mating success until the improvement in mating success is balanced by costs in other fitness components; then the response will stop. That explains why males take risks to mate and why juvenile males develop their secondary sexual characters only on maturation. If they developed them before maturation, they would suffer the costs without enjoying the benefits, which come when mating begins. Many sexually dimorphic traits that reduce the survival of the individuals that carry them can be explained by a trade-off between mating success and survival. Thus sexual selection involves a trade-off, as does selection on life-history traits. However, because the trade-off is determined by interactions between two or more individuals, it has some special features, illustrated in the following example.

Why get up before dawn to make love in the snow at the risk of being eaten?

Every morning before sunrise, from January to May, male sage grouse gather on their traditional display grounds, called **leks**, in eastern Oregon. As dawn comes, the males fan their tails, puff their chests, and arch their wings, strutting and emitting penetrating popping sounds, cooing between the pops. They fight one another for the central display sites, and they remain on the lek despite strong winds and subzero temperatures. Unlike the females that lurk in the bushes around the lek, occasionally entering to mate with a central male, they have large, elaborately ornamented tails that they display in a fan, and large chest

Fig. 9.1 Male sage grouse displaying on a lek, one female in the foreground. (By Dafila K. Scott.)

pouches connected to their lungs from which they release air explosively, producing popping noises that can be heard far away (Fig. 9.1). Their extravagant displays are expensive and dangerous. It takes a lot of energy to display vigorously in cold weather, coyotes and golden eagles are common in that habitat, and displaying males are conspicuous and relatively immobile. They get up before dawn to make love in the snow at the risk of being eaten. Why would such behavior evolve?

Mating success can be improved by competing for mates or being chosen by mates. The two mechanisms have different consequences

The sage grouse lek illustrates the puzzle for which Darwin suggested the solution of sexual selection: some sexually dimorphic traits, traits that differ between the sexes, reduce the survival probability of the organisms that carry them. Darwin thought sexual selection acted through male–male competition and female choice. Those remain the principal mechanisms. In the case of the sage grouse lek, the problem is, what are the females choosing?

Note that the word 'choice' in the context of sexual selection does not necessarily refer to a conscious mental event. It refers to anything intrinsic to an individual that makes that individual more likely to mate with some partners than others. Usually one conceives of this as a signal–receiver system: one partner makes a signal, the other partner receives it and reacts in a way that depends on the information contained in the signal.

Individuals of the sex that is limiting can be choosy, while individuals of the other sex compete for matings.

Mate choice is one of two types of interactions that will lead to sexual dimorphism. The other is competition for mates. Competition for mates can occur both between males competing for females and between females competing for males. Mate competition will be stronger in the sex with the greater reproductive potential, which competes for the sex with the lesser reproductive potential. Usually males have greater reproductive potential than females; thus males usually compete and females usually choose.

While males are competing with each other for females, females are competing among themselves for the most attractive males. Similarly, while it is usually females that choose males for some reason, males may also choose females. Competition for mates is usually more important in males, and mate choice is usually more important in females, but both processes can occur at the same time in both sexes.

Which mate to choose: for direct benefits,

The consequences of mate choice are more subtle and surprising than those of competition for mates. One reason is that choice involves a signal sent by one sex and received by the other sex. The signal communicates characteristics of the potential mate, and choosing a mate is one of the most important decisions a sexually reproducing organism makes. On what criteria should the choice be based? Some are straightforward. Does the potential partner control a superior feeding territory? When both parents care for the offspring, or when one partner feeds the other during part of the reproductive period, is the potential partner a superior forager?

good genes,

Other criteria are not so easily detected. Is the potential partner healthy? Does it carry parasites that might infect oneself or one's offspring? All populations contain some individuals infected with diseases or parasites, but not all infections are obvious. (Parasites may conceal their presence to enhance their probability of transmission.) If a partner uses a reliable signal to advertise that it is healthy and free of parasites, that should create a mating advantage, for mates would be selected to recognize and prefer such signals. Reliable, honest signals must be costly. If they were not costly, sick cheaters could imitate the signal, deceive the partner, and produce mistakes that would eliminate the reason for choosing partners on this basis. Thus potential mates should evolve signals that honestly indicate their health and cost something to produce, and potential partners should choose their mates on the basis of a costly trait.

or good prospects for the mating success of sons?

Sexual selection by mate choice can also lead to surprising results because the genes that determine mate preference and the genes that determine the traits preferred come together in the offspring. The genes for the preferred trait and for the preference then increase in frequency together, in self-reinforcing fashion. This sets in motion an evolutionary process with special features, including the possibility of arbitrariness in preference.

Research on sexual selection has exploded in recent years. Here we only touch on the main points. This chapter addresses the following questions:

The main questions about sexual selection.

1. How did sexual selection originate? It is a consequence of anisogamy.
2. How does sexual selection work? The main ideas are competition for and choice of mates.
3. Is there good evidence for sexual selection? We discuss the spermatophores of katydids, the epaulettes of blackbirds, the long tails of male widowbirds, and the colorful spots and large tail fins of male guppies.
4. What determines the strength of sexual selection? Parental care and

mating systems influence which sex has the greater potential reproductive rate. That influences the **operational sex ratio**, the local ratio of males ready to mate to females ready to mate, which largely determines the strength of sexual selection.

5. How does sexual selection work in plants? Pollen scramble for fertilization and flowers compete for pollinators. Competition among pollen to fertilize ovules is analogous to competition among sperm to fertilize eggs, but pollinator choice is not analogous to mate choice, for the genes for flower morphology (in plant genomes) do not become associated in the offspring with the genes for flower choice (in pollinator genomes).
6. What are alternative explanations for sexual dimorphism? The principle alternatives to sexual selection are ecological sex differences and primary sexual characters.

How did sexual selection originate?

First there was asex, then isogamous sex,

The road from the origin of sexual selection to the sage grouse lek has been a long one. When life originated about 3.8 billion years ago, reproduction was asexual, and it remained so for about 2 billion years. Sexual reproduction with meiosis and recombination originated with the eucaryotes about 1.5–2.0 billion years ago. The first sexual organisms were single celled and produced gametes of equal size: they were isogamous.

then isogamous sex with mating types,

The first step in the differentiation of sexes was the origin of mating types, which occur today in fungi, algae, and ciliate protozoa. In a species with mating types, an individual can only mate with individuals that are not of its own type. Selection for mating types is driven by inbreeding avoidance and frequency-dependent selection. The evolution of mating types can proceed in two directions. In the first, the number of mating types present in a population increases without limit, creating a situation in which almost every individual encountered is a potential mate. This may have happened in ciliates, in which 40 or more mating types have been found. When evolution takes the second direction, the number of mating types is reduced to two. The reduction to two mating types was necessary for the evolution of anisogamy, the critical step on the path to sexual selection.

then anisogamous sex: the origin of males and females.

Once some isogamous population reached an equilibrium of two mating types, selection began to change those mating types into organisms that differed in the size of their gametes. Larger gametes were selected in one mating type because they resulted in zygotes with greater energy stores and better survival probabilities, and because they produced more of the pheromones that attract gametes. Smaller gametes were selected because they could be produced in greater numbers, were more motile, and could actively find the large, pheromone-producing gametes. This led to specialization either on large or on small gametes, for intermediate forms paid the costs of being mediocre at both functions without realizing the full benefits of either. Thus anisogamous populations evolved with two types, one which produced large gametes, eggs, the other which produced small gametes, sperm. The individuals producing eggs were females, those producing sperm were males, by definition.

Eggs are expensive, but sperm are cheap.

Now a key principle of sexual selection came into play. Eggs are expensive, but sperm are cheap. The lifetime reproductive success of females is limited by the number of eggs they can produce, that of males by the number of eggs they can fertilize. Females became a limiting resource for males, setting off competition among males for mates and allowing females to choose their partners. Both primary and secondary sexual characters then evolved. (Primary sexual characters have functions necessary for reproduction but not directly involved in mating success; for example, differences in the biochemistry of the ovaries and testes. Secondary sexual characters have been shaped by sexual selection for success in mating.) Males and females with striking secondary sexual characters are the product of a long history of sexual selection in anisogamous organisms.

Competition for mates

The main forms of direct competition for mates are contests, scrambles, and endurance rivalries.

Contests, scrambles, and endurance rivalries.

1. In a contest, the rivals display or fight directly with one another over mates or over the resources needed to attract mates. Bull elephant seals fighting for a stretch of beach, red deer stags fighting with their antlers for control of a harem, and male great tits fighting to defend their territories are all engaged in contest competition for mates.
2. In a scramble, rapid location of a mate is crucial for success, which is often given to the first to arrive. Pollen germinating on a flower style scramble to reach the ovum first.
3. In endurance rivalry, persistence brings rewards. In frogs and toads with an extended mating season, the ability to keep calling night after night for many weeks can strongly affect reproductive success. Persistence also affects mating success in species with direct male–male contest competition. It can determine the amount of time that a male can display without leaving for food or water, and it can determine the length of a reproductive season during which a male can maintain his top rank in repeated fights with other males.

Females could choose males for their competitive ability.

Mate competition explains large, well-armed males.

Mate competition co-occurs with mate choice whenever females use competitive ability as the criterion for choice.

Mate competition is the main explanation for the evolution of sexual size dimorphism and the weapons used in fights over mates. The most striking examples of **sexual size dimorphism** in mammals occur in pinnipeds—seals and their relatives (Fig. 9.2).

Males can be up to six times as heavy as females, and there is a strong

Fig. 9.2 Southern elephant seal bull and cows at the top; Antarctic fur seal bull and cows to the left; a pair of harbor seals in the right foreground. (By Dafila K. Scott.)

relationship between the degree of sexual size dimorphism and the number of females controlled by a breeding male (Fig. 9.3).

Elephant seals exemplify contests and endurance rivalry,

In elephant seals the reasons for the striking size dimorphism appear to be male contests and endurance rivalry. Unlike whales, pinnipeds cannot give birth in the water. During the 3 month long reproductive season, females haul out on beaches where they give birth and mate before returning to sea. The large concentrations of females are a resource that a male can defend, and males fight with each other for stretches of beach with females on them. Large males win most fights and access to females. Less than one-third of all males copulate at all during a breeding season, and most matings are achieved by just a few males.

and some female choice.

When a large bull controls a harem of more than 50 females, a small bull can sometimes sneak a copulation without being noticed, making

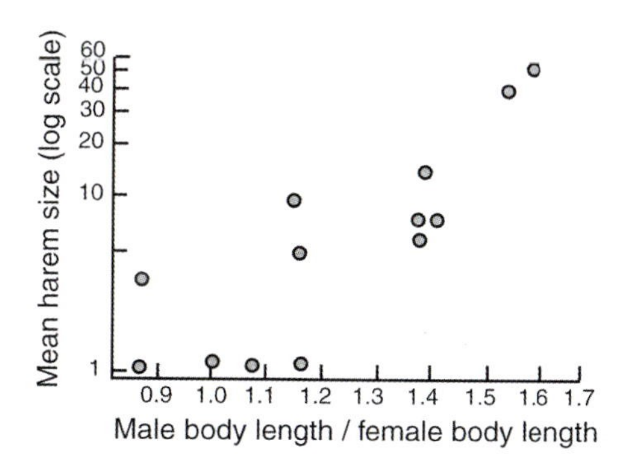

Fig. 9.3 Sexual size dimorphism in pinnipeds increases with the number of females controlled by a breeding male. Each point represents one species. Elephant seals are at the upper right; harbor seals are at the lower left. (From Alexander *et al.* 1979.)

possible some female choice. Female northern elephant seals protest against copulation by emitting loud calls and moving their hind quarters, attracting dominant males who chase off smaller males. Females protest less when approached by dominant males. While female choice has some role, male contests and endurance rivalry explain most of the size dimorphism in pinnipeds.

Weapons have evolved in many species where males fight for mates, including deer and antelope; scarabid, lucanid, and cerambycid beetles; the lips of certain fish, used in wrestling; and narwhales, whose males carry a tusk up to 2.5 m long that can inflict serious injury on rivals. Experimental removal of antlers reduces reproductive success in male red deer and reindeer, for these males use their antlers as both offensive and defensive weapons in contests over mates. Male beetles use their horns to pry up rivals and push them off mating sites or to grasp, lift, and throw their rivals to the ground.

'Reversed' size dimorphism (larger females): explained by direct selection on fecundity and weak sexual selection.

In some species, females are larger than males, probably because sexual selection is weak, and large females gain more fecundity with size than do males. Such 'reversed' size dimorphism is probably explained by selection for fecundity rather than sexual selection through mate competition.

Mate choice

Choosing mates can increase fitness, but choice has costs.

Organisms should be careful in choosing a mate, but not too careful. Excessive carelessness will result in hybridization with other species and offspring of low fitness; excessive discrimination will take so much time that the opportunity to mate will disappear before the choice is made.

Choice based on immediate phenotypic benefit can explain a lot,

Mates should either be chosen because they provide immediate phenotypic benefits or because they provide genes that increase offspring fitness. Some immediate phenotypic benefits are these: the mate has higher fecundity; or is a better provider of food to the partner or care to the offspring; or defends a breeding place that is safer, richer in food, or both; or offers better protection against predators or other potential mates that might harass, than do other mates. For example, in scorpion flies, females choose males based on the quality of the food that the males bring to their mates; in mottled sculpins, which have male parental care, females prefer large males to small ones, for large males are better able to defend nests.

but not extravagant male morphology or leks.

Choice based on immediate phenotypic reward can explain size dimorphism and some exaggeration of traits that produce particular benefits, but it cannot explain the extravagant morphologies and displays of peacocks, sage grouse, and birds of paradise that led Darwin to suggest sexual selection in the first place. In lekking species, the basis of the choice must be primarily the partner's genes, for there is no male parental care. Mating occurs swiftly on the lek, after which the female leaves to raise her offspring on her own. Except for temporary protection on the lek, the only things that a lekking male gives to the female are his genes. In lekking species females will be selected to choose indicators of male genetic quality that predict higher offspring fitness, including the ability of male offspring to attract mates.

Female choice based on male genetic quality: indicator mechanisms,

There are three main ideas on how female preferences for male indicators of genetic quality should evolve. The first is that females should prefer males displaying honest, costly signals that they contain genes for superior survival ability. This idea has been labeled either as the 'good genes' or as the 'handicap' hypothesis, the first stressing the fitness advantages, the second emphasizing the costliness of the honest signal. Andersson (1994) uses the neutral term 'indicator mechanisms' for it.

Fisherian reinforcement,

The second idea is that when females prefer males with higher fitness, their preference genes will be united in their offspring with the male's genes for higher fitness. The female preference genes will then spread in the population because they 'hitch-hike' on the success of the male's fitness genes. Once female preferences are established, they work on male traits that are otherwise neutral or even disadvantageous, except that they are preferred by females. This idea is called the 'Fisherian self-reinforcing' hypothesis after Fisher, who had the idea, and it is also called the 'sexy-son' hypothesis, because the mothers gain in fitness by selecting fathers with heritable traits that make their sons attractive to females in the next generation.

and sensory bias.

The third idea is that females inherit sensory capacities from ancestors that bias the traits that they select in males. For example, color-blind females might select striking black and white patterns but not striking colors, and females incapable of hearing low-frequency sounds will not choose males that emit such sounds. This idea is called the 'sensory bias hypothesis'.

Indicator mechanisms

Advertising the ability to resist parasites and pathogens:

If females select males because males indicate that they can sire offspring with superior fitness, then a male trait that should interest females is resistance to infection by parasites and disease (Hamilton and Zuk 1982). Structures that might have this function are the red belly of the male stickleback, throat wattles in turkeys, and eye color in pheasants. If males produce ornaments to advertise their resistance to disease, then:

(1) male fitness should decrease with increased parasite infection;

(2) ornament condition should decrease with increased parasite burden; maintaining a parasite-sensitive ornament in good condition must be costly;

(3) there must be heritable variation in resistance;

(4) females should choose the most ornamented and the least parasitized males.

These conditions for the evolution of female preferences for males with superior resistance to diseases and parasites appear to be met in guppies, pheasants, and barn swallows, but they have not yet been demonstrated in careful studies of several other species. Females do sometimes select males for parasite resistance, but they also select males for other reasons.

tail length in barn swallows indicates resistance to blood-sucking mites.

Barn swallow nests are infested by a blood-sucking mite. By manipulating levels of parasitism in natural nests, Møller (1994) showed that both nestling growth and adult tail size are reduced by parasites. Fathers with longer tails have sons that live longer. Female barn swallows select males with longer tails, and since males with more parasites have shorter tails, females that select males with longer tails have offspring that inherit superior resistance to blood-sucking mites. Females also prefer symmetrical tails. By choosing males with symmetrical tails, females may be selecting less inbred, more genetically heterozygous males—an indirect genetic benefit. They may also simply be selecting males that can maneuver more adroitly in the air, an important skill in a species that captures its insect food on the wing—a direct, phenotypic benefit.

Attractive male blue tits are bigger and live longer.

In blue tits female choice may produce genetic benefits in the offspring. Kempenaers *et al.* (1992) identified the fathers of blue tits in a Belgian forest by DNA fingerprinting. About one-third of the nests contained offspring sired by more than one male. Some females visited neighboring territories and solicited copulations; some males were often solicited; others were never solicited. Females paired with attractive males (males that were often visited by other females) rarely visited neighboring territories; those paired with unattractive males (males rarely visited by other females) often visited neighboring territories for extra-pair copulations. Unattractive males were smaller and died younger than attractive males, and the offspring of attractive males lived longer than the offspring of unattractive males. This could be explained by direct phenotypic benefit if attractive males had better territories and were better parents. If the extra-pair offspring of attractive males also had better survival, which is not known, it would be hard to explain the results only in terms of direct phenotypic reward.

His review of a large literature led Andersson (1994, p. 79) to evaluate ideas on indicator mechanisms as follows: 'Firm empirical evidence for genetic indicator processes is still lacking. There is better support for indicator mechanisms based on direct material benefits to the female or offspring, where males provide parental care or other resources.'

The Fisherian process

Preference for good genes will select for the preference itself,

Fisher suggested another idea for how female preferences evolve. The Fisherian process can start with mate choice for any reason—direct benefits, good genes, or sensory bias—and once it gets started, it can take on a life of its own that produces mate choices for new reasons. The beginning of a Fisherian process started by the choice of good genes can be illustrated with guppies—small, freshwater fish that live in clear streams in Trinidad and Venezuela. The males have brightly colored spots. Suppose that guppy females start to select males with a trait, such as orange spots, that varies genetically and is an indicator of the ability to acquire food and resist parasites. Females choosing males with larger spots will have sons with better survival. The genes for large orange spots in males will spread in the population because they are associated with better survival, and the genes that make females prefer large orange spots will also spread. So far, the argument is strictly in terms of the evolution of preferences for good genes.

Once this process starts, it makes a new effect possible, for males with large orange spots now not only have better survival, they are also preferred by females. Both because the male offspring survive better, and because they are preferred by females and therefore have superior fecundity, genes for orange spots and for the preference for orange spots spread through the population. The process is self-reinforcing. Now every time a mutation arises that increases female preference for orange spots, it will spread, and the stronger the preference in the females for orange spots, the stronger the selection for big orange spots in males. If the preference becomes strong enough, it can exaggerate the male trait enough to reduce male survival. Thus an initial preference for males with superior survival could evolve to the point where it caused a reduction in male survival.

making the preferred trait an object of selection in itself, regardless of the consequences for survival—up to a point,

Fisher argued that this process would exaggerate male ornaments progressively until the sexual preference was balanced by a reduction in male survival. In fact, if the female preferences are strong enough, then reductions in male survival are not a strong enough cost to stop the process before the population is driven to extinction. Some other factor must be involved to explain how populations with strong female preferences survive. That factor appears to be the cost of having a preference, the cost of taking the time and the risks necessary to choose a male (Bulmer 1989). When the female preference is costly, a Fisherian process can exaggerate the male trait, but not without limit, for it will be stopped by the costs of choice. Given costly preferences, a Fisherian process could not exaggerate female preferences and male ornaments without limit, but it could produce impressive ornamentation.

and explaining the evolution of impressive ornamentation.

The theory of the Fisherian process is more complete than are the experimental tests of its assumptions and predictions. The problem is to find genes for female preferences and male ornaments and to document the dynamics of their coevolution under experimental conditions. Good evidence for one assumption was provided by Bakker's (1993) study of sticklebacks. Stickleback males guard nests into which females lay eggs; the male develops red coloration on his belly in the breeding season. Both the red coloration of the males and the preference of females for males of different degrees of coloration varied genetically, and the sons' intensity of red coloration was genetically correlated with the daughters' preference for red (Fig. 9.4). Such a genetic correlation of male trait and female preference, a necessary condition for the Fisherian process, could also result from selection for indicators of parasite resistance. In fact, by choosing males with brighter red bellies, females avoid parasitized males

In sticklebacks, intensity of male color is genetically correlated with female preference for that color. And by choosing bright males, females avoid parasitized males.

(Milinski and Bakker 1990). Red bellies may be an indicator mechanism, but the Fisherian process may also be involved. Andersson (1994, p. 52) summed up as follows: 'No critical test has been performed that supports Fisherian sexual selection and excludes the alternatives, or estimates their relative importance.'

Sensory bias

Mate choice may depend on pre-existing features of the female sensory system—the preference may evolve before the thing preferred:

The third hypothesis for the evolution of female preferences focuses on their origin. Which male trait is exaggerated by sexual selection may depend on its fit to pre-existing features of the female sensory system (Ryan 1985). The cases that best support this notion are the mating calls of male tungara frogs and the preference of female platyfish for males with sword-like tails.

Tungara frogs

Male tungara frogs call from pools visited by females that select their mates by touching them. Males emit two calls, 'whines' and 'chucks'. The whine appears to be a species-identification call and is given by isolated males. When a male begins to hear competition from other males, he starts to emit 'chucks', which are attractive to females. Isolated males do not emit 'chucks' because they attract the fringe-eared bats that eat displaying males, making 'chucks' costly, and females that select deep 'chucks' have offspring with better survival (Ryan 1985). So far, it appears that female tungara frogs prefer deep 'chucks' because they indicate good genes, and that may well be the mechanism that now maintains the trait. However, the neurobiology of the female ear in the tungara frog and its close relatives suggests that another explanation is plausible for the origin of the preference.

The female ear is biased toward the low-frequency components of the 'chuck', and females of a closely related species have a similar sensitivity to the 'chuck', even though males of that species do not produce it. There are two ways to interpret this evidence. We can postulate that the

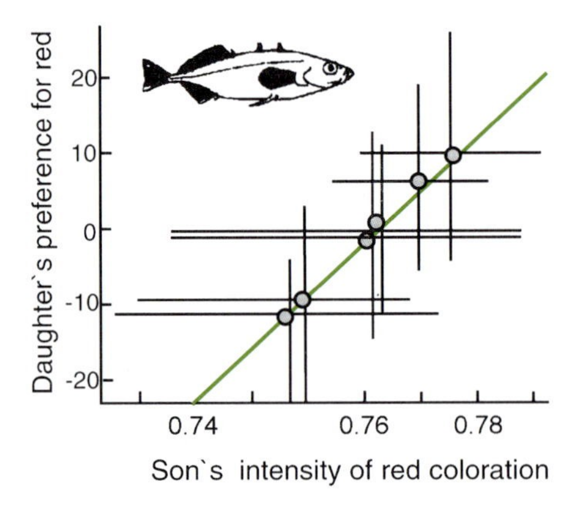

Fig. 9.4 The correlation among stickleback fathers between the intensity of red coloration in sons and the preference of daughters for red males. (From Bakker 1993, reproduced by kind permission of the author and *Nature*.)

two species had a common ancestor whose ears were biased to hear the 'chuck' but whose males did not yet produce it. In that case, the male 'chuck' first evolved in the tungara frog in response to a pre-existing female bias. Or, we can postulate that the common ancestor of both species had both the male 'chuck' and the female sensory bias, but that the related species lost the male 'chuck'. Because closely related species do not have the 'chuck' but do have the preference, the first scenario is more plausible.

Genetic variation for fitness

There is a potential logical problem with the evolution of female preferences for indicators of male genetic quality. It requires that males vary heritably in their fitness. If they did not, there would be no reason to choose among them. However, fitness is certainly under directional selection to increase, and under continued directional selection the genetic variation for fitness should be reduced to near zero. Either the theory makes some assumption that does not hold, or female preferences for indicators of male genetic quality could not evolve because there would not be enough genetic variation among males to give females any benefit from their choice.

Female choice requires male genetic variation in fitness, whose existence is a puzzle, but not an insurmountable problem.

In fact, the simple theory makes several assumptions that do not hold in general. Three factors could maintain enough genetic variation for fitness to allow sexual selection to work: spatial variation, temporal variation, and mutations affecting fitness. If conditions vary from place to place, or from time to time, so that the most fit genotype at one time or place has lower fitness at another time or place, and if gene flow and environmental variation cause changes in fitness rank frequently enough, then considerable genetic variation for fitness can be maintained in a population. Also, if there is a steady flow of favorable mutations into a population, which will happen if the population is large enough, then natural selection will be continually pulling some favorable mutations through to fixation. At any time we would always find some genetic variation for fitness, for some favorable mutations would be increasing in frequency and would not yet have gone to fixation.

Thus this logical objection to the theory of sexual selection is not fatal. How much genetic variation for fitness can be maintained by these mechanisms is a question not yet answered; it needs to be enough to allow female choice to work as a mechanism of sexual selection.

Evidence for sexual selection

Mate choice is common, and both sexes can choose.

Andersson (1994, p.131) recently reviewed studies of 186 species, mostly insects, fish, anurans, and birds, in which sexual selection was demonstrated 232 times. The most common mechanism, female choice, occurred in 167 cases. In 76 of those 167 cases any influence of male contests was excluded. In 30 cases male choice, usually of large females, was documented, and in 58 cases males contested access to females. The trait most commonly selected was male song or display, followed by male body size, male visual ornaments, female body size, male territory, then other material resources. Because of the theoretical interest in mate choice, results on mate choice are probably over-represented and results on male contests are probably under-represented in the literature. Little is known about the genetics of most traits studied.

Several case studies of sexual selection have been mentioned: barn swallows, elephant seals, sage grouse, tungara frogs, and sticklebacks. Four other studies report further evidence for important general principles, or exemplify particularly clever experiments: katydids, guppies, redwing blackbirds, and African widowbirds.

Katydids

The limiting sex is the one with the lower maximum reproductive rate:

The katydid study exploited an unusual feature of katydid reproduction to demonstrate that the limiting sex is the one with the lower maximum reproductive rate. In katydids, or bush crickets (family: Tettigonidae), the sex in which mate competition is stronger depends on the food available. Males transfer their sperm to the female in a large, nutritious spermatophore. The female eats the nutritious part of the spermatophore (but not the sperm that fertilize her), and the nutrients in it increase her fecundity. When food is scarce, females are reproductively limited by the availability of male spermatophores. When food is abundant, males are reproductively limited by the availability of females. Changing food supply should therefore change the sex that must compete more strongly for mates.

the sex that chooses changes with the sex that is limiting.

As predicted, where food was limited, males engaged less in courtship, females fought over males, males were more discriminating in their choice of mates, preferring large, fecund females, and males invested more per reproductive attempt than did females. In contrast, when extra food was supplied, many males courted, females were more discriminating in their mate choice, and females invested more than males. When food is scarce, the male reproductive rate declines because they cannot produce spermatophores rapidly. When food is abundant, spermatophore production is rapid, and females have less reason to compete for males. Thus the sex that experiences greater competition for mates is the one that invests less in each reproductive attempt (Gwynne and Simmons 1990), and the limiting sex is the one with the lower maximum reproductive rate (Clutton-Brock and Vincent 1991).

Guppies

Traits involved in sexual selection evolve to a state in which the costs and benefits of sexual and natural selection balance.

Guppies illustrate the balance of natural and sexual selection on the same trait. The natural enemies of guppies in their native streams are two species of predatory fish (see Chapter 8) whose density increases as one goes from the headwaters of the streams down into larger rivers. As the density of predators increases, the number and size of the colored spots on male guppies decreases. Females prefer males with large orange spots. The orange color in the spots comes from a pigment acquired by eating crustaceans; thus larger spots indicate better foraging ability. Females also prefer males with color patterns that make them stand out from the complex, pebbly background of the stream bottom, with its shifting patterns of light and shade.

Males from populations with low predation pressure engage in more complex courtship maneuvers than do males from populations with high predation pressure. When predators are introduced to previously predator-free populations in the field, male ornamentation and courtship behavior rapidly change in the expected direction (summary in Andersson 1994). Thus natural selection and sexual selection through female preference exert opposing pressures on male coloration and courtship behavior in guppies.

Redwing blackbirds

An extravagant ornament may be maintained by male–male interactions.

Redwing blackbirds are strikingly dimorphic, the females being brown and streaked, the males mostly black with bright red shoulder patches with lower yellow borders. These patches, called epaulets, are displayed in territorial contests with other males. The epaulets can be painted black, with clear paint used on controls, or presented in dummies with epaulets of different sizes. Males with epaulets painted out lose their territories to males with normal patches, and dummies with larger epaulets provoke stronger aggression in territory-holding males than do dummies with smaller epaulets. Males that are just passing through a territory conceal their epaulet, but those prepared to fight for a territory expose it. Thus it appears to function as a threat display. These studies (summary in Andersson 1994) show that an extravagant ornament may be maintained in a population by male–male interactions. Female choice is not excluded but is not necessary.

African widowbirds

Field manipulation experiments reveal female preference for an exaggerated male ornament.

The widow bird study demonstrated female preference for an exaggerated male trait through a field manipulation experiment. Male widowbirds establish breeding territories on the grasslands of East Africa. They have tails that are often more than half a meter long. The way that males use their tails suggests that it functions in female choice rather than male contests, for they do not expand it during territorial contests with males, but they do expand it into a deep keel during the advertising flights they perform when females visit their territories. Andersson (1994) tested the function of these long tails by experimentally shortening and lengthening them in the field. He set up nine groups of four males each, matched for territory quality and tail length. He took one male at random from within each group and cut its tail to 14 cm. The piece removed was then glued to the tail of another of the four males in that group. The other two males served as controls. One was not manipulated; the other had its tail cut at the midpoint and reglued. To measure the mating success of the males, Andersson counted the number of nests on each territory. Before the manipulations (Fig. 9.5), all males had on average 1.5 nests on their territories. After the manipulations, and sufficient time for females to build new nests, the males with shortened tails averaged only half a nest per territory and the males with lengthened tails averaged nearly two nests per territory. Thus female widowbirds prefer to build their nests on the territories of males with longer tails, and natural selection may be preventing further increases in male tail length, for females preferred much longer tails than are found in natural populations. This case demonstrates female choice for an exaggerated male ornament.

What determines the strength of sexual selection?

The strength of sexual selection is determined by differences in the potential reproductive rates of the two sexes, the operational sex ratio, the mating system, and the pattern of parental care.

Darwin saw a relationship between patterns of parental care, the mating system, and the strength of sexual selection. For analysis in depth we refer you to books by Clutton-Brock (1991) and Andersson (1994). Here we mention only key points: Bateman's principle, differences in the potential reproductive rates of the two sexes, the operational sex ratio, mating systems (especially harem and lek polygyny and monogamy), and sex role reversal.

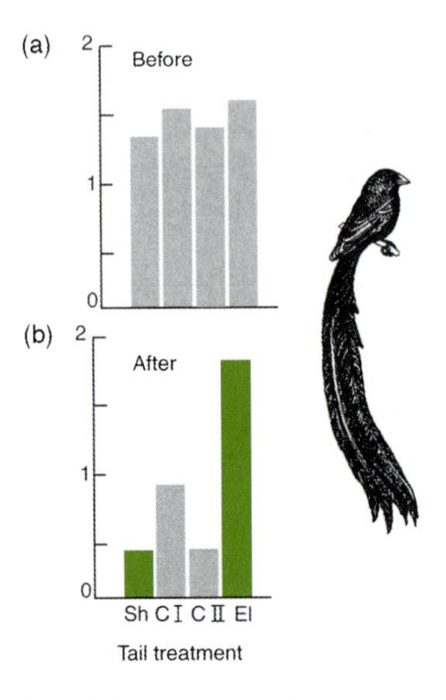

Fig. 9.5 Experimental changes in the length of the tails of male long-tailed widowbirds cause changes in the numbers of nests that females build on their territories. (a) The number of nests per territory before tails were artificially shortened or lengthened; (b) the number of nests after the manipulation. Sh, tails shortened; El, tails elongated; CI, first control; CII, second control. (From Andersson 1982, reproduced by kind permission of the author and *Nature*.)

Potential reproductive rates and operational sex ratios

Bateman (1948) demonstrated that male fruit flies (*Drosophila*) can increase their lifetime reproductive success by increasing the number of females that they mate with, whereas female *Drosophila* achieve most of their reproductive success with a single mating. Bateman concluded that this was why males competed for mates and females were the scarce resource. Bateman's principle holds for species that have no parental care and no male provisioning of mates. For more complex mating systems, we need a more general principle: the sex with the higher potential reproductive rate will compete more strongly for mates, and it will be the sex subject to stronger sexual selection (Clutton-Brock 1991).

The sex with the higher potential reproductive rate will compete more strongly for mates, and it will be the sex subject to stronger sexual selection.

The principal factors determining the potential reproductive rates of the two sexes are the mating system and the pattern of parental care. How these factors interact to influence the strength of sexual selection can be seen in the determinants of the operational sex ratio, the ratio, at any given place and time, of receptive females to sexually active males. This ratio depends on how individuals of the limiting sex form groups, on differences between the sexes in survival rates, and on differences between the sexes in the time that it takes to find a mate, breed, and care for the offspring. The operational sex ratio is the main immediate determinant of the opportunity for sexual selection. It varies strikingly with mating system and type of parental care.

The operational sex ratio—the ratio of receptive females to sexually active males—determines the opportunity for sexual selection. It varies with mating system and type of parental care.

Mating systems

Mating systems have strong effects on the operational sex ratio. They can be classified by the sex in which more individuals have offspring or by which sex has more mating partners. The categories that result are useful but rough tools, often concealing beneath their labels as much interesting biology as they reveal by their contrasts. In monogamy, equal numbers of males and females have offspring, and each sex has one partner, e.g. moorhens, albatrosses, penguins, geese and swans, dung beetles, and some cichlid fish. In some species, the pair mates for life; in others, for a season, with 'divorce' possible. In polyandry, more males than females have offspring, and each female has two or more male partners, e.g. pipe fish, some plovers and sandpipers, and jacanas. In polygyny, more females than males have offspring, and each male has two or more female partners, e.g. species where males hold harems, such as gorillas, elephant seals, horses, and red deer, and species where males display on leks, including grouse, cotingas, birds of paradise, eight species of mammals, including Uganda kob and fallow deer, and at least one insect, a Hawaiian *Drosophila*. In polygynandry, each sex may have several partners, e.g. the dunnock and the blue tit, where individuals appear to be monogamous but in fact frequently engage in extra-pair copulations.

Mating systems: which sex has more offspring or more mating partners?

These labels describe rough patterns. Precise comparisons require quantitative measurements of the mating success of males and females of different ages and sizes to establish the effect of the mating system on the operational sex ratio.

Monogamy: each sex has one partner and equal numbers of each sex have offspring.

Polyandry: more males than females have offspring, each female has two or more male partners.

Polygyny: more females than males have offspring, and each male has two or more female partners

Patterns of parental care are closely associated with mating systems. Species in which both parents care for the offspring are usually monogamous (where both sexes have only one mate) or polygynandrous (where both sexes have more than one mate). Species with female-only parental care are usually polygynous—males mate with several females, but each female mates with only one male. Those with male-only parental care are either polyandrous, where one female mates with several males (pipe fish and phalaropes), or polygynandrous (in sticklebacks, the males guard the eggs of several females, and females can deposit eggs in several nests). However, there are striking exceptions. Sea horses are monogamous but have male parental care. The male sea horse carries the eggs in a modified stomach pouch with placenta-like functions. It is thought that in sea horses there was a transition through polyandry to monogamy.

Sexual selection: strongest in species with harems or leks, with interesting exceptions.

Sexual selection has apparently been strongest in species with harems or leks. In harem-holding pinnipeds, the more females a male can control, the more intense the fights among males over females, and the greater the difference in the sizes of the sexes. Striking examples of sexual dimorphism in plumage ornaments are found in lekking birds: grouse, peacocks, and birds of paradise. Not all species that hold harems are strikingly size dimorphic: horses and zebras are examples. Not all species with leks have males with exaggerated ornamentation: one-quarter of all lekking birds are sexually monomorphic. And not all monogamous species are sexually monomorphic, although sexual monomorphism is more common among monogamous species.

The match between mating system and sexual dimorphism is not precise. There are several possible explanations. The classification of the mating system could be wrong (this can be checked with genetic fingerprinting of offspring to determine the real parents). Opportunities for mate choice, and reasons for choosing mates, may not match the classification of mating systems precisely. And the strength of sexual selection can vary strongly among the species within a mating system category.

Sex-role reversal

The rule confirmed by an exception with an obscure origin.

In some species—giant water bugs, some other hemipteran insects, some cichlid fish, pipefish, and some sandpipers—only males care for offspring. They provide a test case for sexual selection theory: males are the limiting resource, so females should compete for males and males should choose mates. In the spotted sandpipers studied by Oring in Minnesota, each male rears one clutch of four eggs per season, but a female can lay up to five clutches per season. This reverses Bateman's principle: here female reproductive success increases with the number of mates, but male reproductive success does not. Females arrive earlier than males on the breeding grounds and fight among themselves for breeding territories, for the number of males that a female can attract increases with the size of her territory and the length of foraging beach that she controls. Thus males make greater parental investment, have the lower potential reproductive rate, and are the limiting resource for the females, which compete more strongly for mates (Andersson 1994). How such a mating system originates is not clear.

Sexual selection in plants

Sexual selection in plants: scramble competition among pollen grains to reach the ovary, and competition among flowers for pollinators.

Many plants are cases of sex-role reversal: female functions compete for scarce male services.

Sexual selection in plants occurs through competition among pollen grains to reach the ovary and through competition among flowers for pollinators. Sexual selection differs in plants because many plants are hermaphrodites, with male and female functions combined in a single individual, and because many plants use insects as pollinators. Sensory bias operates in pollinator–flower coevolution, influencing flower morphology by restricting the types of signals that result in an improvement in pollination success. And a new element enters the process through the introduction of the pollinator genome to the coevolutionary game, in which the Fisherian process is then not possible because the genes determining flower preference in the pollinator are never associated in the same genome with the genes that determine morphology in the flower. Showy floral displays are exaggerated ornaments used to compete for scarce pollen—they are cases of sex-role reversal, where female functions compete for scarce male services. Sometimes the exaggeration of female traits is driven by the avoidance of nectar-stealers that do not pollinate, rather than by sexual selection; for example, the Madagascar star orchid with its 30 cm floral tube and pollinator with a 30 cm long tongue (Fig. 2.5).

Sexual selection on gametes: sperm competition and choice by eggs

Both male–male competition and female choice can occur with gametes.

Up to now we have discussed sexual selection as a process operating on the large, diploid stage of the life cycle. It also operates on gametes. In organisms where multiple insemination occurs, there will be competition between the sperm donated by different fathers. The females may be passive, but it is more likely that females will actively choose sperm if the sperm carry information on male quality. Female choice can be exercised either by the eggs themselves, through molecular mechanisms, or by the inseminated females, through morphological structures and physiological mechanisms.

Alternative explanations of sexual dimorphism

When males and females have different ecologies, their different morphologies are caused by natural, not sexual, selection.

Sexual selection need not be the only cause of differences between males and females. When males and females have different ecologies, their different morphologies are caused by natural, not sexual, selection. Female mosquitoes suck blood from vertebrates, males feed on flower nectar, and the sexes have different mouth parts. Sexual dimorphism may also, of course, reflect primary sex differences, the differences in morphology that are directly associated with reproduction rather than mating success, such as differences in male and female genitals. However, even genitals may be subject to sexual selection if they are used in sperm competition, as is the case in many insects or when females can express a preference for a particular genital morphology by controlling ovulation or insemination (Eberhard 1996). There the distinction between primary and secondary sex characters breaks down.

Sexual selection was invented to explain sexual dimorphism, but mate choice can operate in sexually monomorphic species. We only need to assume that the mates are looking for the same sorts of characteristics in each other. It is then simply harder to see that sexual selection is occurring, for there is no difference between the sexes to suggest what might be selected by mate choice.

Summary

This chapter considers the evolutionary consequences of competition for and choice of mates—sexual selection.

- Sexual selection is a component of natural selection in which mating success trades off with survival.
- Sexual selection accounts for many of the attractive ornaments of plants and animals.
- Contest competition for the mate that is the scarcer reproductive resource explains much of sexual size dimorphism, particularly in polygynous species. Active choice of mates of the nonlimiting sex has also been well documented.
- Whether the complicated, indirect mechanisms of the Fisherian process are needed to explain ornaments in lekking species remains to be seen. It is preferable to begin with simple explanations that can be easily tested.
- Alternatives for sexual dimorphism should be considered in the following order:
  (1) primary sex differences;
  (2) ecological sex differences;
  (3) choice for direct phenotypic benefit;
  (4) choice for good genes;
  (5) the Fisherian process.

The mechanisms of sexual selection produce conflict within and among the sexes. This is one kind of evolutionary conflict. The general causes and consequences of evolutionary conflicts are the topic of the next chapter. When conflicts exist, all participants may suffer, and when conflicts are resolved, interesting adaptations result. Mate choice is involved both in sexual selection and in speciation. Speciation is discussed in Chapter 11.

Recommended reading

Andersson, M. (1994). *Sexual selection.* Princeton University Press, Princeton.

Clutton-Brock, T. H. (1991). *The evolution of parental care.* Princeton University Press, Princeton.

Darwin, C. (1871). *The descent of man, and selection in relation to sex.* John Murray, London.

Eberhard, W. G. (1996). *Female control: sexual selection by cryptic female choice.* Princeton University Press, Princeton.

Møller, A. P. (1994). *Sexual selection and the barn swallow.* Oxford University Press, Oxford.

Questions

9.1 Wild turkeys are dramatically sexually dimorphic; domestic turkeys are less so. Suppose that in the wild, female turkeys chose males on the basis of expensive traits that indicated disease resistance. Furthermore, suppose that artificial selection for rapid weight gain in domestic turkeys destroyed female choice. What would you predict about the evolution of disease resistance in domestic turkeys? If we observe that domestic turkeys are less resistant, does that necessarily mean that wild females had been choosing more resistant males, or are other hypotheses equally plausible?

9.2 Butterfly fish are sexually monomorphic, apparently monogamous, and strikingly colored. Can their striking coloration be explained by sexual selection involving mate choice? An alternative explanation is that the striking colors are species recognition mechanisms, that they help butterfly fish to avoid choosing the wrong mate. If that is the case, then what is the difference between sexual selection by mate choice and species recognition?

Chapter 10
Multilevel selection and genomic conflict

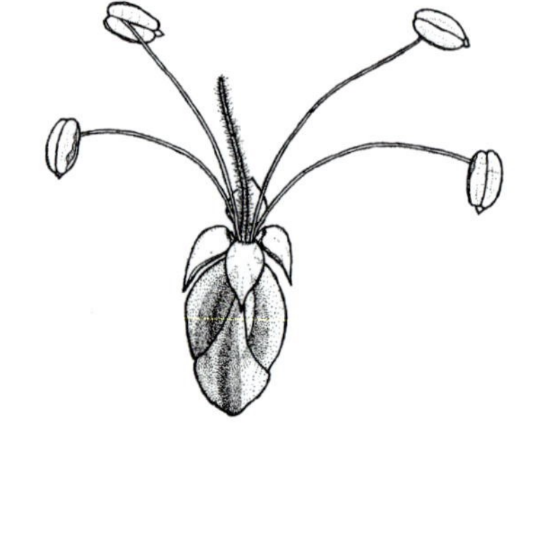

Introduction

In previous chapters, adaptive evolution was described as the product of the differential reproductive success of genetic variants. Some organisms contribute more offspring to the next generation than others, and traits correlated with reproductive success become more common if they are, to some extent, heritable. Thus adaptive evolution requires variation, reproduction, and heredity. Any collection of 'things' that exhibit these three characteristics will respond to natural selection. Note that this description lacks any genetic detail, except the very general statement that the variation must be in part heritable. That is why Darwin could describe adaptive evolution in essentially this way although he did not know about genes, how they are transmitted, the nature of genetic information, or where it is located.

The relationship between adaptive evolution and the genetic transmission system is special: genetic systems have evolved and influence further evolution.

In this chapter we concentrate on some consequences of genetic details for adaptive evolution. Those consequences are sometimes very important, for properties of the genetic system can make the response to natural selection surprisingly subtle and complex. The relationship between adaptive evolution and the genetic system is special, for natural selection has shaped genetic systems, while genetic systems shape the response to natural selection.

Multilevel natural selection

A multicellular eucaryotic organism (Fig. 10.1) contains cells; a typical cell contains a nucleus and mitochondria; a nucleus contains chromo-

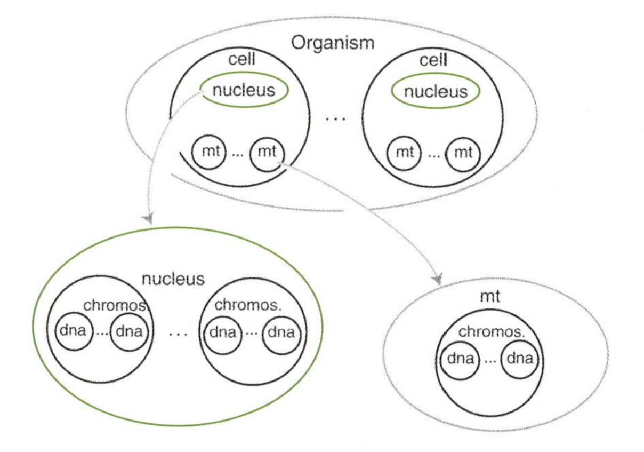

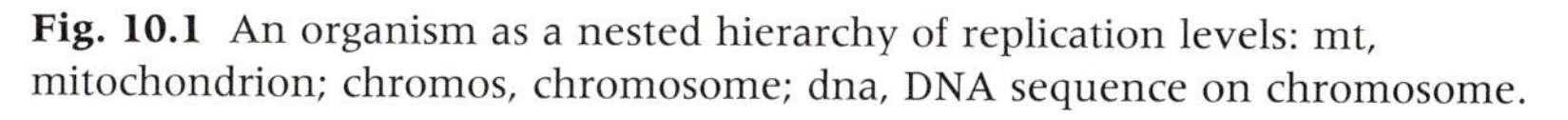
Fig. 10.1 An organism as a nested hierarchy of replication levels: mt, mitochondrion; chromos, chromosome; dna, DNA sequence on chromosome.

somes; and a chromosome contains coding DNA sequences (genes) and noncoding sequences. Each mitochondrion contains several genomes in the form of circular DNA molecules; these DNA molecules contain coding and (sometimes) noncoding sequences.

Many nested structures—nuclear genes, genes in organelles, cells, and organisms—obey the three conditions for natural selection: variation, heredity, and reproduction.

Many of these structures satisfy the three conditions for adaptive evolution: variation, heredity, and reproduction. Individual organisms vary, part of their variation is heritable, and they can reproduce: thus natural selection operates on individual organisms. But cells also show variation: there are different types, such as epithelial cells, lymphocytes, and liver cells; they reproduce by mitotic division; and they show heredity—liver cells, for example, divide to produce liver cells. Cells are also subject to natural selection. So are mitochondrial genomes and repeated DNA sequences on nuclear chromosomes.

A hierarchy of replicating units experiences multilevel evolution. When a trait is favored at one level, but selected against another, genomic conflict occurs.

Thus a multicellular organism is a hierarchy of replicating units, many of which are undergoing adaptive evolution. This is **multilevel evolution**. One important consequence of multilevel evolution, **genomic conflict**, occurs when a trait is favored at one level but selected against at another, or when different genes affecting the same trait experience contradictory selection pressures because they follow different transmission rules. Male sterility in flowering plants is discussed here briefly to introduce the essential characteristics of genomic conflict. It is treated in greater detail below.

An example: natural selection simultaneously favors hermaphroditic flowers coded by genes in the nucleus and male-sterile flowers coded by genes in the mitochondria.

Several hermaphroditic plant species, including those in the genus *Plantago* (plantain), contain some individuals that produce no viable pollen. These male-sterile plants are effectively females, in contrast to normal individuals that have flowers with both male and female reproductive structures (Fig. 10.2). That the mutation for male sterility is in the mitochondrial genome gives us a clue to its evolutionary success, which otherwise would be a puzzle. Because mitochondria are transmitted only in the female line, mitochondrial mutations that enhance female fitness are favored by natural selection, irrespective of their effect on

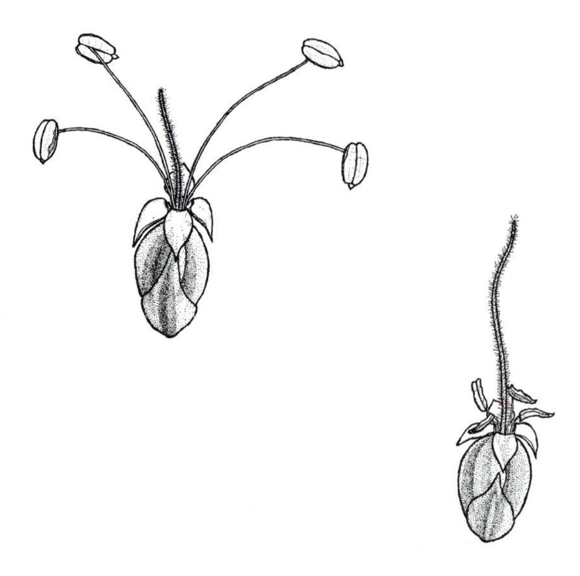

Fig. 10.2 Normal hermaphrodite (upper) and male-sterile (lower) flowers in *Plantago coronopis.* (From Koelewijn 1993.)

male fitness. A mitochondrial mutation that suppresses male function and allocates more resources to female function will be selected.

Nuclear genes, in contrast, are transmitted with equal probability through pollen and ovules, and nuclear genes affecting reproductive organs are selected for equal allocation to male and to female function (Chapter 8). Male-sterile plants do have higher fitness through seed production than hermaphrodites, but not enough to compensate the nuclear genes for the fitness lost in male function. Thus there is genomic conflict: natural selection favors hermaphroditic flowers coded by genes in the nucleus and male-sterile flowers coded by genes in the mitochondria. The outcome of such a conflict can be hard to predict. In this case, both types occur in frequencies that vary.

The concept of multilevel selection can be extended to higher levels. Individuals occur in groups and groups make up a species. Groups and species also satisfy the criteria for adaptive evolution in some sense: they vary, they reproduce by division, and they show heredity (Fig. 10.3).

Another example: multilevel selection on virulence—myxoma virus in rabbits.

Just as conflicts can occur between genetic levels within an individual organism, so do conflicts arise between the effects of natural selection operating at higher levels. Consider, for example the evolution of the virulence of a pathogen infecting a host population. The myxoma virus infects rabbits. Its replication rate within a rabbit determines the virulence of the virus—fast replication leads to more severe disease and more rapid death of the rabbit than slow replication. When there is genetic variation among viruses within a rabbit, selection within rabbits increases virulence. However, selection on the virus for transmission between rabbits favors viruses that have the greatest probability of being transmitted to new

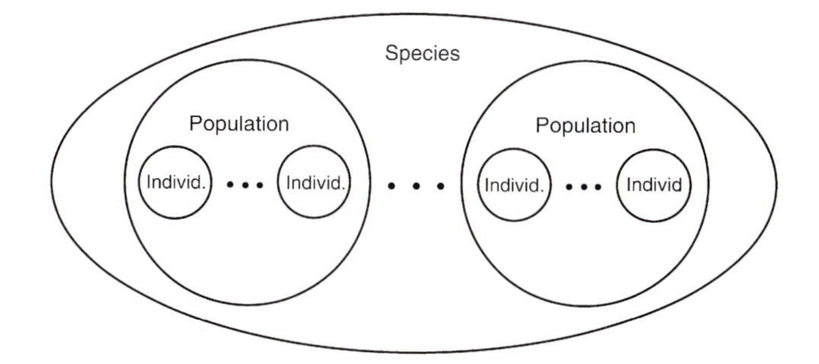

Fig. 10.3 A nested hierarchy of replication levels upwards from individual to species.

hosts, i.e. those living in rabbits that survive for a long time. Evolution at this level reduces the virulence of the virus. Thus there is a conflict between two levels: selection among viruses within hosts increases virulence, and selection between hosts reduces virulence. The outcome is an intermediate level of virulence. Which process dominates depends on the amount of variation among the viruses within rabbits, which is enhanced by a high viral mutation rate and by multiple infections; on the variation among rabbits in the severity of the disease, caused by infections by different viral strains; on the transmission efficiency of each viral strain; and on the evolution of resistance in the rabbits.

Two-level selection and genomic conflict

An example of conflict with two-level selection: mitochondria multiplying within cells. Any systematic bias in transmission opens the door to genomic conflict.

We now describe the simplest type of multilevel selection, two-level selection, to see how it gives rise to genomic conflicts. Figure 10.4 shows a two-level hierarchy. The outer circle represents the higher level, for example a unicellular organism or a cell in a growing multicellular organism. The inner circles represent the lower level, for example mitochondria within the cell. Two types of mitochondria, genetic variants marked *A* and *a*, occur within the same cell. Mitochondria replicate within the cell and are distributed at cell division to the daughter cells.

Three outcomes are possible:

1. If both mitochondrial types replicate at the same rate and are fairly distributed to the daughter cells (Fig. 10.4a), there is no natural selection at the lower, mitochondrial level.
2. If *A* replicates faster than *a* (Fig. 10.4b), the proportions of *A* and *a* mitochondria change in the next cellular generation, and natural selection does operate on mitochondria. Whether it is in conflict with selection at the cellular level depends on the effects of *A* and *a* on the cell's phenotype. If *A* contributes more to the cell's fitness than *a*, there is no conflict, for selection works in the same direction at both levels. But if *a* benefits the cell more than *A*, there is genomic conflict because selection favors *a* at the level of the cell but *A* at the level of the mitochondria.

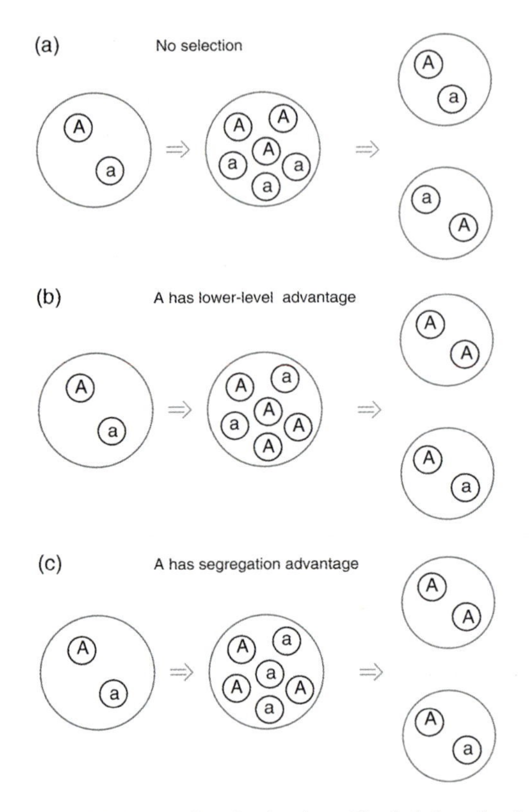

Fig. 10.4 Two-level selection. The left-hand column represents newly formed cells; each, for example, with two mitochondria (*A* and *a*). The middle column represents mitochondrial replication. The right-hand column represents the result of cell duplication. (a) No selection at lower level; (b) lower-level selection favoring *A*; (c) biased segregation favoring *A*.

3. If *A* and *a* mitochondria replicate at the same rate but are distributed unequally—with biased segregation—to the daughter cells (Fig. 10.4c), selection also operates on mitochondria. As in case 2, genomic conflict arises if the mitochondrial type favored at segregation is selected against at the cell level.

In cases 2 and 3 the ratio of *A*:*a* changes from one cellular generation to the next because mitochondrial transmission is biased in favor of *A*, because of faster replication or segregation. This illustrates the key role in the causation of genomic conflicts played by the transmission of the lower-level replicating units—the mitochondria—to the descendants of the higher-level replicator—the cell. Any systematic bias in this transmission creates genomic conflict.

Mitochondrial 'petite' mutations in yeast

Two-level selection producing genomic conflict is exemplified by mitochondrial mutations in yeast. The cells of baker's yeast normally form large colonies on a solid culture medium, but occasionally a small, or 'petite', colony is found. Such petite mutants often have defective mitochondria, suffer severe metabolic problems, and grow poorly, which explains their small colony size. Each yeast cell contains many mitochondria, and each mitochondrion contains many genomes. How can so many defective mitochondrial genomes accumulate in a yeast cell? Petite mutations, which are often large deletions, allow faster replication of their mitochondrial genome, which then out-competes the other mitochondria within the cell. All the cells derived in successive divisions from a cell with mutant mitochondria inherit the defective mitochondria and thus form a 'petite' colony. This is a clear case of genomic conflict, for the mutation is favored in mitochondrial replication but selected against among the yeast cells, whose growth it impairs.

What is the fate of petite mutants in a yeast population? Selection at the two levels—mitochondria and cells—occurs on different time scales. The process of mutation, increase in frequency, and fixation at the mitochondrial level takes place in only a few cell generations because there are many mitochondrial divisions between two successive cell divisions. Thus the petite mutation is successful in the short term. What are its chances in the long term?

The 'petite' mutation is favored at the level of mitochondrial replication but is selected against among yeast cells.

First consider yeast cells reproducing by asexual division, as is normal under natural conditions. Because the mutation lowers the fitness of the yeast cells, petites are selected against in competition with normally growing yeast, and in time they must disappear. Occasionally new petite mutations will occur, have short-term success, and be eliminated.

However, yeast cells can also become sexual. Then two cells of different mating type fuse, and the resulting diploid zygote undergoes meiosis to produce haploid asexual cells. Some petite mutations, called suppressive petites, show biased segregation in crosses between a petite strain and a normal strain: these crosses yield only petite offspring. The long-term success of suppressive petite mutations is more difficult to predict than the fate of petite mutations in asexual yeast cells. They are favored at the mitochondrial level, selected against at the cell level, and favored in sex with non-petite cells. Sexual fusions enable these petite mutations to 'infect' normal lineages that under asexual reproduction can only become petites by mutation. For the petite mutation, sex resembles the horizontal transmission of a parasite. Whether suppressive petites can get established depends on the frequency of sex and on their negative effect on the cellular growth rate (question 10.1). Sexual reproduction clearly provides more opportunities for suppressive petites than does asexual reproduction.

Genomic conflict in asexual systems

Genomic conflicts arise less easily in asexual systems because variability among the lower-level replicators is rare and because the lower-level replication gene causing a genomic conflict shares the fate of the higher-level replicator in which it finds itself.

The yeast example shows that genomic conflicts arise less easily in asexual systems for two reasons. First, a genomic conflict presupposes selection acting at two or more levels. This implies that at the lower level there has to be genetic variation, in our example among the mitochondria, which in an asexual system can only arise by mutation. Because mutation rates are low, variation among the lower-level replicators is limited. Secondly, the long-term fate of a lower-level gene that causes a genomic conflict is linked in an asexual system to the fate of the higher-level replicator in which it finds itself. It cannot 'escape' by moving into another replicator, and because genomic conflict lowers fitness at the higher level, such a gene will be removed by selection.

Genomic conflict in sexual systems

Sex generates genetic diversity upon which natural selection can act at several replication levels. It also creates chances for genetic elements to change hosts.

Neither reason why asexual systems are well protected against genomic conflicts applies to sexual systems, for sex combines genomes from different lineages, and it can generate genetic diversity upon which natural selection can act at several replication levels (the 'input' aspect of sex, see Fig. 10.5). Fusion is followed by a meiotic division that restores the original ploidy of the reproductive cells with genetic recombination and segregation (the 'output' aspect of sex). At this stage genetic elements can become associated with replicators different from those of their original hosts. For example, in Figure 10.5 the lower-level variants *A* and *a*, originally associated with higher-level variants *B* and *b*, have changed their association after recombination. Furthermore, the segregation process itself offers a chance to gain a transmission advantage. Some genes violate the Mendelian transmission rules by distorting segregation. We examine an example of **segregation distortion** (also called **meiotic drive**) below, but first we consider a particularly instructive example that shows the crucial importance of 'lineage mixing', characteristic of sexual systems, for the generation of genomic conflict.

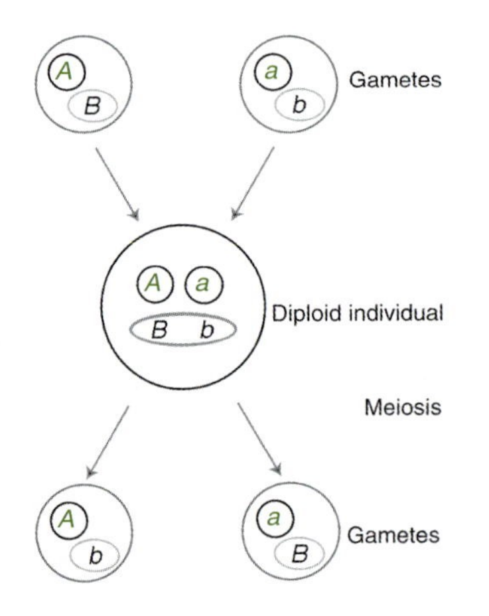

Fig. 10.5 Gamete fusion and subsequent segregation and recombination. *B,b* is a pair of alleles on chromosomes in the nucleus. *A,a* are alleles at a mitochondrial locus in the cytoplasm. Here we depict what happened before uniparental inheritance of mitochondria evolved.

Transmission of bacterial plasmids

Many bacteria carry not only a chromosome but also smaller, circular DNA molecules called plasmids. These plasmids usually have some genes important for their own propagation, but they may also contain genes useful to the host cell, like those coding for antibiotic resistance. If plasmids were only transmitted vertically (i.e. at bacterial cell division to the daughter cells), their long-term fate would be strictly coupled to that of their bacterial host. If they lowered the cell's fitness, natural selection would remove them from bacterial populations by favoring cells that did not have such plasmids. However, many of these plasmids induce their host to conjugate with another, uninfected, bacterial cell. During conjugation the recipient acquires a plasmid while the donor cell retains a copy of the plasmid.

Horizontal plasmid transmission, one form of bacterial sex, creates genomic conflict between plasmids and their host bacteria.

This horizontal transmission of plasmids is one form of bacterial sex, for bacterial chromosomal genes are sometimes also transferred and recombine with the recipient's chromosome. The consequences for genomic conflict are serious: the long-term fate of the plasmids is no longer coupled to that of their host, and if infection by horizontal transfer is frequent enough, plasmids can establish themselves in a bacterial population despite an average negative effect on bacterial fitness. Then there is genomic conflict because natural selection favors both maintenance of a plasmid via vertical and horizontal plasmid transmission and its loss via the fitness difference between plasmid-bearing and plasmid-free bacteria.

Several plasmids have evolved a remarkably clever mechanism to enhance their stable maintenance in the bacterial host population. They make the host cells addicted to their presence by producing both a toxin and its antidote. The antidote molecule is less stable than the toxin molecule. As long as the plasmid is present in a cell, there is no problem, for continuous production of both molecules guarantees that the toxin is ineffective. But if by chance a copy of the plasmid fails to get transmitted vertically to a daughter cell, this cell dies because the antidote has disappeared while toxin molecules are still present. Thus once a cell has acquired a plasmid, its descendants remain addicted to it. Although this addiction mechanism helps the plasmid to maintain itself, it would not guarantee maintenance in strictly asexual bacteria with only vertical transmission, where the host cell could lose in competition with uninfected cells. Sex with horizontal transmission overcomes, for the plasmid, the consequences of negative fitness effects on the bacterial cell.

Some plasmids addict their host cells to their presence by encoding both a toxin and its antidote.

Segregation distortion in eucaryotes: the *t*-haplotype in mice

A basic principle of genetic transmission, discovered by Mendel, is that an individual heterozygous for a pair of alleles, *A* and *a*, produces 50% *A* gametes and 50% *a* gametes (the law of equal segregation). This is a direct consequence of the separation of each chromosome pair during meiosis. However, the law of equal segregation is sometimes violated (Lyttle 1991).

Segregation distorters are genes, regions of chromosomes, or chromosomes that are in conflict with the rest of the nuclear genome over segregation: the *t*-allele system in mice is a well-studied example.

One of the best-studied cases concerns the *t*-haplotype in mice. The name derives from the tailless phenotype that occasionally occurs in some species, including *Mus musculus* and *M. domesticus*. Crossing experiments in the 1930s seemed to show that at the short tail, or *T*, locus, three alleles were segregating: *T*, *t*, and +. The homozygotes *TT* and *tt* died before birth, while heterozygote *Tt* and *T*+ individuals were short-tailed or had no tail. Remarkably, the *t* allele was transmitted much more frequently than the normal 1:1 ratio in crosses involving heterozygous *t*+ or *Tt* males.

Let us consider a simplified situation with only the *t* and the + alleles, which is enough to communicate the essence of meiotic drive. Heterozygous *t*+ males produce between 90% and 100% *t*-bearing sperm, depending on the particular *t*-haplotype. Several loci within the *t* complex produce a 'toxic' effect on + chromosomes in heterozygous males, to which they themselves are largely immune—a poison/antidote system. All *t*-haplotypes produce male sterility in the homozygous condition, and *tt* homozygotes are also lethal or have severely reduced viability, both in males and females.

As demonstrated by a simple population genetic model (question 10.2), a stable polymorphism can be expected for *t* and + chromosomes under broad conditions. The genomic conflict, caused by opposing selection pressures at the level of the gametes (favoring *t*) and the individual (favoring +), results in both types of chromosomes being maintained. This polymorphism may have been present in mouse populations for 3 million years, predating the speciation events leading to the present species in the *Mus* subgenus (other trans-species polymorphisms are discussed in Chapter 15).

The genomic conflict does not, however, necessarily lead to stable coexistence of *t* and + chromosomes. For example, if the fitness reduction in *tt* homozygotes was small enough, the segregation distortion in heterozygotes would be strong enough to drive *t*-haplotypes to fixation. (question 10.3).

This genomic conflict reduces adaptation at the individual level: mean individual fitness is lower with the distorter than without it. The *t*-haplotypes in mice illustrate how segregation distortion can produce negative effects for individual fitness. Understanding the evolutionary stability of fair Mendelian segregation is therefore one of the basic challenges in evolutionary biology.

Segregation distortion can only be discovered if it is associated with a phenotype. Otherwise it may be there, but we cannot 'see' it. In mice, it was discovered through the associated tail abnormalities, which may have nothing to do with the cause of distortion. Another easy phenotype through which segregation distortion can be discovered is an abnormal male/female ratio in the offspring of a cross. This can indicate meiotic drive caused by a gene on a sex chromosome, although other explanations are possible. Several sex chromosomal segregation distorters are known, such as the SR (sex ratio) trait in *Drosophila melanogaster*, which has been investigated in great detail.

The cytoplasm as battleground for genomic conflicts

Segregation distortion appears to be a rare exception. Most chromosomal genes have normal meiotic behavior; their transmission follows the Mendelian patterns. Because meiosis and mitosis are strictly regulated processes that prevent a biased distribution of chromosomes over the daughter cells, the evolution of a successful distorter is unlikely. In fact, even for *t*-haplotypes and other well-investigated nuclear distorters, meiotic segregation seems to be normal, and the apparent segregation bias in heterozygous males is probably caused by post-meiotic killing of nuclei with the normal chromosome. Thus it appears to be very difficult to create effective distortion. All chromosomal segregation distorters investigated so far consist of two genetic components: a gene producing a harmful effect in the homologous chromosome and a gene conferring resistance to the distorting chromosome. These two genes must be closely linked, for recombination would destroy their effect. Therefore, it is hard to see how segregation distorters evolve. Two simultaneous mutations producing closely linked genes with such functions are highly implausible, and if either mutation occurs separately, it has no selective advantage.

Outside the nucleus there are more opportunities for biased segregation of genetic elements. Cytoplasmic genes, for example in mitochondria and chloroplasts, are not distributed at cell division by a process as precise as meiosis. They occur in many copies per cell, and if a cell contains tens or hundreds of mitochondria, a precise segregation mechanism is not only difficult to evolve but also unnecessary, for if one daughter cell receives fewer mitochondria than the other, subsequent mitochondrial divisions easily compensate for the initial deficit. Other genomes, such as viruses or plasmids, may also occur in cytoplasm. They also occur in many copies per cell and lack a precise distribution mechanism at cell division. Evidence from many organisms suggests that an initial heteroplasmic condition, in which a single cell contains different alleles of a cytoplasmic gene, cannot be maintained through a series of consecutive mitotic cell divisions. After a few cell divisions the daughter cells become homoplasmic (containing copies of just one or the other allele). This suggests that cytoplasmic genes usually differ in their replication rates. Any cytoplasmic mutant that is competitively superior can, like the suppressive petite mitochondria in yeast, gain a segregation advantage in a heteroplasmic situation.

Opportunities for biased segregation are greater in the cytoplasm than among the nuclear genes, for cytoplasmic genes are not distributed to daughter cells with high precision. A competitively superior cytoplasmic mutant can gain a segregation advantage in a cell with several cytoplasmic genomes.

Uniparental transmission of cytoplasmic genomes

In sexual life cycles, transmission of cytoplasmic genes is uniparental. Both the male and the female gamete may have cytoplasmic genomes, but the resulting offspring usually contain the maternal cytoplasmic genes. Very few exceptions to this rule are known. In anisogamous species the maternal cytoplasmic contribution to the zygote outweighs the paternal contribution by so much that this itself may explain why only maternal cytoplasmic genes occur in the offspring. Active degradation of paternal cytoplasmic genomes has also been observed. Active elimination of the cytoplasmic genes from one parent also occurs in the zygotes of isogamous species where both gametes contribute roughly equal amounts of cytoplasmic genomes but only those of one parental mating type are found in the offspring.

The evolution of uniparental cytoplasmic inheritance is probably mediated by genomic conflict.

Why would the inheritance of cytoplasmic genes be uniparental, in contrast to the biparental inheritance of nuclear genes? The evolution of uniparental cytoplasmic inheritance is probably mediated by genomic conflict.

Biparental cytoplasmic inheritance would produce strong competition among cytoplasmic genes to get into zygotes, leading to conflict between the cytoplasmic and the nuclear genome.

What are the consequences of biparental cytoplasmic inheritance? Cytoplasmic genomes from both gametes then occur in the zygote, and no mechanism ensures that after cell division the daughter cells have exactly the same cytoplasmic composition. It is hard to see how such a mechanism could work because of the many copies of cytoplasmic genomes and their scattered location in the cell. Because the cytoplasmic genomes are not under segregation control, intracellular competition and sampling effects determine their distribution after cell division. In a multicellular organism originating from a heteroplasmic zygote, cell lineages will differ in their cytoplasmic genetic content. In particular the genetic composition of the cytoplasm of the cells from which the gametes are derived is crucial, for only those cytoplasmic genes will be passed on to the offspring. Cytoplasmic genes will compete for this position. A cytoplasmic mutation with increased probability of winning the competition has a good chance of being selected, for it enjoys a transmission advantage—even if it lowers the fitness of the individual carrying it. Here again are the ingredients of genomic conflict: natural selection favors a gene at one level (the cytoplasmic genome) but selects against it at another level (the individual organism).

Uniparental inheritance of cytoplasmic genes prevents genomic conflict.

Now consider uniparental inheritance of cytoplasmic genes. If a cytoplasmic mutation occurs, like the one postulated above, its fate will be fundamentally different. If it occurs in a male, it will not be transmitted to the offspring anyway. If it occurs in a female, it can be transmitted to offspring, but only along the female line of descent. Direct competition with other alleles in the zygote and young embryo is prevented, for in half the cases the mutation comes from the mother and is exclusively transmitted irrespective of the paternal allele, and in half the cases it comes from the father and is never transmitted. The genomic conflict that occurs with biparental inheritance is prevented, and the fate of the cytoplasmic mutation only depends on its effect on the individual fitness of its carriers. If it enhances individual fitness, it will increase in frequency; if it lowers individual fitness, it will be eliminated.

Because biparental inheritance of cytoplasmic genes makes organisms vulnerable to genomic conflicts, whereas uniparental inheritance prevents them, it is attractive to suggest that this has been the main reason for the evolution of uniparental inheritance of cytoplasmic DNA from an ancestral pattern of biparental inheritance. Population genetic models suggest that this scenario for the evolutionary transition from biparental to uniparental inheritance can work under certain conditions.

This may be why uniparental inheritance evolved.

Uniparental transmission and the origin of males and females

In discussing the evolution of uniparental inheritance of cytoplasmic genes we assumed the existence of males and females. Indeed, sexual differentiation seems to be essential for uniparental transmission. If both parental gametes were identical in type, it would not be clear beforehand which gamete would transmit its cytoplasmic DNA and which gamete would not. Then competition for transmission would arise and cytoplasmic mutants enforcing transmission would spread. Uniparental inheritance would not be stable, and the door would be open to genomic conflict. When the two gametes that form a zygote are always of different mating type, transmission or nontransmission can be coupled to mating type. So it seems that male–female dimorphism, or at least mating types, evolved before the evolution of uniparental cytoplasmic inheritance.

The interesting suggestion has been made that it may have been the other way round: that male–female or mating-type differentiation evolved to regulate uniparental cytoplasmic inheritance. In other words, selection to improve an imperfect mechanism of uniparental cytoplasmic inheritance triggered the evolution of two sexes or mating types. A supportive observation is that some organisms that do not have a system of two sexes, namely hymenomycetes (mushrooms) and ciliates, accomplish sex without cell fusion, exchanging nuclei without cytoplasm. Because cytoplasms of the two parents are not mixed, cytoplasmic inheritance is automatically uniparental. Therefore—so the proponents of this hypothesis argue—there is no need for the regulation of uniparental cytoplasmic inheritance, and consequently no selection for sexual differentiation: in their view, an exception that suggests the rule.

Did gametic mating type asymmetry evolve before or after the evolution of uniparental cytoplasmic inheritance?

Uniparental transmission and male–female differences create the potential for new genomic conflicts

Uniparental cytoplasmic inheritance prevents the genomic conflicts that easily arise in a system of biparental transmission without precise segregation control. However, this evolutionary solution is actually a

Uniparental inheritance creates an asymmetry in transmission favoring cytoplasmic genes that enhance female fitness at the expense of male fitness. Nuclear genes should counteract such effects.

Trojan horse. Once established, uniparental inheritance creates a fundamental asymmetry in transmission that constantly favors any genetic element in the cytoplasm that enhances female fitness at the expense of male fitness, or shifts the sex ratio towards more females, for by doing so it automatically increases its own fitness. If this produces genomic conflict, nuclear genes will be selected to counteract the effects of the cytoplasmic gene.

On top of the asymmetry in cytoplasmic transmission, sexual selection (Chapter 9) has shaped additional asymmetries between males and females. Many cases of genomic conflict generated by the male/female asymmetry have been discovered. We discuss two examples: male sterility in plants, and genomic imprinting in mammalian development.

Male sterility in plants

About 5% of angiosperm species are **gynodiecious**, meaning that some individuals are male sterile. Male sterility can be detected in phenotypes by the occurrence of normal hermaphroditic plants and plants that produce seeds but no viable pollen in the same population (see Fig. 10.2).

This trait has been investigated in natural populations of *Plantago* and *Thymus* and in several agricultural crops where the presence at appreciable frequencies of plants deficient in male function is puzzling. These plants can only reproduce as females and should have lower fitness than conspecifics that can function both as males and as females. Selection should eliminate the genotypes that cause male sterility.

Male sterility is explained by genetic conflict initiated by mitochondrial genes. Nuclear 'restorer genes' resist the effects of the mitochondrial genes.

The apparent paradox disappears once we know that in most cases the male-sterile phenotype is caused by a mitochondrial mutation. For mitochondrial genes, which are inherited exclusively along the female line, a male is a 'dead end' from which no transmission to offspring is possible, and elimination of male function does not affect the transmission prospects of mitochondrial genes as long as there is enough pollen in the population. Indeed, a mitochondrial mutation that eliminates male function while enhancing the female fitness of its carriers will be favored, for the fate of mitochondrial genes is affected only by their carriers' female fitness.

In response (see Chapter 8), selection on nuclear sex-ratio genes will act to establish equal investment in offspring of both sexes. When the frequency of male steriles in a population increases because selection favors a mitochondrial gene causing male sterility, that increase in turn selects any nuclear mutation that suppresses cytoplasmic male sterility and restores a more equal sex ratio. Such nuclear 'restorer genes' are found in gynodiecious species, where selection for mitochondrial male sterility conflicts with selection for nuclear restorers of male fertility.

Indirect evidence from interspecific hybrids suggests that the nuclear genome sometimes 'wins' the conflict.

The long-term outcome of conflict between mitochondrial male sterility genes and nuclear restorer genes is not easy to predict, for it depends on the precise relations between the fitnesses of the different genotypes. Indirect evidence suggests that the nuclear genome sometimes 'wins' the conflict, for male-sterile individuals can result from hybrid crosses between related species within which no male sterility occurs. This can be interpreted as evidence for complete within-species restoration of male fertility involving different mitochondrial mutations and nuclear restorers in the two species. In hybrids, a nuclear restorer from one species encounters a mitochondrial male sterility gene from the other species, against which it is ineffective. The evolutionary dynamics of male sterility can be complex, for natural populations often contain several mitochondrial male sterility mutations together with corresponding nuclear restorer genes.

Genomic imprinting in mammalian development

In mammals a remarkable phenomenon called genomic imprinting has been discovered, in which the sex of the parental genome determines the expression of genes in the offspring. For some genes the paternal copy is expressed and the maternal copy is inactive. For other genes the pattern is reversed: they are expressed if derived from the mother, not expressed if derived from the father. Most genes are not subjected to imprinting; imprinted genes often affect growth. At loci where imprinting occurs, individuals are effectively haploid because only one allele is active. Moreover, imprinting prevents parthenogenesis in mammals: a parthenogenetic mutant would possess only maternal alleles, so that development would be disrupted by the lack of expression of genes coming from the male germ line (see Chapter 1).

In genomic imprinting the sex of the parental genome determines the expression of genes in the progeny. For some genes only the paternal copy is expressed; for others only the maternal copy. Only a few genes are imprinted. They usually control growth.

Several hypotheses have been suggested for the evolution of imprinting, including Haig's (1992) interesting idea that it may be the product of genomic conflict resulting from the different interests of the male and female parent. A gene enhancing the rate at which a mammalian embryo extracts resources from the mother may be selected in males but not in females. In the mouse, insulin-like growth factor 2 (Igf2) is expressed in the fetus and promotes the acquisition of resources from the mother across the placenta. The paternal copy of the *Igf2* gene is expressed, but the maternal copy is inactive. There is another gene, the insulin-like growth factor 2 receptor (*Igf2r*) which appears to inhibit the action of *Igf2*. This gene shows the reverse pattern: the maternal copy is expressed and the paternal copy is inactivated.

Genomic imprinting may have evolved because of genomic conflict resulting from the different interests of the father and mother. Paternal genes are expected to manipulate the mother to provide more nutrients to the fetus. Maternal genes are expected to resist.

According to Haig, this makes evolutionary sense as follows. If females can have offspring from several males, then it is not in the interest of a male to let the female hold resources in reserve for future offspring fathered by other males. Thus paternal genes are expected to manipulate the mother to supply more nutrients to the fetus, while maternal genes are expected to protect her from overprovision that would compromise future reproduction. Only with strict life-long monogamy would the interests of the father and mother coincide, and conflict would not be expected. If Haig's explanation is correct, selection for these two genes acts in opposite directions in males and females, thereby creating a remarkable genomic conflict between paternal and maternal genes whose effects occur in the offspring.

Evidence from mice provides support.

The long-term outcome of this conflict may well be the situation observed. Mice with an inactivated paternal copy of *Igf2* are only 60% of normal size at birth, while mice with an inactivated maternal copy of *Igf2r* are born 20% larger than normal. This situation may be a stable reconciliation of paternal and maternal interests—but it could also be interpreted as a tense stand-off, a balance of interests in constant and enduring conflict.

Importance of genomic conflicts in evolution

Genomic conflict may have been the driving force in many evolutionary transitions.

What is the evolutionary significance of genomic conflict? Has it been a major force of evolutionary change, or has it played a modest role? It is too early for a definitive answer, for genomic conflicts have been recognized and studied only recently. Hurst *et al.* (1996) suggest a role for genomic conflicts in the origin of chromosomes, mating types, sex, meiosis, sexual selection, diploidy, genome size, and speciation. Only future work can reveal whether these suggestions can be supported by evidence.

Summary

This chapter concentrates on multilevel selection within organisms that causes genomic conflicts because selection at different levels works in opposite directions.

- Organisms consist of a hierarchy of replication levels, at each of which natural selection may occur simultaneously.
- Organisms occur in groups, and under some conditions groups are also subject to natural selection, for they may form new groups and disappear at different rates, depending on their composition.
- Replicating units that occur in few copies and whose replication and segregation are strictly controlled, such as cell nuclei and their chromosomal genes, do not easily cause genomic conflicts.
- Replicating units that occur in many copies and whose replication and segregation are not strictly controlled, such as cytoplasmic genetic elements, more easily cause genomic conflict.
- Sexual organisms are more prone to experience genomic conflicts than asexual organisms.
- Genomic conflicts can generate evolutionary change and may have been involved in several key evolutionary events, such as the evolution of the male–female distinction.

Genetic conflicts can also account for uniparental inheritance of cytoplasmic genomes, male sterility in plants, and genomic imprinting of growth genes in mammals. Whether they can also account for speciation, the subject of the next chapter, is not yet clear.

Recommended reading

Haig, D. (1992). Genomic imprinting and the theory of parent–offspring conflict. *Seminars in Developmental Biology*, **3**, 153–60.

Hurst, L. D., Atlan, A., and Bengtsson, B. O. (1996). Genetic conflicts. *Quarterly Review of Biology*, **71**, 317–64.

Mock, D. W. and Parker, G. A. (1997). *The evolution of sibling rivalry.* Oxford University Press, Oxford.

Lyttle, T. W. (1991). Segregation distorters. *Annual Review of Genetics*, **25**, 511–57.

Questions

10.1 Is two-level conflict, as exemplified by petite mutants in yeast, analogous to cancer? Just as mutant petite mitochondrial genomes spread within a yeast cell, out-competing normal mitochondria, mutant cancer cells may spread within the body, out-competing normal cells. Cancer seems to be a case of genomic conflict because selection at the level of the somatic cells is in conflict with selection at the individual level. Is there an essential difference between petite yeast and cancer where the analogy breaks down?

10.2 The *t*-haplotype polymorphism in mice involves segregation distortion favoring *t*-bearing chromosomes in heterozygous, *Tt*, males and lethality of male and female *tt* homozygotes. Consider a theoretical case in which segregation distortion occurs both in males and females: suppose that of the gametes produced by A_1A_2 heterozygotes a fraction k has genotype A_2 and a fraction $1 - k$ has genotype A_1 ($k > 0.5$). Assume that A_2A_2 homozygotes are unviable or sterile. Adapt Table 4.2 to show that these conditions lead to a stable allele frequency equilibrium $\hat{q} = 2k - 1$.

10.3 If in question 10.2 the A_2A_2 homozygotes are normal (fitness 1), the distorting allele A_2 is expected to become fixed (A_1 will disappear from the population). When this has happened, will the segregation distorting effect of A_2 still be observed?

10.4 In Chapter 1 it was argued that group and species selection are not likely to have shaped many patterns and certainly not many adaptations. Compare individual to group selection from the point of view of genomic conflict. What determines the power of selection at each level in a conflict?

10.5 Male sterility occurs in about 5% of the flowering plants and appears to be due to mitochondrial mutations in almost all cases that have been investigated. In hermaphroditic animals, however, male sterility is very rare. Can you think of plausible reasons for this difference between plants and animals?

Chapter 11
Speciation

Introduction

Speciation is the bridge from micro- to macroevolution.

Chapters 2–10 described the microevolutionary processes that occur within populations: natural selection, drift, changes in gene frequencies, the origin and maintenance of genetic variation, the lineage-specific expression of genetic variation, the evolution of sex, life histories, and sex allocation, sexual selection, and genetic conflict. In this chapter we start to make the transition to macroevolution, the patterns in fossils and phylogenies above the species level. The bridge between micro- and macroevolution is speciation, which is responsible for the diversity of life.

About 10 million species currently exist; many more are extinct.

Life forms are incredibly diverse. They range in size from viruses that can be seen only with electron microscopes, through hundred-ton whales, up to the clones of some trees that may cover a square kilometer. Some live under the Antarctic ice, others in hot thermal vents on the ocean floor at temperatures well above the boiling point of water. Present evidence indicates (see Chapter 15) that all organisms now alive—with the possible exception of viruses—shared ancestors that lived more than 3500 million years ago (earlier forms may have left no descendants). Thus there has been enormous diversification during evolution. The basic unit in which most life forms are classified is the species, of which about 10 million currently exist. Only about 1.4 million species have been described and named. The fossil record shows that many species have existed that are now extinct. Understanding how this diversity came about is a central question in evolutionary biology. How are species defined, and how do they originate?

What is a species?

Identification of species in practice

Living things tend to occur in groups, such that those within a group resemble each other more than those from different groups. Anyone can discriminate between a leopard and a tiger, although the individuals of these related species have many traits in common. This means that some traits—including coat pattern—vary less within each species than between them (Fig. 11.1a).

Thus the species unit seems to be a natural one, not just an arbitrary invention of biologists who need a classification system to communicate about the organisms they study. But species differences are not always so clear-cut. Taxonomists use all sorts of differences—morphological, behavioral, and genetic—to identify species. Occasionally they have serious problems deciding how much a group must differ to be classified as a separate species. Sometimes the discriminating traits between two species partially overlap (Fig. 11.1b). For example, *Drosophila melanogaster* and *D. simulans* are **sibling species** whose females cannot be distinguished and whose males can be distinguished only by experts. Populations of the same species may also differ, particularly when they live far apart. Across their geographic range species are often subdivided into a mosaic of subspecies or races, and where two subspecies meet and hybridize a so-called hybrid zone occurs. For example, the crow *Corvus corone* has two subspecies in Europe. The all-black carrion crow *Corvus corone corone* occurs in much of western Europe, and the hooded crow *Corvus corone cornix*, which is gray with a black head, wings, and tail, inhabits eastern and northern Europe, northern Scotland, and western Ireland (Fig. 11.2).

Taxonomists use all sorts of differences—morphological, behavioral, and genetic—to identify species. Sibling species and hybrids pose some problems.

Where the subspecies meet there is a hybrid zone varying in width from 20 to 200 km in which they freely interbreed. All sorts of intermediate phenotypes can be found within (but rarely outside) the hybrid zone, which is thought to be old and stable, having existed since the last ice age. We do not know why it is stable. Hybrid zones are much studied

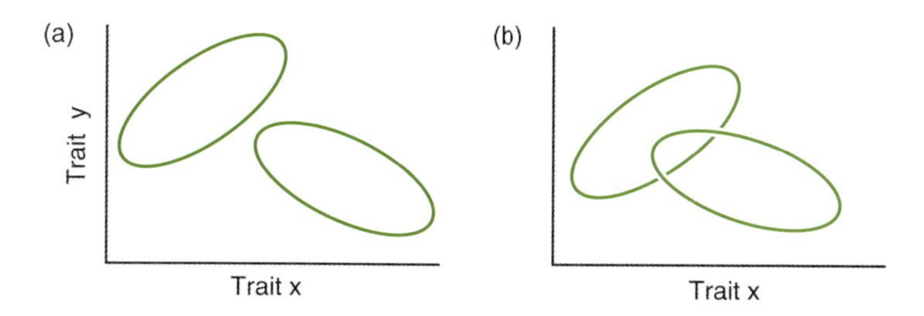

Fig. 11.1 Hypothetical distributions of the values of two traits in two different species. In both diagrams the within-species variation is smaller than the between-species variation. In (a) the trait value distributions are completely separated, while in (b) there is some overlap.

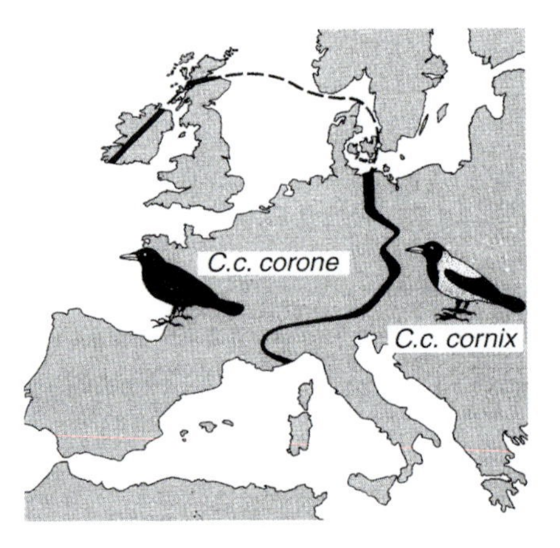

Fig. 11.2 Map of Europe with the distribution of the carrion crow *Corvus corone corone* and the hooded crow *C. c. cornix* and the hybrid zone between them (from Skelton 1993).

because they may provide insights into processes that contribute to speciation, as we will see later in this chapter.

Species concepts

Evolutionary biologists disagree on species definitions.

How should one define species? What characteristics do species have in common? Evolutionary biologists still disagree on this issue. Remarkably, judging from the extensive scientific literature on this topic, abstract species concepts are more problematic than the practical identification of species.

The biological species concept

According to the biological species concept the decisive criterion is the ability to interbreed.

The most influential species concept has been the biological species concept (BSC), advanced by Dobzhansky (1937) and propagated by Mayr: 'species are groups of actually or potentially interbreeding natural populations which are reproductively isolated from other such groups' (Mayr 1963).

Thus, according to the BSC the decisive criterion is successful sexual reproduction: the ability to produce fertile offspring. The popularity of the BSC stems from two biological insights. First, sexual reproduction promotes uniformity within a species by genetic recombination. All individuals that can interbreed share a common gene pool; copies of genes in one individual may end up in future descendants of any other conspecific individual, but not in descendants of an individual of a different species. Genetic recombination within the common gene pool thus prevents strong divergence of any subgroup of individuals. Secondly, if two groups do not interbreed, there is no gene flow between the gene pools, allowing further genetic divergence between the groups by natural selection and genetic drift. The inability to interbreed prevents species from merging together at a later time when conditions have changed. Such species remain distinct in sympatry, and this criterion is often used in practice to define 'good species.' Thus—according to the BSC—the number of 'good' species cannot decline through hybridization because species that can fuse would, by definition, not be 'good species.'

Problems with the BSC Not all organisms occur in groups within which there is sexual interbreeding and between which gene flow is prevented. We consider two types of exceptions: asexual organisms and interspecies hybrids.

Asexual organisms Absence of sexual recombination is rare but does occur in some groups of organisms. For example, the imperfect fungi appear to have lost the sexual life cycle and reproduce exclusively through spores formed by mitosis. They are classified into species because within-group variability in 'diagnostic' traits is small compared to between-group variability. Thus the species *Aspergillus niger* is recognized by the traits it shares with other *Aspergillus* species and by a few special characters, of which the black color of its spores is the most important. This species consists of a very large collection of asexual clones, and each clone is reproductively isolated from all other clones. Strict application of the BSC would force us to consider each clone a separate species, which is neither practical nor meaningful. Because the BSC refers by definition to sexual reproduction, it is best not to apply this species definition to the imperfect fungi. Another example is *Taraxacum officinale,* the common dandelion. The species consists of sexual diploid plants and asexual triploid clones. Experts can distinguish some clones or groups of clones from each other and from diploids on the basis of morphological details. Thus one could argue that, in accordance with the BSC, the asexual triploids and the sexual diploids should not be grouped together and each clone (or group of similar clones) should be given a separate name. Indeed, some systematists have constructed long lists of so-called *Taraxacum* microspecies. A similar situation occurs within the genus of water fleas, *Daphnia,* where molecular systematics has revealed a complex network of hybridization events and where some clones are obligately asexual while other groups of clones engage in intermittent sexual reproduction.

Asexual organisms also cluster into things that look like species. The BSC would make each clone a separate species, which is not helpful.

Interspecies hybridization The BSC is also problematic when species are sexual, but barriers to interspecies breeding are not strong. Interspecies matings producing fertile hybrids are not common among related animal species (although they do occur), but they are frequent in plants and fungi. The difficult question then arises, with how much gene flow between gene pools can the BSC retain its meaning? For the carrion crow and the hooded crow, the current opinion is that both belong to the same species and the BSC would apply, but wolves and coyotes are considered separate species, despite the fact that they can and do hybridize, and some wolf populations are threatened with extinction through hybridization.

Fertile hybrids are common in plants and fungi. With how much flow between gene pools can the BSC retain its meaning?

A special situation occurs in bacteria and viruses. Bacteria are classified

Even bacteria have localized sex.

into species on the basis of their pathology, morphology, and antigenic properties. Comparisons of multilocus genotypes suggested that little recombination was going on and that bacterial populations consisted of independent clones, as in the imperfect fungi discussed above. However DNA sequencing of bacterial genes has revealed that individual genes may be composed of bits and pieces of different—sometimes of very different—origin (Maynard Smith *et al.* 1991). Such mosaic gene structure is probably a consequence of 'localized sex,' the exchange by homologous recombination of small stretches of DNA transferred into recipient cells by conjugative plasmids, transformation, or transduction. Evidence of mosaic gene structure has also been obtained in viruses. The implication is that pieces of DNA can travel between related (and sometimes between unrelated) species. Thus bacterial and viral species are not genetically isolated from each other, and gene pools are generally wider than named species.

Darwin considered a species definition like the BSC, based on sterility barriers, but rejected it because so many recognized species can hybridize in nature (Mallet 1995).

The phylogenetic species concept

The phylogenetic species concept (PSC) defines a species as a monophyletic group composed of 'the smallest diagnosable cluster of individual organisms within which there is a parental pattern of ancestry and descent'.

The BSC lacks a historical dimension, for it can only be applied to contemporary organisms. Indeed, since every living sexual organism is linked via an uninterrupted chain of successive ancestors to totally different life forms hundreds of millions of years ago, there must have been a continuity of sexual fertility between ancient ancestors and the present organism. The BSC provides no criterion for where to draw the lines between successive species along such a line of descent. Such a phylogenetic perspective is provided by the phylogenetic species concept (PSC), defined by Cracraft (1983) as a **monophyletic** group composed of 'the smallest diagnosable cluster of individual organisms within which there is a parental pattern of ancestry and descent'.

Within such a group organisms share certain derived characters that distinguish them from other such groups. This species concept avoids the problems associated with the strict reproductive isolation required by the BSC.

High-resolution molecular characterization could split an established species into many small groups.

Problems with the PSC First, it is not clear how many shared derived characters a monophyletic group of organisms should have to be classified as a separate species. If one searched hard enough with high-resolution molecular methods, an established species could be split up into many very small groups of individuals that each shared a common derived character. Clearly, giving species status to all such small groups is not meaningful, for in this way any newly derived trait would produce a new species and the number of species would explode. Recently the PSC has started to be modified to avoid extreme division of species, which may well lead to an attractive alternative or addition to the BSC.

The Procrustean fallacy

In Greek mythology, the hero Theseus was born and raised by his mother and grandfather on the Peloponnese, south of Athens. When he reached manhood, he learned that his father was Aegeus, king of Athens, and marched off to meet him. On the way to Athens, he had to cross the Isthmus of Corinth and overcome various threats. One was the robber Procrustes. When Procrustes captured a victim, he placed him on a bed. If the victim was too long for the bed, Procrustes cut off the head and feet. If the victim was too short, Procrustes stretched him to fit the bed. Theseus killed Procrustes, freeing the local people from his cruelty, and proceeded to Athens, where he eventually became king.

In science we can commit mistakes that resemble Procrustes' crimes. We invent some concept, like the species concept, then try to cut and stretch nature to fit it, ignoring or explaining away what does not fit. However, nature is often too diverse to be described by a single concept, and that is the case with species. In some groups, species are genetically well-separated entities. In others, there are no clear boundaries. It is best to accept the natural diversity, to apply concepts only when they are appropriate, and to be sensitive to situations where they do not apply. Because it is so useful, for many reasons, to be able to identify organisms, biologists tend to push a species concept as far as it can go, but do not forget that they all have their limits. Do not be misled by the fact that a name exists; it may not describe a natural object. The groups in which one must be particularly sensitive to this problem include many plants, fungi, and bacteria, but such errors can also occur with animals. The assumption that species can be identified in a meaningful way underlies the phylogenetic methods of Chapter 12.

Nature is often too diverse to be described by a single concept, and that is the case with species, particularly among plants, fungi, and bacteria.

The origin of species

Speciation is one of the central processes in evolution. A theory of biological evolution that could not explain speciation would be seriously flawed. Do species originate as an inherent consequence of microevolution, or must different, additional processes be involved? Is speciation caused by natural selection acting on variations produced by mutation and recombination and by neutral genetic drift? If so, speciation would be caused by the same processes that drive microevolutionary change within populations. There are at present no reasons to think that speciation requires mechanisms beyond those that generate change within species: speciation appears to be a by-product of intraspecific evolution.

Speciation is a by-product of intraspecific evolution.

A general speciation scheme

A broad look at biological diversity reveals two things:

1. Many species have unique combinations of features that we can use to distinguish them. When such differences are hard to detect, the species involved are usually closely related. (The rare exceptions, the

distantly related sibling species, such as the *Tetrahymena pyriformis* complex (Chapter 6), pose intriguing puzzles.)

2. Almost all species are reproductively isolated from each other. Again, closely related species are exceptions when they can hybridize.

Copies of genes that occur in one species usually cannot reach descendants of individuals from other species, although some 'leakage' may occur. Eucaryotic speciation involves at least two processes: genetic separation and phenotypic differentiation.

Thus, generally speaking, copies of genes that occur in one species cannot reach descendants of individuals from other species: each species has its own gene pool (although some 'leakage' may occur between the gene pools of closely related species). While the correspondence between species and separated gene pools is not correct for bacteria and viruses, in eucaryotic species speciation involves at least two processes: the splitting up of one gene pool into two or more largely separated gene pools, and the diversification of one biological form into two or more phenotypically different forms. We call these two processes **genetic separation** and **phenotypic differentiation**.

Populations can be genetically separated geographically or reproductively.

Populations can be genetically separated because they are either geographically or reproductively isolated. With geographical isolation interpopulational matings do not occur because of the physical separation, but matings might occur and be fertile if the populations did mix. Upon secondary contact interpopulational matings could also be sterile because the formerly isolated populations had diverged genetically, or would be avoided because of behavioral divergence. Then reproductive isolation would be definite and the gene pools would be completely separated. Reproductive isolation may also originate within a population when differences in mating behavior between subpopulations in the same area lead to isolation.

The two processes characterizing speciation can drive each other in a positive feedback loop.

Many scenarios of speciation have been suggested. They differ in their assumptions about the relative importance of the processes that contribute to genetic separation and phenotypic differentiation. Moreover, genetic separation and phenotypic differentiation can interact during speciation (Fig. 11.3). Some degree of genetic separation facilitates diversifying selection because it weakens the homogenizing effect of gene flow, and differentiation may enable mate recognition and assortative mating to enhance genetic separation by reducing the frequency of matings between different types. Thus the two processes that characterize speciation can drive each other in a positive feedback loop.

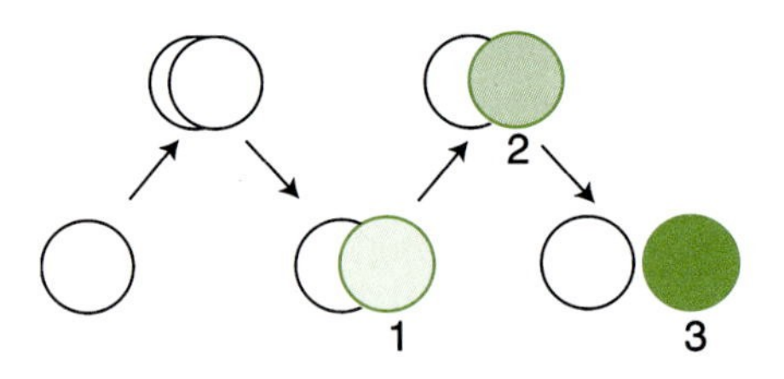

Fig. 11.3 Scheme showing the interaction of genetic separation and differentiation in speciation. Genetic separation is represented by the separation of circles (gene pools); morphological differentiation is represented by increasing color contrast.

Two particularly controversial issues have generated much discussion: the role of geographical isolation and the origin of reproductive isolation.

The role of geographical isolation

Is complete geographic isolation of populations, with no gene flow, necessary to start speciation? Some think that most speciation events occurred in **allopatry**, with different populations completely isolated in space. Others hold that speciation in **sympatry**, where the subpopulations diverge while continuing to live in the same place, has also played a significant role.

Is complete geographic isolation necessary to start speciation?

Allopatric speciation

It is easy to imagine that speciation starts when populations become geographically isolated, are exposed to divergent selection, and evolve independently. After enough time they will have accumulated so many genetic differences that they will be reproductively isolated if they come into contact again. If isolation is complete, speciation has occurred. This is the allopatric model of speciation (Mayr 1963), represented in its simplest form by Figure 11.4. According to this model, the first step in speciation is for one population to split into two or more completely isolated subpopulations.

Geographically isolated populations evolve independently and after enough time have accumulated enough differences to isolate them reproductively if they come into contact again.

Such a distribution pattern may be caused by migration, by local extinctions of intervening populations, or by geological events. The barriers separating the populations may be geographical or ecological. Examples are populations on islands separated by water, in lakes, on mountain tops, in patches of forest surrounded by savannah, or in fields surrounded by forest.

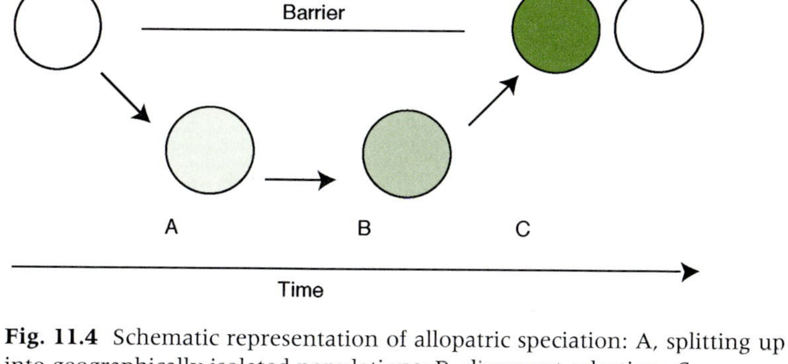

Fig. 11.4 Schematic representation of allopatric speciation: A, splitting up into geographically isolated populations; B, divergent selection; C, reproductive isolation at secondary contact.

Speciation in Darwin's finches probably involved geographic isolation, ecological specialization, and secondary dispersal and reinforcement.

Darwin's finches: an example of allopatric speciation
The finches on the Galápagos Islands suggested to Darwin the idea of descent with modification from a single ancestor species. They are thought to have speciated allopatrically (Lack 1947; Grant 1986), for the Galápagos Islands formed within the last 5 million years, as volcanoes emerged from the sea. The islands have never been connected to a continent or to each other. Thirteen different species are recognized, with up to 10 on a single island. The initial stages of the process were presumably as follows. About 3 million years ago a small group of birds from South or Central America colonized at least one of the islands. After the population had established itself, dispersers colonized other islands. Because ecological conditions varied among the islands, the genetically isolated populations encountered different selective forces and differentiated. The next stage was the establishment of secondary contact, through dispersal, between the differentiated populations. If birds from two populations did not interbreed, or if their offspring were inviable or sterile, speciation was completed in allopatry. If the populations were only partly isolated, so that some successful interbreeding was still possible, the situation is not so clear. This aspect—**secondary reinforcement**—is further considered below, where we discuss the origin of reproductive isolation during speciation.

Sympatric speciation

Sympatric speciation involves speciation in the presence of continued gene flow between the diverging groups. It has been controversial.

Often geographic and ecological conditions cause partial isolation of subpopulations, reducing gene flow between subpopulations but not stopping it completely. Does incomplete isolation between subpopulations under divergent selection allow speciation? This model is called sympatric speciation (speciation occurring within a single population; Fig. 11.5), in contrast to allopatric speciation, which assumes complete geographic isolation at the start of the speciation process.

While allopatric speciation is undisputed, whether sympatric speciation is likely and common, or can only occur under restrictive conditions, has been controversial (Bush 1994). The main problem is whether subpopulations can become reproductively isolated and differentiate in the presence of some gene flow.

Host shifts in phytophagous insects: a plausible example The most likely form of sympatric speciation involves selection to use different re-

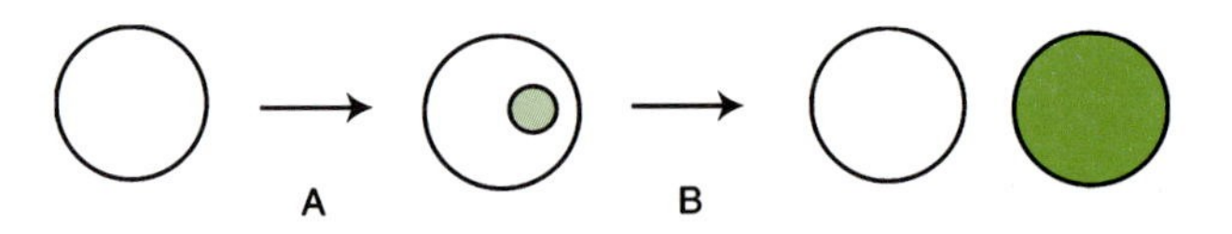

Fig. 11.5 Schematic representation of sympatric speciation: A, small divergence leading to some degree of genetic separation within a single population; B, further differentiation and genetic separation produce complete reproductive isolation.

sources. For example, parasitic insects may shift hosts: some individuals start to lay their eggs in a new species of host, perhaps because it has recently arrived or increased in numbers. Such a situation has been studied by Bush and coworkers in *Rhagoletis polmonella* flies. These flies used to lay their eggs in hawthorn, whose fruits are eaten by the larvae. In 1864 *Rhagoletis* was found in apples, later it parasitized other fruits, and now one can discriminate between an 'apple race' and a 'hawthorn race' that differ genetically, being characterized by different frequencies of enzyme variants.

Rhagoletis polmonella flies living on hawthorn and apples are a good example. Genetic differentiation and partial reproductive isolation have evolved in sympatry.

These flies share with many parasitic insects a behavior that strongly favors differentiation in sympatry: females prefer to lay eggs in the type of fruit from which they came themselves. Females that grew up in apple prefer to lay eggs on apples, and females from hawthorn prefer hawthorn. Males are true to the type of fruit they came from in their mating behavior: males that emerged from apple tend to mate on apple, and the same holds for males from hawthorn. Thus matings are mostly between males and females from apple and between males and females from hawthorn.

The breeding times of the two races have also diverged, contributing to their sympatric reproductive isolation. In the laboratory, however, flies from both races still interbreed freely. Therefore, although in nature considerable reproductive isolation has evolved, the potential for full interbreeding remains. Speciation is not yet complete, but the genetic differentiation and partial reproductive isolation observed in *Rhagoletis* did occur in sympatry. Whether speciation will be completed under these conditions, given sufficient time, remains an open question.

Fig wasps may also have speciated sympatrically.

The phylogenetic distribution of some plant-eating insect taxa suggests sympatric speciation by host shifts. For example, there are hundreds of species of fig wasps, each breeding on its own species of fig. Allopatric speciation seems implausible here, for it would require many geographic isolation events, whereas many fig species and their wasps are now sympatric.

Sympatric speciation can also occur through shifts in the flowering times of plants on a very local scale.

Divergence of flowering time in plants In plants partial reproductive isolation may occur following the evolution of different times of flowering. The grass species *Agrostis* and *Anthoxanthum* are able to grow near tailings from mines despite the high concentrations of copper, lead, and zinc in the soil (see Chapter 2). Particular types within these species are more tolerant of metals than the normal types. On the tailings only the tolerant types can survive, while the normal types dominate the vegetation just outside the mine tailings. The tolerant and normal types show signs of incipient reproductive isolation because they flower at slightly different times and the tolerant types have a higher degree of self-fertilization. The differences in flowering time are genetic and may represent adaptations to the local conditions. Thus this example parallels the *Rhagoletis* case discussed above: adaptation to different conditions may produce sympatric reproductive isolation as a by-product.

Sudden sympatric speciation by polyploidization

Sympatric speciation by polyploidization is undisputed.

A special and undisputed form of sympatric speciation is caused by a change in the genetic system that produces sudden reproductive isolation. Several mechanisms are known, most involving major chromosomal alterations. The most common is polyploidization, a doubling of the complete set of chromosomes, which arises due to irregularities in cell division. If the chromosomes double, but the cell fails to divide, then the cell contains twice the normal number of chromosomes. This is called **autopolyploidy**. Further normal divisions transmit the polyploid condition to all descendant cells. In somatic tissue this leads to a polyploid region. In reproductive tissue the outcome depends on details; one possibility is the production of gametes containing twice the normal haploid number of chromosomes.

More common than autopolyploidy is **allopolyploidy**, resulting from hybridization between related species followed by doubling of the chromosomes (Fig. 11.6). Hybrids often suffer because quasi-homologous chromosomes do not pair properly during meiosis. When the complete hybrid chromosome set doubles, every chromosome has a perfectly homologous partner, and meiosis can proceed normally.

Once a polyploid individual has arisen, by whatever mechanism, it is reproductively isolated from the parental types because a mating with a parental diploid individual would yield triploid offspring. Triploids are almost always sterile because of severe problems in meiosis. They may, however, be able to propagate asexually, and in fact many asexual clones in plants (and some animals) are triploid.

70–80% of plant species probably originated as polyploids.

Speciation by polyploidization has been particularly important in flowering plants, where 70–80% of the species are thought to have originated as polyploids. Many crop plants became polyploid recently, perhaps as part of the process of domestication. These include wheat, oats, potato, tobacco, cotton, and alfalfa. Further in the past, apples, olives,

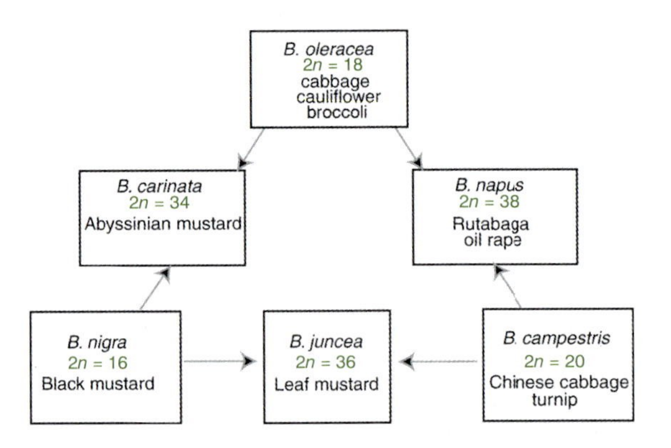

Fig. 11.6 Allopolyploidy in the genus *Brassica*. Three different parent species have hybridized in all possible pair combinations. Subsequent chromosome doubling in the hybrids has produced three new polyploid species.

willows, poplars, and many genera of ferns may have originated as polyploids.

Stable, viable polyploidy is most likely to arise in organisms capable of self-fertilization and asexual reproduction. Self-fertilization enhances the probability that two unreduced gametes unite to form a polyploid individual; two diploid gametes would fuse to form a tetraploid, for example. Asexual reproduction avoids problems in chromosome pairing and segregation during meiosis. Speciation by polyploidy is less significant among animals, because animal mating systems often do not allow self-fertilization and vegetative reproduction. Exceptions are found in some taxa of fish, earthworms, crustaceans, and a few others, where polyploidization has been common, and in groups with parthenogenetic females, including beetles, moths, and pill bugs.

Speciation by polyploidy is less frequent among animals. Exceptions include some fish, earthworms, crustaceans, beetles, and moths.

The origin of reproductive isolation

The physical separation of allopatric populations guarantees zero gene flow between them and thus their reproductive isolation. If they come into contact again they may or may not be reproductively isolated. According to the allopatric speciation model, the divergence built up during allopatry will have caused intrinsic reproductive isolation as a by-product that only becomes apparent in secondary sympatry. This reproductive isolation may be prezygotic or postzygotic. Prezygotic reproductive isolation occurs when individuals from different populations do not mate because of behavioral differences or because they mate at different times. Postzygotic reproductive isolation occurs when the different populations do interbreed but without success: the matings are sterile or the offspring inviable. Allopatric speciation leading to complete prezygotic or postzygotic reproductive isolation is unproblematic. How frequently speciation has followed this course remains to be seen.

Reproductive isolation may be prezygotic or postzygotic.

But what happens when allopatric populations show only partial reproductive isolation in secondary sympatry? Or when sympatric populations have developed partial reproductive isolation, like the apple and hawthorn races of *Rhagoletis*? How much gene flow between the different populations or races will stop the speciation process? Under what conditions will reproductive isolation become more complete and when will it break down?

What happens in secondary contact? When will reproductive isolation become more complete?

Molecular aspects of reproductive isolation

Genetic recombination causes the exchange of genetic information between lineages, and the amount of genetic recombination determines the extent of genetic isolation. Genetic recombination among enterobacteria depends on the difference in their DNA sequences (Vulic *et al.* 1997). The more genomes diverge in sequence, the less likely it is that they will recombine. More precisely, genetic recombination appears to require enough blocks of sequences that are identical in the two mating partners, and these blocks must be large enough to allow recombination to begin. The extent of genetic isolation increases exponentially with

In bacteria and yeast sequence divergence is the basis for genetic isolation.

increasing sequence divergence. Thus in bacteria sequence divergence offers a structural basis for genetic isolation. Similar results have been found in yeast, which has a typical eucaryotic sexual cycle with meiosis, and some data on recombination in mice also point in this direction. Thus sequence divergence could be the general barrier to recombination that characterizes fully completed speciation.

Reinforcement

Secondary reinforcement of reproductive isolation in hybrid zones is problematic.

Because postzygotic isolation has high costs—matings are sterile or wasted on hybrid offspring of poor viability—Dobzhansky thought that mechanisms would be selected to convert postzygotic into prezygotic isolation. This process is known as **secondary reinforcement**. Reinforcement is particularly relevant in hybrid zones, where hybrid progeny often have low fitness. Here reinforcement mechanisms would help to keep species separate, whereas hybridization destroys the distinction between them. Although reinforcement seems plausible at first sight, many experts doubt that it has an important role. One problem is that hybrid progeny may backcross to either of the parental types, so that within a few generations a whole range of hybrid types will be formed, some of them very similar to one parental type, others resembling the other parental type. Then hybridization between the two parental types will be rare because they will be separated by the hybrid zone, and reinforcement would be very weak. We return to the role of reinforcement below when we discuss experimental evidence on speciation.

The role of sexual selection

Sexual selection may cause prezygotic reproductive isolation.

Two well-investigated and spectacular cases of rapid and profuse speciation suggest an important role for sexual selection in causing prezygotic reproductive isolation.

The radiations of *Drosophila* in Hawaii . . .

Speciation of drosophilids (fruit flies in the genus *Drosophila*) in the Hawaiian archipelago has been studied since the 1960s by Carson and Kaneshiro and their coworkers. About 800 *Drosophila* species are endemic to the Hawaiian islands (compared to about 2000 species in the rest of the world). Speciation among the Hawaiian drosophilids has probably followed a course similar to that of Darwin's finches on the Galápagos islands. A few fruit flies must have arrived on one of the islands and dispersed from there to other valleys, mountains, and islands, with allopatric diversifying selection followed by secondary sympatry. The males of different species often differ greatly in body and wing patterns and have unusual modifications of mouthparts and legs, but females of different species are often similar (Fig. 11.7). Furthermore, related species show little genetic divergence in allozymes or DNA sequences.

What drove speciation in this group and how can we explain the rapid morphological differentiation of the males of related species? Kaneshiro believes that many of the remarkable morphological fea-

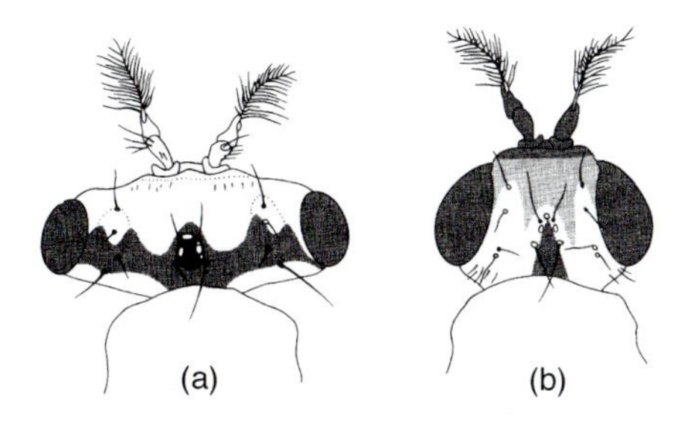

Fig. 11.7 *Drosophila heteroneura* and *D. silvestris* are closely related sympatric species with similar ecology but dissimilar courtship behavior. (a) The hammer-shaped head of male *D. heteroneura*; (b) the head of male *D. silvestris*, which closely resembles the heads of females of both species. (From Skelton 1993.)

tures of the males are the by-products of genes controlling courtship behavior (Kaneshiro and Boake 1987). He suggests that prezygotic reproductive isolation originates through allopatric changes in courtship behavior, possibly as a chance event in a small founder population, and has often initiated speciation in Hawaiian *Drosophila*, preceding adaptation to different food sources and divergence in other characters.

and cichlid fishes in Lake Victoria suggest that sexual selection can drive speciation.

The explosive speciation of cichlid fishes belonging to the genus *Haplochromis* in Lake Victoria in Africa is another example of sexual selection driving speciation. The lake contains an estimated 500–1000 haplochromine cichlid species that evolved since the last ice age when the lake was completely dry, perhaps less than 13 000 years ago. Males of related sympatric species always differ in color (red, blue, or yellow); intraspecific male color variation also sometimes occurs. Females do not have bright colors. Seehausen *et al.* (1997) have shown that female choice of male color causes reproductive isolation between sympatric species and between intraspecific color morphs. In laboratory experiments, females of a sympatric red/blue species pair preferred males of their own species over those of the other species under broad-spectrum illumination. Under monochromatic light, where color differences are masked, mating preferences disappeared. Water pollution caused by humans may actually threaten species diversity. The increased turbidity impairs the visibility to such an extent that in some areas females can no longer distinguish males of related species from those of their own species. Since hybrids are fully fertile, species diversity may decline as a consequence—although this effect on cichlid diversity in Lake Victoria is small compared to the impact of introduced Nile perch and anoxic conditions.

As in the Hawaiian drosophilids, sexual selection for particular types of males (here males with striking colors) may be the first stage in speciation. This could occur in sympatry given some spatial heterogeneity. For example, depth affects color perception, and new mate preferences may

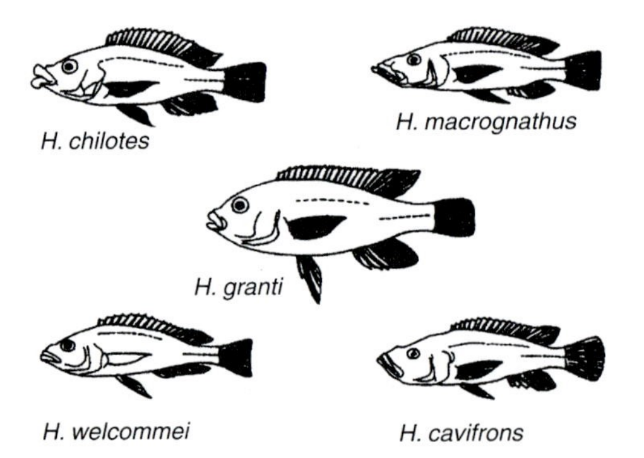

Fig. 11.8 Some *Haplochromis* species from Lake Victoria, illustrating rapid evolution of morphological diversity. As many as 1000 species may have evolved in Lake Victoria in as little as 12 000 years. (From Greenwood 1974.)

stabilize in a subpopulation that stays long enough at a different depth in the lake. In the first stage of speciation many incipient species could coexist, separated by mate choice. The second stage in the speciation of haplochromines in Lake Victoria is thought to be diversifying selection on feeding habits and other specializations. The resulting differentiation can be rapid because the jaw apparatus of the cichlids consists of a large number of bony elements that allow independent modifications (see Fig. 11.8). Indeed, the species in the lake had extremely diverse feeding habits, specializing on fish scales, eggs and larvae sucked out of the mouths of mouth-brooding cichlids, phytoplankton, snails, fish, or insects.

The experimental evidence

Non-zero gene flow (sympatry) in combination with divergent selection can produce reproductive isolation.

It is often impossible to reconstruct the speciation processes that have led to presently existing species. Even though much circumstantial evidence might point to sympatric speciation, how could we be sure that there never was an allopatric stage? For this reason, experiments designed to duplicate part of the speciation process under controlled laboratory conditions help to obtain information about the feasibility and relative importance of the various aspects of speciation. When Rice and Hostert (1993) reviewed the results of laboratory experiments on speciation, they came to two important conclusions:

1. Complete geographic isolation (allopatry) is not necessary to produce reproductive isolation: diminished but non-zero gene flow (sympatry) in combination with divergent selection can also produce reproductive isolation

Many experiments (mostly with *Drosophila*) confirm that divergent selection applied to allopatric populations can produce pre- and postzygotic reproductive isolation as a by-product. Reproductive isolation is probably caused by pleiotropic effects of genes that were selected during adaptation to the different environmental conditions used in the experiment. When postzygotic isolation was observed, it appeared to be environment-dependent. Only rarely did unconditional postzygotic isolation evolve.

There is, however, strong experimental support for the evolution of reproductive isolation between sympatric populations connected by gene flow, provided the divergent selection is strong relative to the gene flow. Evolution of reproductive isolation in sympatry was particularly successful when divergent selection was applied to several characters simultaneously.

2. Reinforcement is probably unimportant in generating prezygotic isolation

Reinforcement may not happen very often.

Prezygotic isolation between a pair of related species is often stronger when the individuals tested come from an area where the species are sympatric than when they are collected in allopatry. This has been thoroughly confirmed for *Drosophila* species by Coyne and Orr (1989); the pattern has also been observed in other species, including fish and frogs. These observations are consistent with the idea that prezygotic isolation has evolved to prevent the production of hybrids with low fitness (reinforcement), but they do not prove it. An alternative explanation is 'reproductive character displacement', which assumes that speciation was already complete when the species came into secondary contact, but that prezygotic isolation mechanisms diverged in secondary sympatry simply to reduce the amount of time spent mistakenly courting partners that were already reproductively isolated. Reinforcement, in contrast, presupposes ongoing gene flow while prezygotic isolation evolves. Of many experiments designed to demonstrate reinforcement, only one or two were successful. The great majority failed. Thus the importance of reinforcement in producing prezygotic isolation is doubtful.

Summary

This chapter discusses the definition of species and how species originate.

- Several species concepts have been developed. The biological species concept, which groups organisms on the basis of interfertility, has been the most influential. Species concepts are problematic, for none fits all groups of organisms.
- Speciation is characterized by genetic separation and morphological differentiation. These processes can interact to enhance each other's effects.
- Several scenarios have been proposed for the start of speciation.
- Some believe that allopatric speciation between populations that are

physically separated is the rule, and sympatric speciation between populations connected by gene flow is an exception.

- Those who think that sympatric speciation need not be rare find support in the results of experiments that mimic speciation processes in the laboratory.
- Also controversial is the significance of reinforcement, or selection for the prevention of crossbreeding between two populations that produce hybrids of low viability. Reinforcement has encountered theoretical objections and is not supported by the results of experiments that were designed to demonstrate it.
- Evidence on two spectacular and profuse speciations—Hawaiian fruitflies and cichlid fishes in Lake Victoria—suggests that sexual selection has an important role in bringing about reproductive isolation.

Speciation produces branches in phylogenetic trees. How we can infer the branching patterns in those trees, the patterns that document the relationships of species, is discussed in the next chapter.

Recommended reading

Howard, D. J. and Berlocher, S. H. (ed.) (1998). *Endless forms: species and speciation.* Oxford University Press, Oxford.

Mayr, E. (1963). *Animal species and evolution.* Harvard University Press, Cambridge, Massachusetts.

Otte, D. and Endler, J. (ed.) (1989). *Speciation and its consequences.* Sinauer, Sunderland, Massachusetts.

Questions

11.1 There are well-studied examples of 'half-way' speciation in the form of partially reproductively isolated populations, and cases of completed speciation demonstrated by closely related sibling species, but very few documented examples of the complete process of speciation. What is required to demonstrate speciation, and why is this apparently so difficult?

11.2 The main processes involved in speciation are genetic separation and phenotypic differentiation. List plausible causes for both processes.

11.3 The following type of experiment has been used to support the reinforcement model of speciation. Equal numbers of male and female virgins were collected from two strains of *Drosphila* and mixed to allow mating. Genetic markers allowed offspring to be classified as coming from homotypic or heterotypic matings. Only males and females from the homotypic matings were used for the next generation, in which the procedure was repeated. After 20 generations increased prezygotic isolation was observed. Do you think this is a proper way of testing the possibility of speciation by reinforcement? If not, how would you modify the experiment?

Chapter 12
Systematics

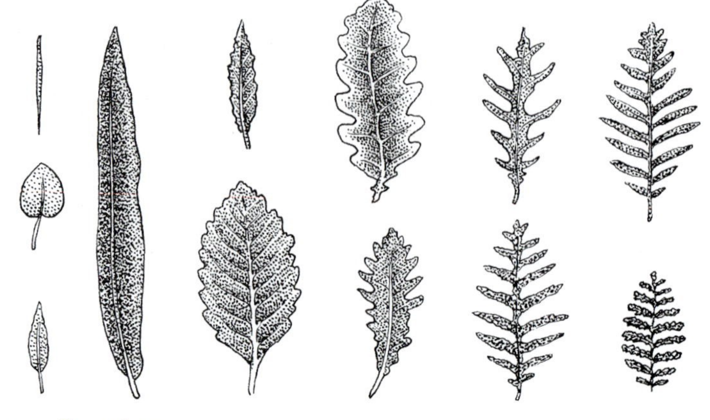

Introduction

The goal of systematics: to understand the relationships of all living things and thereby infer the history of life since its origin.

Chapter 11 described what species are and how we think they originate. Speciation creates a branching tree of relationships that describes evolutionary history. The goal of systematics is to discover the structure of that tree—the relationships among organisms that result from their having common ancestors. In this chapter we first describe the problems that systematics tries to solve and some surprises that have emerged from the solutions. Then we discuss the main methods and their key assumptions, introducing enough terminology to keep concepts clear. The chapter ends with case studies that illustrate the power of modern systematics.

All organisms have a history; that history is reflected in their relationships, and knowledge of their history and relationships affects our interpretation of their biology. Because the characters within a single individual attained their state at different times in the past, individuals are mosaics of ancient and recent features. Only a phylogeny—a tree describing relationships—can tell us the order in which traits evolved, which parts of the mosaic are older, which are newer, and approximately when each evolved. For example, a hand with five fingers is an ancient character state, about 300 million years old, that we share with all tetrapods. It originated within the amphibia before the origin of the reptiles. Our chin is a recent innovation, less than 5 million years old, unique to the human lineage, and not shared with chimpanzees.

Thus relationships are important. That they are not always clear at first glance is illustrated in the next example.

Are the two pandas each other's closest relatives?

The giant panda, a bamboo-eating specialist now found only in Western Szechwan, looks like a bear, but its genital anatomy and vocalizations are unlike those of any bear, and it has an extra digit, derived from a

Fig. 12.1 The giant panda and the lesser panda, two mammals that both eat bamboo and live in Asia. Molecular systematics has shown that the giant panda is related to the bears, whereas the lesser panda is the sister group to all canoid carnivores. Thus 'pandas' is not a natural group. (By Dafila K. Scott.)

wrist bone, found in no other animal, that it uses to grasp bamboo shoots. Its closest relative was long thought to be the lesser panda, which also lives in Asia, eats bamboo, and whose molar teeth resemble those of the giant panda. The lesser panda has a long, ringed tail and resembles a raccoon or coatimundi (Fig. 12.1).

Appearances can be misleading: 'pandas' are not a natural group.

Until analysis of their DNA sequences became possible, it was not clear whether they were closest relatives and to what group each belonged. Molecular systematics has now shown that the giant panda is most closely related to the bears, and the lesser panda is the sister group of all the canoid carnivores (the dogs and their relatives, including the bears and raccoons). The similarity of their molar teeth, adapted to bamboo-feeding in both species, misled us about their relationships, for their teeth resembled each other because natural selection had shaped them to similar tasks, not because that morphology had been inherited from a common ancestor.

The next example shows that systematics is not an abstract, academic discipline. It has social relevance.

Who transmitted the HIV virus to a rape victim?

Suppose you are a forensic scientist asked by the police to identify which of two suspects transmitted the HIV virus to a rape victim who developed AIDS after the rape was committed. You take blood samples from the victim, the two suspects, and an AIDS patient whom you are certain had nothing to do with the rape case. From those blood samples you isolate the RNA genome of the HIV virus and confirm that all four persons are infected. From the RNA you prepare a DNA copy that you sequence. The critical part of the four sequences is 30 bases long and looks like this:

Unrelated person	AAGCTTCATAGGAGCAACCATTCTAATAAT
Suspect 1	AAGCTTCACCGGCGCAGTTATCCTCATAAT
Suspect 2	GTGCTTCACCGACGCAGTTGTCCTTATAAT
Rape victim	GTGCTTCACCGACGCAGTTGCCCTCATGAT

Using a computer program, you prepare a phylogenetic tree based on these four sequences. The program examines all the possible trees and delivers the one that implies fewer mutations in nucleotides than any other (Fig. 12.2).

Systematics can be used to identify criminals.

This tree clearly suggests that the second suspect infected the rape victim. The conclusion is supported by four changes in sequence that are shared by the second suspect and the victim: at position 1, A→G; at position 2, A→T; at position 12, G→A; and at position 20, A→G. Since the rape occurred the virus has continued to evolve in both suspect and victim. In the suspect, there has been a change at position 25, C→T; and in the victim there have been two changes, at position 21, T→C and at position 28, A→G.

This example is artificial, but similar methods were used to show that a dentist transmitted HIV to his patients (Hillis *et al.* 1996, p. 524).

The clarification of puzzling relationships is one goal of systematics. Another is answering one the biggest questions in biology: What does the tree of life look like?

The tree of life

The tree of life describes the relationships of all organisms on the planet that share a common origin. That may include everything except viruses.

Systematics infers historical relationships from similarities. Those relationships are described as branching patterns in the tree of life, which can be used as a framework in testing evolutionary hypotheses. Systematics aims to describe reliably the tree that relates all organisms

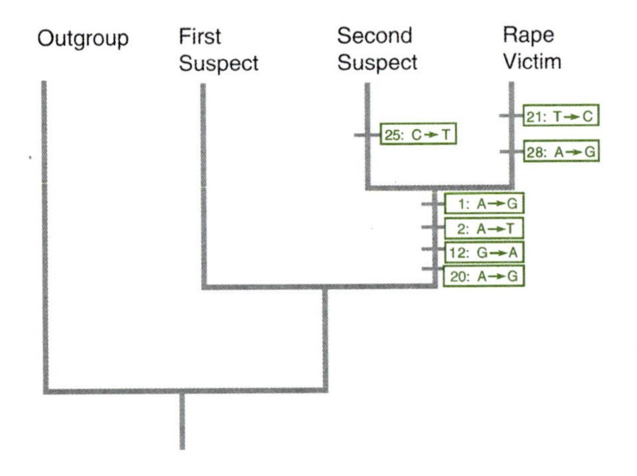

Fig. 12.2 The phylogenetic tree based on the nucleotide sequences in the text that implies the fewest changes overall. The tree is rooted using the unrelated person, and it clearly identifies the second suspect as the probable rapist. The character changes that support each branch unambiguously are listed. (Prepared by C. Baroni-Urbani.)

on this planet (except the viruses, which may have had an independent origin). Research is continually revising that tree. One recent description is given in Figure 12.3.

Most branches in the tree that formed since the origin of the sexual eucaryotes resulted from speciation events (Chapter 11). All branches prior to the origin of sex, and some branches since then, especially those within the procaryotes, describe the divergence of asexual clones.

Not all evolutionary divergences were clean branching events. When horizontal gene transfer or hybridization were involved, there was a period when the diverging groups had net-like rather than branch-like relationships. Most methods of systematics impose branching patterns on relationships that are actually net-like, which happens when hybridization is frequent, as in the plants. The origin of the eucaryotes, when several procaryotic genomes combined into one eucaryotic organism, was also net-like (colored arrows in Fig. 12.3).

Convergence and divergence

Why appearances deceive:

Systematics infers historical relationships from similarities, and thus it seems reasonable to start by assuming that things that look similar are related. However, appearances can deceive, and methods have been developed to extract the real tree of life from misleading similarities and misleading differences. Two common reasons for deceptive appearances are **convergence** (things that look similar have different ancestors) and **divergence** (things that look different have the same ancestor).

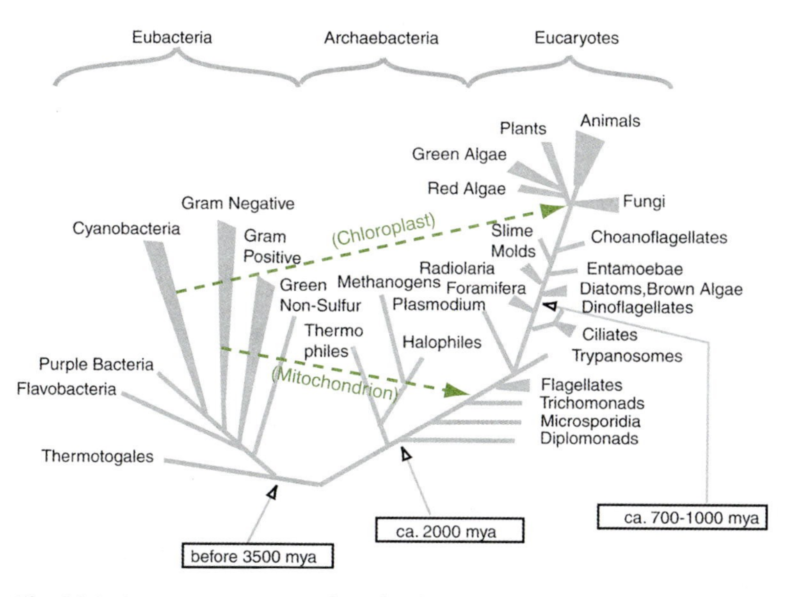

Fig. 12.3 A recent attempt to describe the tree of life. This hypothesis could be changed by the analysis of whole-genome sequences, which are currently being accumulated. (From Gerhart and Kirschner 1998.)

convergence—in DNA sequences,

Convergence occurs, as in the molar teeth of giant and lesser pandas, when selection adapts unrelated organisms to similar conditions. Convergence is one cause of **homoplasy**, a term that covers all types of similarity not due to ancestry. It is probably the most common cause of morphological homoplasy. In DNA sequences, where there are only four possible states (A, T, G, C), homoplasy arises frequently due to random mutations, for two sequences can share a nucleotide at a given position not because it was inherited from a common ancestor, but because it mutated to that state.

in cactuses and euphorbs,

The Old and New World succulent plants, which have adapted to arid conditions with similar morphology, are a spectacular example of convergence. Some New World cacti (Family Cactaceae) are so similar in shape to some Old World euphorbs (Family Euphorbiaceae) that only an expert can tell them apart (Fig. 12.4). In other groups flowers and fruits have converged because plants with different ancestors are pollinated by similar insects or dispersed by similar birds.

in birds that pollinate and vertebrate predators that swim,

When they encounter phenotypic convergence, biologists must rely on molecular divergence or on additional morphological data to discover the underlying relationships. In animals, striking examples of convergence include the bills of Old World sunbirds, New World hummingbirds, and Hawaiian honey creepers, adapted to extract nectar from deep flowers; the fusiform shape of porpoises, tunas, sharks, and ichthyosaurs, adapted to fast swimming; and the wings of birds and bats, which have independently evolved similar adaptations to rapid long-distance flight or to hovering in front of flowers.

and in the hemoglobin of high-flying birds;

Adaptive convergence also occurs in molecules, such as the hemoglobin of birds that fly at high altitudes. The bar-headed goose lives at

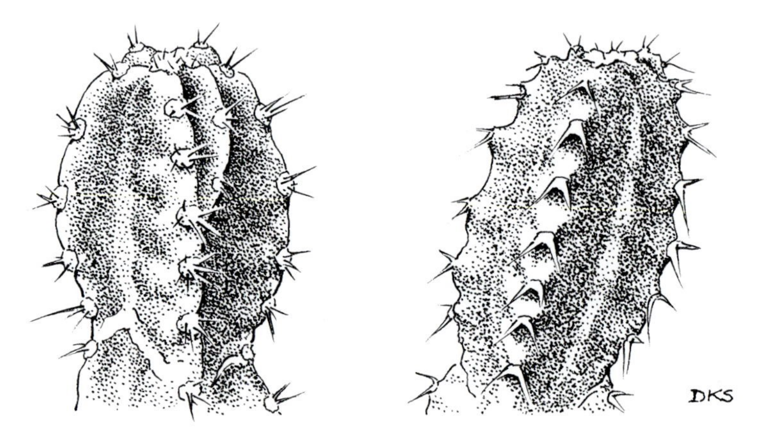

Fig. 12.4 Convergence, which is one reason for homoplasy, is one important reason why simple resemblance is not a reliable indicator of systematic relationships. The cacti of the New World and the euphorbias of Africa have evolved similar morphologies to deal with similar ecological problems. Left: the cactus *Cereus validus* from Argentina, half size. Right: the euphorb *Euphorbia resinifera* from Morocco, double size. (By Dafila K. Scott.)

altitudes of 4000 m and flies up to 9200 m in the Himalayas; the Andean 'goose', a relative of the mallard duck, lives at 6000 m in the Andes. Both species have hemoglobin molecules with higher oxygen affinity than those of their lowland relatives, and one mechanism of increased oxygen affinity is precisely the same in both species—disruption of a contact between the α- and β-chains through the same amino acid substitution in the β-chain (Gillespie 1991). Thus exactly the same modification of the hemoglobin molecule occurred at least twice, in species whose nearest relatives have hemoglobin molecules with different structures.

and divergence—in Hawaiian lobelias,

Divergence occurs when selection or drift modifies species descended from the same ancestor so that originally similar species become dissimilar.

Divergence is well illustrated by the Hawaiian lobelias (Givnish *et al.* 1995). One of the six native genera of Lobeliaceae, *Cyanea,* contains 55 species that constitute 6% of the endemic flora of Hawaii. They are restricted to particular islands or parts of islands. All *Cyanea* descend from a single ancestor; many have undergone striking changes in growth form, leaf size and shape, and flower morphology (Fig. 12.5). They vary from 1 to 14 m in height with leaves that can be simple, compound, or doubly compound and that range from 0.3 to 25 cm in width and up to 1 m in length. The flowers, which coevolved with endemic birds, have corolla tubes that range in length, among species, from 15 to 85 mm. The genus includes shrubs, trees, and a vine. Any classification based solely on growth form or leaf morphology would not reflect phylogeny, placing members of this genus into several unrelated families. The traits that do group them are flower structure, fruit color, some other morphological traits, and DNA sequence.

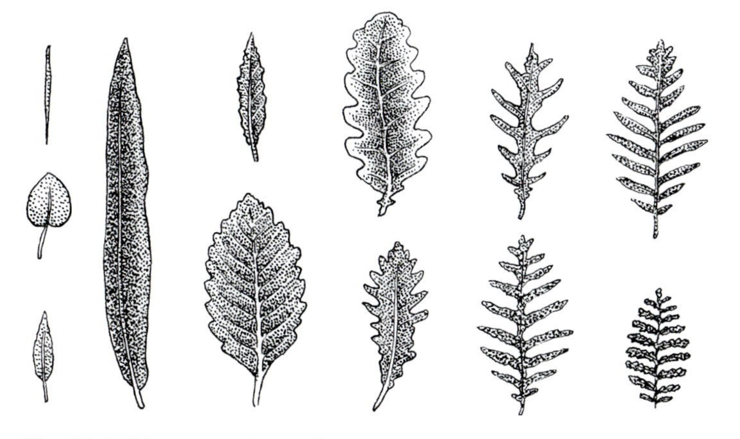

Fig. 12.5 Divergence is another important reason why simple resemblance does not indicate relationship. The Hawaiian lobelias have undergone a dramatic radiation in which their leaves have evolved many different forms. Their flowers and DNA sequences continue to indicate that they are closely related. (By Dafila K. Scott.)

and in all other adaptive radiations.

Among animals, the radiations of cichlid fish in the great lakes of Africa, of finches in the Galápagos and drepaniid birds in Hawaii, of land snails in Polynesia, and of amphipods in Lake Baikal all show striking recent divergence. On a larger scale, divergence is the reason for the diversity of life.

The homology problem

The many cases of convergence and divergence make clear that it is not easy to see when similar traits in two species are similar because those species shared an ancestor in which that trait occurred. However, that is precisely what we need to find if we want to build reliable phylogenetic trees: we need to establish that morphological traits and DNA sequences in two or more species are homologous. If homology were not a problem, systematics would be simple.

Homology is an *hypothesis* that similarity of a trait in two or more species indicates descent from a common ancestor.

Two morphological structures are called homologous by morphologists if they are built by the same developmental pathway and share the same relative position to other structures, such as nerves and blood vessels. The hypothesis for their similarity is derivation from a common ancestor from which similar developmental mechanisms were inherited. Two genes are called orthologous if they have similar DNA sequences; orthology suggests common ancestry but not necessarily morphological homology. Such molecules are similar by inspection, homologous by hypothesis. The determination of orthology is more reliable for long DNA sequences (the DNA coding for an enzyme often contains about 3000 nucleotides), where it is very improbable that random mutation would yield similar states in two organisms, as is homology for complex morphological structures. They are less reliable for short sequences and simple structures.

Morphology and molecules: homology and orthology.

There can be a connection between molecular orthology and morphological homology, but the connection is not necessary. During evolution genes can acquire new roles in new structures, and cases are known where structural homology has been preserved over long periods of time while both DNA and protein sequence homology have been destroyed (recall the case of *Tetrahymena*). When DNA sequence similarity and morphological homology have different phylogenetic patterns, the difference tells us something interesting about the evolution of development (Wagner 1989).

Recent surprises

As systematics solves the puzzles posed by misleading similarities and dissimilarities to uncover reliable ancestral relations, new phylogenetic surprises are being published all the time. For example:

Myxozoans—single-celled parasites—are highly modified cnidarians.

1. Myxozoa, or myxosporidians, are single-celled parasites of invertebrates and fishes, causing serious damage. After penetrating the skin they have a complex fungus-like life cycle in the cells of their hosts: a multinucleate plasmodium ingests host cytoplasm, then forms sporo-

gonts that fuse, make zygotes, undergo meiosis, and form spores. The spores resemble the stinging cells (nematocysts) of coelenterates, not fungal spores (Fig. 12.6). Myxozoa were traditionally classified with another group of single-celled parasites, the microsporidia, and in 1989 they were placed in a phylum of their own. Now molecular systematics strongly suggests that they really are highly modified cnidarians, and thus what was thought to have been a phylum has now become a family related to jellyfish and sea anemones.

2. The 'worms' and their relatives have long been an obscure group whose muddled relationships are now being clarified. For example, the nemerteans, or ribbon worms, appear to be celomates, not acelomates, and the chaetognaths, or arrow worms, appear to be primitive pseudocelomates, not deuterostomes. In both cases, phyla have been moved from one major division of the animal kingdom to another.

3. The closest relatives of whales appear to be ungulates: pigs, antelope, deer, and their relatives.

4. The social hymenoptera (bees and ants) appear to be the closest relatives of the ichneumonid and braconid wasps, which are parasitoids. If so, bees and ants had parasitoid ancestors.

Chaetognaths are primitive pseudocelomates, not deuterostomes. Whales are most closely related to the ungulates. Bees and ants may have had parasitoid ancestors.

The revolution in systematics

Systematics has been revolutionized and invigorated by powerful new methods and rich new sources of data. Major methods are cladistics, also called phylogenetic systematics, and maximum likelihood, a statistical method for estimating the reliability of candidate trees. The ready availability of DNA sequences now contributes an almost unlimited number of characters to the analysis of relationships.

Powerful new methods: cladistics, statistics, and computers. Rich new sources of data: molecular sequences.

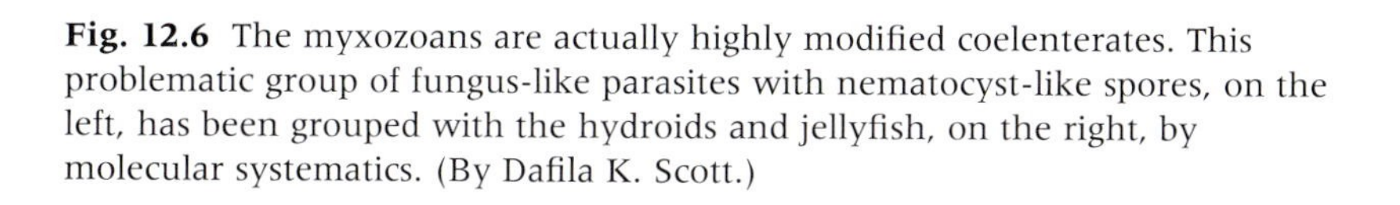

Fig. 12.6 The myxozoans are actually highly modified coelenterates. This problematic group of fungus-like parasites with nematocyst-like spores, on the left, has been grouped with the hydroids and jellyfish, on the right, by molecular systematics. (By Dafila K. Scott.)

What is a phylogeny?

A phylogeny is an hypothesis about relationships expressed as a tree.

A phylogeny is an hypothesis about relationships expressed as a tree diagram, such as Figure 12.3. All parts of a tree are inferred except for the tips of the branches, which are the observed and described species, living or fossil. (Incidentally, the only illustration in Darwin's *Origin of species* is a phylogenetic tree.) The usual phylogenetic methods assume that there is no hybridization or horizontal gene transfer through microbial vectors once speciation is complete (this is certainly not the case for many bacteria). Fossils, which help to test, support, and calibrate phylogenetic trees, are usually interpreted as tips of dead branches, not as direct ancestors of living species.

Types of phylogenetic trees

Rooted and unrooted trees, plesiomorphy and apomorphy.

Phylogenetic trees may be rooted or unrooted (Fig. 12.7). To root a tree, one uses an outgroup that is clearly not in the group being analyzed; a sister group is best if it can be identified reliably. Thus the outgroup for a phylogeny of the tetrapods should be a fish, probably a lungfish, not an echinoderm. The connection to the outgroup defines the root of the tree. If the tree is unrooted, then one cannot identify the outgroup or—as in the tree of life—there is no outgroup because cellular life is thought to have originated only once. Rooting a tree is an important step, for rooting introduces the notion that some character states are ancestral, or **plesiomorphic** (plesio = near, related; morph = form), and others are derived, or **apomorphic** (apo = derived). This has important implications for interpreting character evolution.

The tips of phylogenetic trees are described taxa; everything else in such trees is inferred.

Trees have terminal nodes and internal nodes, peripheral branches and interior branches. The terminal nodes represent described groups whose characters are known from observation; the internal nodes represent hypothetical ancestors whose character states are inferred.

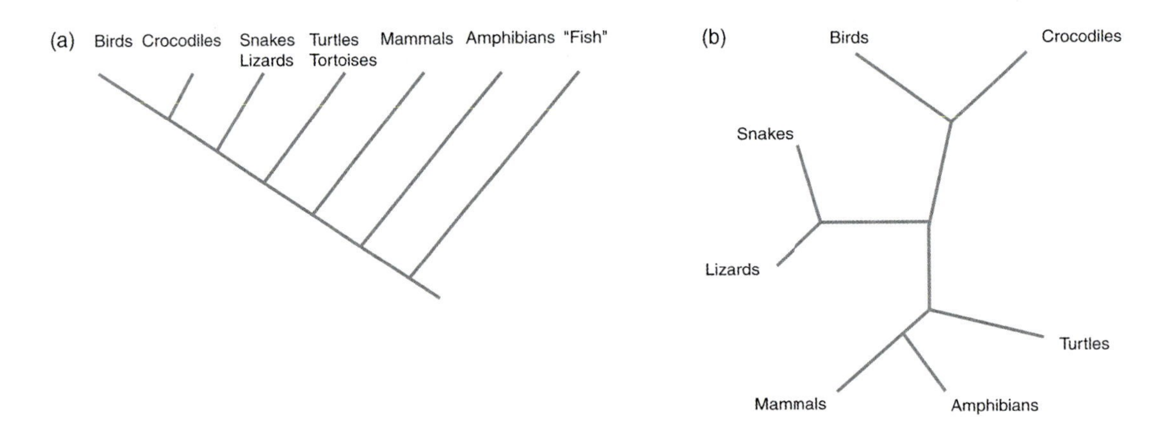

Fig. 12.7 (a) A rooted tree of the living tetrapods. The evolutionary sequence is clear, and the outgroup ('fish') is specified. (b) An unrooted tree. There is no outgroup, and the evolutionary sequence is obscure.

Peripheral branches terminate in described taxa; interior branches connect hypothetical ancestors.

Trees based on distances may be **additive** or **clocklike**. If the tree is additive, then the branch lengths add up to yield the distance between any two species on the tree. In clocklike trees, branch length is supposed to be proportional to elapsed time. In both cases the rate of branch divergence does not have to have been smooth and linear—just the same along all branches.

Types of data

Phylogenetic trees can be built from data that are either qualitative, discrete characters (such as the presence or absence of a morphological structure, or the presence or absence of four nucleotides at a particular location in a DNA sequence) or quantitative traits (such as ratios of the lengths of body parts, or percentage similarity in two DNA sequences).

Cladistics

The approach to systematics now known as 'cladistics' was formulated by Hennig (1966) and is now widely used. It is distinguished by its clearly stated assumptions: only shared derived character states—**synapomorphies**—are phylogenetically informative (see below), and the best tree requires the fewest changes in character states: it is the most **parsimonious** tree (e.g. Fig. 16.3a). Hennig introduced parsimony as a criterion for judging trees; the development of parsimony methods for tree building came later. Cladistics brought into systematics some technical terms that are now widely used.

The analysis of shared derived character states (synapomorphies) yields a tree with the fewest changes in character states—the most parsimonious tree.

Terminology

Plesiomorphy, synapomorphy, and homoplasy

These are the three possible kinds of relationships between character states in two species (Fig. 12.8). Primitive similarity, where homologous traits share the same character state because they were inherited from a distant ancestor, is called **plesiomorphy**; it is associated with **paraphyly** (see below). Derived, or advanced, similarity, where homologous traits are in the same character state because that state originated in their immediate common ancestor, whom they share with no other species, is called **synapomorphy**, or shared derived similarity. It is associated with **monophyly** (see below).

Plesiomorphy and paraphyly, synapomorphy and monophyly.

According to cladistics, only synapomorphy is informative about relationships because only it provides evidence of exclusive common ancestry (strict monophyly). For example, if we try to classify horses, lizards, and humans by number of digits, the character state five fingers, which would group humans with lizards rather than horses, is not relevant because it is ancestral to all three. If the problem were to group horses,

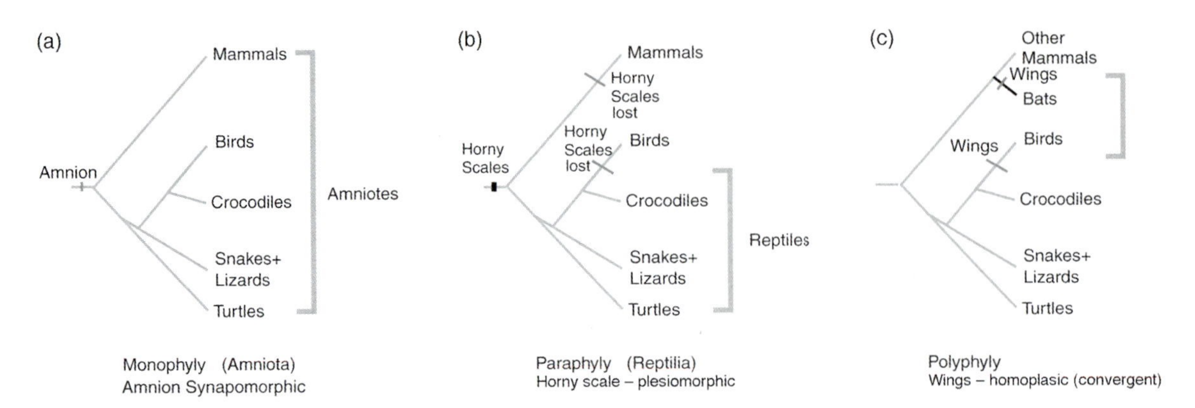

Fig. 12.8 (a) The amniotes, a monophyletic group, are defined, in part, by the synapomorphic trait, amnion. (b) The reptiles, a paraphyletic group, are characterized, in part, by the plesiomorphic trait, horny skin. (c) The winged extant tetrapods, a polyphyletic group, are characterized by the combination of a plesiomorphic trait—four limbs—and a homoplasic (convergent) trait, wings.

zebras, and humans, finger number would become relevant, for it is in a shared derived state in horses and zebras relative to humans.

Homoplasy and polyphyly.

The third relationship between characters is **homoplasy**, which describes similarities that may mislead us about relationships. Such similarities, which result from convergence, parallelism, and reversal, are associated with **polyphyly** (see below). The parsimony assumption states that the best tree is the one that minimizes homoplasy. Homoplasy is revealed in the optimal tree by conflicts among character states that indicate either that the character has undergone an evolutionary reversal, or that it has originated independently more than once, or a combination of both.

These relations—plesiomorphy, synapomorphy, and homoplasy—change depending on the taxonomic problem being investigated. A trait such as the possession of a chitinous exoskeleton, which is a valuable synapomorphy at a higher systematic level (such as the relations of phyla within the animal kingdom) can be an uninteresting plesiomorphy at a lower systematic level (such as the relations of species within a genus of butterflies). A chitinous exoskeleton helps to distinguish the arthropods from the annelids and the mollusks, but it is of no use in distinguishing one species of butterfly from another, for they all have it.

Monophyly, polyphyly, and paraphyly

These terms describe situations that arise when we attach names to parts of a phylogenetic tree—when we classify organisms by creating taxonomic groups. The tree itself is the central hypothesis, the base for reference; special problems arise when we describe the tree with a limited set of categories.

A group is **monophyletic** if all species in it are descended from a common ancestor that is not the ancestor of any other group and if no species descended from that ancestor is outside it (Fig. 12.8a). One goal of classification is to ensure that all named groups are monophyletic. The Amniota are monophyletic. Monophyletic groups are defined by synapomorphies.

Monophyly: a natural group with everything in it.

Paraphyly and polyphyly describe mistakes in taxonomy that result either from a defective tree or from a mistaken description of a correct tree.

A group is **paraphyletic** if it does not contain all species descended from the most recent common ancestor of all the members of the group (Fig. 12.8b). The most famous paraphyletic group is the reptiles, which does not include the birds and mammals. At the time the term 'reptiles' was coined, the true relations of the major vertebrate groups were not known, but the term has survived. Other paraphyletic groups include the fish, the invertebrates, the algae, the gymnosperms, and the dicots. The identification of paraphyly as a frequent mistake in taxonomy was perhaps Hennig's major contribution. Paraphyletic groups are characterized by plesiomorphies.

Paraphyly: a natural group with some members missing.

A group is **polyphyletic** if the species in it are descended from several different ancestors that are also the ancestors of species classified into other groups (Fig. 12.8c). Polyphyly results from the mistaken use of convergent characters (homoplasies) in constructing the tree. If we were to put the birds and the bats into one group of winged tetrapods, that group would be polyphyletic.

Polyphyly: an unnatural mixture of origins.

The main issue: homology or homoplasy?

Given some taxa, some traits, and the character states for each trait, how do we know which traits and which character states are homologous? Homology indicates shared ancestry and is phylogenetically informative. Homoplasy indicates similarity independent of ancestry and is phylogenetically misleading. Thus the analysis of homology is crucial, for it determines the reliability of all that follows. The homology of DNA sequences is determined by aligning the sequences being compared (see Fig. 12.2). The homology of morphological traits is determined by analyzing developmental and functional relationships.

The first step in phylogenetics is to analyze homology.

Given an hypothesis of homology, the next step is to code the characters so that they can be used in an algorithm, manual or automated. (Technical issues in coding get careful attention from experts.) If there are not too many species and characters, all possible trees can be compared on a computer. One looks for the simplest tree, the one that contains the fewest changes in character states. Every branch in a phylogenetic tree should be supported by at least one synapomorphy, one derived character state shared by the species on that branch.

If there are many species and many characters, then the number of possible trees is so great that even the fastest computers cannot examine them all in a reasonable time, and compromises must be made. For example, with five species the number of possible rooted trees is 105, but

For many species and traits, not all trees can be compared—there are too many.

with 10 species it is 34 459 425 (the number of trees is a factorial function of the number of species). For some types of data, one may prefer trees based on distance measures or maximum likelihood methods.

The tree itself reveals homoplasies defined by parsimony.

Given a candidate tree, we can see the main problem—distinguishing homoplasy from homology—in a new light. If that distinction cannot be made prior to the construction of the tree, then the tree itself determines which character states are homoplasies. The simplest tree has the fewest homoplasies and contains the fewest character conflicts. Often several trees are equally simple. Then one takes as reliable the features shared by all equally simple trees and does not put much faith in the rest.

No *group* should be described as primitive or derived, for if you approach the problem having made assumptions about ancestral states, you may simply recover in the analysis the assumptions you built in at the outset. However, every tree implies that some *character states* are primitive and that others are derived.

Molecular data and homoplasy

Sequence alignment is critical; an alignment is an implicit assumption about homology and phylogeny.

Because descendants inherit traits from their ancestors through genes, the history of descent is recorded by changes in genes, by changes in the genetic molecules themselves. Molecular data on sequences in genes are a simple form of character data: the characters are positions in the sequence, and the character states are the nucleotides or amino acids at those positions. This sounds simple but assumes that the positions compared are homologous, that they derive from the same positions in a common ancestor. There are two problems with this assumptions. First, with four nucleotides, the probability that two nucleotides are the same simply because of mutation is 25%. Secondly, in all but the most highly conserved sequences, we must assume that insertions and deletions have occurred to make the sequence homology believable. This means inserting gaps into one or more sequences to bring positions inferred to be homologous into the same column. Thus alignment is a critical step that involves implicit assumptions about homology and phylogeny. There are algorithms that align sequences automatically, introducing some objectivity, but the selection of an algorithm can itself be subjective, and such algorithms are not always reliable. In practice many alignments are performed manually.

Homoplasy remains a problem in sequence data.

Even after sequences have been aligned, homoplasy remains common. Homoplasy can be reduced but not eliminated by selecting and weighting characters. Because of the problem of homoplasy in sequence data, strictly cladistic methods cannot extract all the information available in sequences. That is one reason that molecular systematists also use maximum likelihood methods and distance methods.

The neutral model and the molecular clock

The dating of lineages starts with a fossil whose age marks, at least approximately, the divergence of the lineages. Many methods in molecular systematics assume that mutations are then fixed at the same overall rates in each lineage. This assumption connects evolutionary genetics (Chapters 3, 4, and 5) to systematics. The important part of the assumption is the regular rate of substitution of nucleotides. Molecular evolution does not have to be neutral for the methods to work, but it does need to have a strong form of statistical regularity that is most plausibly supplied by neutrality. Mutations occur independently and different nucleotides are fixed in each lineage, and as time goes by, differences in sequences accumulate. That is not controversial.

A key assumption: mutations are fixed at the same rates in each lineage.

What is controversial is the assumption of a **molecular clock**—the claim that each lineage accumulates changes in sequences (substitutions, or mutations that have been fixed) *at the same rate*. The number of changes that have accumulated then estimates the time elapsed since the lineages shared common ancestors.

A molecular clock: the number of changes estimates the time elapsed since the lineages shared common ancestors.

There are several problems with this idea. First, the genomes of all organisms within a group share a similar structure that determines both the mutation rate and which parts of the genome are exposed to selection. Thus the rate of nucleotide change should be similar within large groups—eucaryotes, procaryotes, RNA viruses—but not between them. The clock should tick at different rates in groups with different genomic structures.

The clock should tick at different rates in different lineages because of differences in genome structure and generation time.

Secondly, lineages differ in generation time. Small organisms usually have shorter generations than large ones, and body size varies dramatically among lineages. Since the mutation rate is a rate per generation, not a rate per year, one would expect changes to accumulate more rapidly in lineages with short rather than long generations. This effect has been demonstrated in several cospeciating groups, such as gophers and their lice.

Thirdly, if the group being analyzed met all the assumptions required for a molecular clock, then the phylogenetic tree produced by the analysis would be clocklike, meaning that the distance from the common ancestor to the tips of each of the branches would be the same: each path would have the same number of codon or amino acid substitutions. However, this is rarely the case. One might be willing to accept a rough correlation between divergence times and number of differences in substitutions between lineages, rather than a precise fit to a clocklike tree. However, even there problems arise, for the confidence limits that one can place on such relations are so broad that in practice the resulting 'clocks' are imprecise.

Molecular trees rarely have a structure that suggests a clock.

For all these reasons many studies try to avoid the assumption of a molecular clock.

Nevertheless, the best predictor of the amount by which orthologous sequences in two species have diverged remains the time that has elapsed since they shared a common ancestor. A longer branch in a molecular phylogeny suggests that more time has elapsed along that branch than along a shorter one—except, of course, when they are sister-branches (this does happen!).

Time to common ancestor and sequence divergence are, however, related—if only roughly.

Using the right molecule for the problem at hand

Each type of molecule and method of analysis is best suited to a certain range of problems. Hillis *et al.* (1996) and Avise (1994) discuss in depth the advantages and disadvantages of various molecules and methods. Here we simply make the point that one should choose the right combination for the problem at hand.

Molecules are like radioisotopes: they change at different rates. Mitochondrial DNA is useful out to several hundred million years.

Molecules are like radioisotopes: they change at different rates. For example, uranium-238 has a half-life of 4.5×10^9 years, which makes it useful for dating objects of about the age of the Earth or the Moon, whereas carbon-14 has a half-life of 5600 years, which makes it useful for dating archeological objects from a few hundred to about 20 000 years old. The genomes of RNA viruses such as HIV change so quickly that every person infected soon carries an identifiably different strain. Mitochondrial DNA, which is haploid, has a relatively fast substitution rate. It evolves rapidly enough to be useful for comparisons of lineages that diverged recently, but it can also be used to establish relationships among groups that are several hundred million years old. Beyond that point it becomes so altered by repeated mutations that the useful information becomes obscured by noise.

For deep time, we need highly conserved genes, such as those coding for ribosomal RNA.

To get good molecular information on events that occurred in deep time, we need highly conserved genes, genes that change very slowly, such as the DNA that codes for the small subunits of ribosomal RNA. Such genes contain useful information about events that occurred 500–1500 million years ago. They can be used, for example, to test the idea (Margulis 1970, 1981) that mitochondria and chloroplasts are intracellular symbionts derived from procaryote ancestors, an idea now strongly supported by sequence data:

1. The nucleotide sequences of the 16S RNA gene from chloroplasts indicate that chloroplasts are more closely related to photosynthetic cyanobacteria than to the nuclear genome of maize.
2. Similar analysis suggests that mitochondria are derived from the α subdivision of the purple bacteria.

Such genes were used to find the free-living relatives of mitochondria and chloroplasts.

Comparing the phylogenies of organelles (Fig. 12.9a) and nuclei (Fig. 12.9b) reveals striking differences in their geometry. The nuclear sequences support the traditional view that the plants, fungi, and animals form a group distinct from the protists. The mitochondrial sequences suggest that the plant mitochondria were derived from the purple bacteria independently and much more recently than the mitochondria found in fungi, ciliates, green algae, and animals. Chloroplasts appear to have been acquired more recently than mitochondria, for their sequences have diverged less from those of their bacterial ancestors, and less of the chloroplast genome than the mitochondrial genome has been transferred to the nuclear genome. An alternative is that the symbiotic events occurred at about the same time but that chloroplasts and mitochondria subsequently evolved at strikingly different rates.

Thus highly conserved DNA sequences record the ancient history of

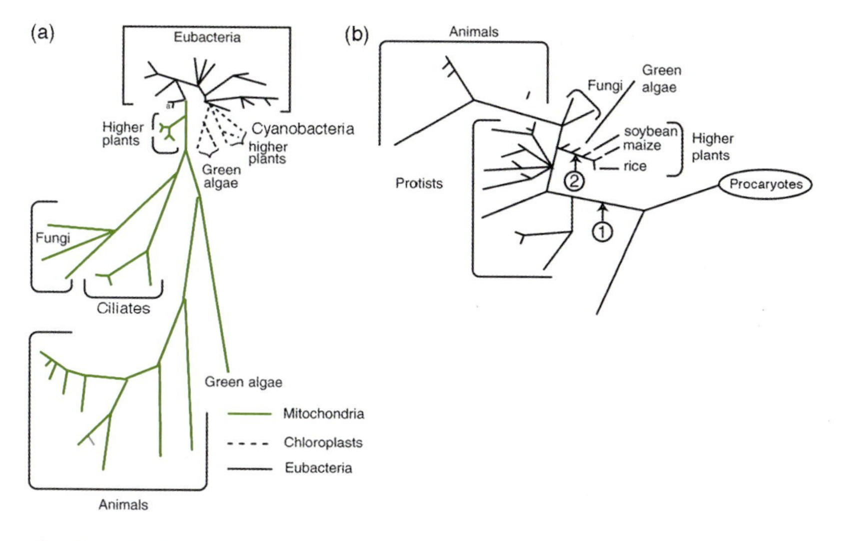

Fig. 12.9 Molecular systematics confirms the symbiotic origin of mitochondria and chloroplasts. (a) The molecular phylogeny of the plastids. Mitochondria are related to purple bacteria and chloroplasts to cyanobacteria. (b) The molecular phylogeny of the hosts. 1, where the mitochondria joined the eucaryotes; 2, where the chloroplasts joined the eucaryotes. (From Li and Graur 1991.)

organisms that have left no fossils, at a time in Earth history when we have very little other information, and allow us to test a major evolutionary hypothesis, the symbiotic theory for the origin of eucaryotic cells.

The theory and rationale of tree building

There are three commonly used methods for building trees: parsimony, distance, and maximum likelihood. Parsimony methods apply the logic of cladistics to molecular data, morphological data, or both together. Distance and maximum likelihood methods are usually applied to DNA sequence data where homoplasy is common.

Parsimony methods

This family of methods is based on cladistics: simpler hypotheses are preferred to more complicated ones, and shared derived similarity is decisive. As much similarity among taxa as possible is explained by common ancestry. In most data sets, some characters will conflict with this explanation, making necessary assumptions of homoplasy (convergence, parallelism, or reversal). When parsimony is applied to sequence data, evolutionary rates must be small or nearly equal in different lineages if the method is to yield trees that approach the true tree with increasing amounts of data. Otherwise the information needed will be washed

Parsimony methods explain the greatest possible portion of characters shared among taxa by common ancestry.

away by repeated mutations to one of the only four possible nucleotide states (this effect is also called saturation).

Distance methods

A distance is the number of differences in character states between two taxa. All distance methods start by calculating the distances between all pairs of species. Common methods include cluster analysis, neighbor-joining, and optimality methods.

Cluster analysis

Cluster analysis is easy but misleading.

Cluster analysis is simple and fast but flawed. It assumes that changes occur at the same rate along all branches of the tree, which is questionable, and that grouping on the basis of total similarity is logically valid, which ignores the cladistic distinction between informative synapomorphies and uninformative plesiomorphies.

Neighbor joining

Neighbor-joining methods deal with different rates in different branches.

Neighbor joining is the most important of the methods that yield additive trees, trees in which it is assumed that the lengths of the branches between any pair of taxa can be summed to yield the amount of evolutionary change that has occurred between them. It proceeds by linking the closest pair of nodes, which defines a common ancestor; removing the distal branches, which converts the new common ancestor into a terminal node on a smaller tree; then continuing. This method can deal with trees in which changes are occurring at different rates in different branches.

Optimality methods

Optimality methods estimate uncertainty but assume that pairwise distances are independent, which is not generally true.

There are several definitions of optimality and ways to find the best tree. Optimality methods produce estimates of the uncertainty with which we can view the various parts of trees.

Maximum likelihood methods

Maximum likelihood methods find the tree most likely to produce the data observed.

Maximum likelihood methods find the most likely tree given the data. That tree is found by considering a large set of hypothetical, candidate trees—all possible trees or a representative sample of them. For each candidate tree the probability is calculated that a given initial state will yield a given final state at the end of a defined interval of time. The probabilities for all changes in the tree are then multiplied to yield the total likelihood of the tree. In making those calculations, branch lengths play an important role. The best tree is then the one with the maximum likelihood (Felsenstein 1988). The method is logically appealing and computationally expensive. Its range of application is increasing as computers improve.

Finding and evaluating trees

Phylogenetic trees are now calculated routinely with computer software (see, for example, Hillis *et al.* 1996). Here we simply assume that data and software are available and a tree can be built.

There are two basic approaches, exact and heuristic. The exact approach finds the best possible tree. If only a few taxa are being compared, we can work through all possible trees, calculate a score for each, and pick the one with the best score: for example, the fewest number of changes in character states. This gives us a frequency distribution of scores, telling us how many trees are nearly optimal, as well as which one is the best. However, it cannot be used for large numbers of taxa, for the number of possible trees grows factorially with the number of taxa: only 945 for 7 taxa, but over 2×10^{20} for 20 taxa. For this reason, heuristic methods have been developed for estimating optimal trees for large numbers of taxa.

Evaluating cladograms: strict consensus trees

Strict consensus trees show what is supported by all equally parsimonious trees.

With enough characters and a reliable analysis of homoplasy, parsimony methods should yield the single simplest tree. In practice, there may not be enough characters in the data set, or homoplasy may be too frequent to resolve all groups. The result may be many, equally parsimonious trees. What should we make of them? One sensible approach is to ask: 'In how many of the equally parsimonious trees is a particular group supported?' If it is supported in all of the equally parsimonious trees, then it can be included in a **strict consensus tree**. This is the most conservative criterion; there are others (Forey *et al.* 1992).

Evaluating distance and maximum-likelihood trees

Bootstrap values: above 90%, strong support; below 70%, suspicious.

Software packages deliver confidence measures for trees estimated with distance and maximum likelihood methods. One of the most common is the **bootstrap value**. **Bootstrapping** is a method of estimating confidence in the tree by taking random samples of the original data, calculating a tree from the new data set, then repeating the procedure many times. The bootstrap values are then the percentages of cases in which groups appear in the resulting set of trees. Bootstrap values above 90% are regarded as strong support; those below 70% should be viewed with suspicion.

Some tips for evaluating trees

Do not accept phylogenetic trees uncritically.

Here are some questions to ask when judging the reliability of a tree:

1. What are the estimated errors in the distances? If the distance between two groups is less than the estimated error, then other hypotheses of relationship are tenable.
2. What are the bootstrap values? If a branch is optimal in more than 90% of all subsamples, that branch is strongly supported.

Trees with long branches are a problem for parsimony methods applied to sequence data, for the longer the branches, the more likely it is that homoplasies will accumulate by mutation. Lineages evolving at different rates pose problems for all methods. Methods are available for dealing with those problems, and new methods are being published every month.

The genealogy of genes and the phylogeny of species

Different genes in the same organisms can have different evolutionary histories.

Molecular systematics is not only used to build trees relating species; it is also used to construct the history of single genes. The trees constructed from different genes in the same organisms often have different structures because each gene has had a different evolutionary history. For events occurring within a species, the recovery of a reliable gene genealogy must be done in sequences with little or no recombination, such as the mitochondrial genome, because recombination produces nets, not branches. A gene genealogy can differ from a species phylogeny because mutations do not occur simultaneously and are not constrained to occur during speciation (Fig. 12.10). One gene may have diverged prior to a speciation event, another gene may have diverged after that speciation event. Thus genes have different genealogies, and only some genealogies have the same structure as the phylogeny of the species in which the genes occur. Examples are discussed in Chapter 15.

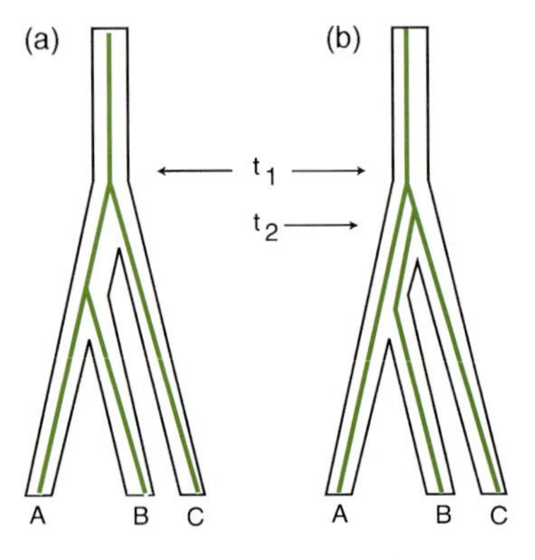

Fig. 12.10 Phylogenies of species and genealogies of genes. The species tree is described by the large outer figure and is the same in both cases. The gene genealogy is described by the lines contained within the tree. Whenever more than one line is present, the gene is polymorphic. In both cases there is a mutation at time t_1, prior to the first speciation event. (a) The gene tree and the species tree have the same branching pattern, i.e. the second mutation occurs between the first and the second speciation event. (b) The genes and species have different branching patterns. The second mutation event occurs shortly after the first mutation event at time t_2, and the genealogical split predates species divergences. (From Avise 1994.)

Summary

This chapter discusses what phylogeny can tell us, what phylogenetic trees are, and what issues arise in constructing them.

- Systematics estimates the relationships among taxa to get working hypotheses in the form of phylogenetic trees. A phylogenetic tree records the history of a group.
- To construct a phylogenetic tree, one first determines which characters are homologous, measures them on all the taxa being analyzed, then codes the measurements for analysis. Different codings can produce different results.
- Trees are built from data using methods that can produce different trees. If all methods yield the same branch of a tree in a large and reliable data set, then that branch can be regarded with confidence. If the data are equally consistent with several branching patterns, judgment about relationships should be suspended.
- Phylogenetic analyses are becoming increasingly reliable. They are producing surprises about both relationships and character evolution.

Improvements in both data and analysis are resolving major questions about the history of life, the subject of the next chapter.

Recommended reading

Avise, J. C. (1994). *Molecular markers, natural history and evolution.* Chapman & Hall, New York.

Brooks, D. R. and McLennan, D. A. (1991). *Phylogeny, ecology and behavior. A research program in comparative biology.* University of Chicago Press, Chicago.

Forey, P. L., Humphries, C. J., Kitching, I. L., Scotland, R. W., Siebert, D. J., and Williams, D. M. (1992). *Cladistics.* Oxford University Press, Oxford.

Hillis, D. M., Moritz, C., and Mable, B. K. (ed.) (1996). *Molecular Systematics,* (2nd edn). Sinauer, Sunderland, Massachusetts.

Questions

12.1 Why is it more reliable to interpret fossils as the tips of dead branches than as the direct ancestors of living species?

12.2 Can a phylogenetic tree ever be anything more than 'just a working hypothesis'?

12.3 If every branch in the tree of life defines a new natural group, and if the tree of life includes everything from bacteria to whales, then how useful is the Linnaean system of taxonomy, which tries to place all organisms into nested categories, species within genera within families within orders within classes within phyla? If that system does not work well, what would you recommend to replace it? A taxonomy should accurately reflect all the information in a phylogenetic tree. Is that goal possible?

Chapter 13
The history of life I: the evolutionary theater

Introduction

Chapters 2–10 discussed microevolution: natural selection and inheritance, and their consequences for adaptive and neutral evolution. Sex, life histories, sexual selection, and conflict were major themes. Chapter 11, on speciation, made the bridge from micro- to macroevolution. Macroevolution consists of patterns at and above the level of the species that document the history of life. Such patterns are described through the two main approaches to macroevolution: systematics and paleontology. Chapter 12 described some of the methods and major insights of systematics. This chapter, where the focus shifts towards paleontology, describes major patterns in the history of life and our planet, the events and structures that have formed the changing theater of the evolutionary play. Its goal is a survey of high points that might motivate further reading in paleontology and geology. The next chapter discusses the major transitions in the organization of life that mark the key events in evolution.

Some insights of history

The history of a maple tree

Landscapes and organisms are mosaics of forms of vastly differing ages.

On a wooded shoreline in south-east Alaska (Fig. 13.1) we see maple trees, bracken fern, moss-covered rocks, orchids, dragonflies, a frog, a mule deer, and a great blue heron. In the intertidal zone, barnacles, mussels, and a starfish are visible. A loon swims on the water surface, beneath which we can see anemones, an octopus, a crab, sea urchins, and algae. In deep water, a family of humpback whales attacks a school of herring.

Fig. 13.1 A shoreline in south-east Alaska. The elements, the organisms, and the landscape itself are all mosaics of parts and forms that originated at very different times in the history of the planet. (By Dafila K. Scott.)

The elements themselves formed at different times following the Big Bang.

The organisms and landscape consist of elements, parts and forms that appeared in the history of the universe and planet at vastly different times. The cells of the maple tree are constructed from molecules, consisting of atoms with nuclei. Hydrogen—the most common nucleus—formed in the first milliseconds of the Big Bang about 15 000 million years ago (mya). Helium and lithium appeared after 3 minutes, but all the heavier nuclei, including the light elements that form the building blocks of most biological molecules (carbon, nitrogen, oxygen, sulfur, phosphorus, sodium, calcium, and the other elements up to and including iron), were created much later through thermonuclear fusion in stars, then expelled into interstellar space in novas or supernovas. Almost all elements heavier than iron were created in supernova explosions (Mason 1992). Most elements found on Earth today originated 5000–15 000 mya, before the solar system formed from recycled starstuff.

Life is built from recycled starstuff.

Some of the elements formed in supernovas, including copper, zinc, selenium, molybdenum, and iodine, have important biochemical roles (Stryer 1988). Copper occurs in plastocyanin, part of the photosynthetic mechanism that transforms light energy into chemical energy. Copper helped to oxygenate the planet. Zinc is an essential part of carboxypeptidase A, an enzyme that breaks down proteins. Zinc aids digestion. Selenium is part of an enzyme that protects cells from poisons. Molybdenum forms part of the nitrogenase enzyme that converts molecular nitrogen into usable ammonium ions in nitrogen-fixing bacteria. Molybdenum helps to fertilize the biosphere. Iodine, part of the vertebrate hormone thyroxine, helps to control the expression of genes involved in growth and energy metabolism. Both the maple leaf and your body are built of recycled starstuff.

The names and dates given by geologists to the major divisions of Earth history are listed in Table 13.1, together with the major develop-

Table 13.1 The history of the Earth and life

Period	Duration (mya)	Comment
Cosmic	15 000/20 000–5000	Big Bang, nucleosynthesis, stellar recycling
Solar	5000–4600	Formation of the solar system, inner planets bombarded by planetoids, original atmosphere of Earth, if any, blown off by solar wind
Archaean	3900–3600	**Origin of life**
	3500	Fossil photosynthetic cyanobacteria
Proterozoic	2500–570	Gradual oxygenation of the atmosphere
Paleoproterozoic	2500–1600	**Origin of eucaryotes, of sex**
Mesoproterozoic	1600–1000	Protist radiation
Neoproterozoic	1000–570	**Origin of multicellularity** algae and sponges, coelenterates and worms
Phanerozoic	570–0	
Paleozoic	570–250	
Cambrian	570–510	Fossil mollusks (amphineurans, gastropods cephalopods), annelids (polychaetes), arthropods (trilobites, chilicerates), brachiopods, echinoderms
Ordovician	510–438	Fossil mollusks now include scaphopods and bivalves, nautiloids, big radiations of ammonites and chelicerate arthropods, bryozoans appear, all major modern echinoderm classes (sea stars, urchins, cucumbers, lilies, brittle stars) appear, first jawless vertebrates (Agnatha), first land plant spores. **Invasion of land**. At end, in a diffuse **mass extinction**, 22% of families disappear
Silurian	438–410	Appearance of ascomycete fungi, of scorpions, radiation of jawless fishes. First land plant megafossils: club mosses. Closest aquatic relatives: Charophyceae (a group of mostly freshwater green algae). First spore-eating arthropods
Devonian	410–355	First forests indicated by rooted soils. Major vascular plant groups radiate. Appearance of mosses, ferns, squid, myriapods, insects, ticks, jawed fishes, lungfishes, first vertebrate tetrapod—an amphibian (*Ichthyostega*), origin of insect flight
Carboniferous	355–290	First modern soils, earthworms, seed plants appear, harvestmen (opilionids), dragonflies, orthopterans, reptiles. First pollen-eating insects
Permian	290–250	Cycads appear, another ammonite radiation. **Plant–pollinator coevolution begins**. At end, the trilobites disappear, along with 83% of all marine invertebrate genera, in the **biggest mass extinction**; at end a single megacontinent, a single ocean
Mesozoic	250–65	
Triassic	250–205	Gymnosperms radiate: first pines, podocarps, araucarias; radiation of odonates and orthopterans, first fossil teleost fish, crocodiles, mammals. Another ammonite radiation. First flying reptiles (pterosaurs). At end, **mass extinction**, 20% of families of marine invertebrates disappear, including some bivalves, cephalopods, gastropods, brachiopods, sponges, marine reptiles. Timing controversial. Mechanism undecided

Table 13.1 Continued

Period	Duration (mya)	Comment
Phanerozoic (Cont.)		
Mesozoic (Cont.)		
Jurassic	205–135	First fossil yews, octopods, decapod crustacea, diptera, hymenoptera, lepidoptera, sharks, rays, frogs, salamanders, birds (avian flight). Teleosts radiate. Yet another ammonite radiation. Reptiles radiate. Vertebrate homeothermy (birds, dinosaurs). Turtles. Origin of modern insect pollinators, intensification of plant–pollinator coevolution
Cretaceous	135–65	Angiosperms radiate at expense of pteridophytes and cycads; conifers persist at about half former level. First fossil walnuts, termites, wasps, honeybees, lizards, snakes, loons, marsupials. At end, **mass extinction**: dinosaurs and ammonites disappear. Brief fern spike, soot deposits, shock quartz, iridium spike. Mechanism: meteorite impact, massive volcanism, or both
Cenozoic	65–0	
Paleogene	65–23	
Paleocene	65–53	First fossil rhododendrons, grasses, maples, willows, herons, falcons, anhingas, avocets, parrots, rollers, placental mammals: lagomorphs, whales, rodents; mosquitoes
Eocene	53–37	First fossil beeches, elms, casuarinas, tilias, primates, ungulates, canids, proboscids, bats (mammalian flight)
Oligocene	37–23	Origin of grasslands, further mammal radiation
Neogene	23–1.6	
Miocene	23–5.3	Grass-processing horse teeth, all subfamilies of grasses
Pliocene	5.3–1.6	First hominid fossils (*Australopithecus*), first orchid fossils, bipedal hominids 4 mya in East Africa
Quaternary	1.6–0	
Pleistocene	1.6–0.01	*Homo erectus* in Africa and Asia (1.5), *Homo sapiens* (0.3), Neanderthals (0.1); widespread burials (0.09–0.04), **language** (0.06?), art-cave painting, sculptures, decoration (0.03). In upper Paleolithic Europe, 50% survive to age 21, 12% survive to age 40
Holocene	0.01–0	End of Wisconsin Glaciation, domestication of cows, goats, sheep, horses; origin of agriculture, **writing**

From IUGS (1989), Briggs and Crowther (1990), Benton (1993), Kenrick and Crane (1997), Labandeira (1998a,b)

ments in the history of life. Please read that table now and refer to it while reading the rest of this chapter.

Angiosperms evolved at least 130 mya, wood perhaps 350 mya.

If we could let evolutionary history unfold before us, how far back in time would we have to go to see the first maple tree? (The fossil record gives a conservative answer, for the dates of first appearances in the fossil record can only be pushed back by new discoveries.) According to the fossil record, we could have seen a maple leaf 60 mya. A maple is an angiosperm, a flowering plant, and traces of angiosperm pollen have been found in the early Cretaceous, about 130 mya. A maple has a tall, upright stem, made possible by an unusually strong composite material, wood. Wood evolved in gymnosperms 350–230 mya.

Maples are terrestrial plants. The earliest traces of terrestrial plants are spores recovered from the late Ordovician and sterile stems from the early Silurian, 440–430 mya, and there are fossil soils from the late Devonian, 360 mya, with traces of vascular roots. Maples are multicellular organisms, which appeared in the fossil record about 1000 mya. They are eucaryotes with endosymbiotic organelles, a spindle apparatus, and true meiosis. The eucaryote ancestors acquired the purple nonsulfur bacterial ancestors of mitochondria between 3000 and 1700 mya, probably several times; mitochondria may be polyphyletic. Later the eucaryotes that became plants acquired chloroplasts from cyanobacteria between 1500 and 1000 mya, probably several times; chloroplasts may also be polyphyletic. Maple trees photosynthesize carbohydrates from water, carbon dioxide, and solar energy, and photosynthesis can be traced to fossil blue-green algae (cyanobacteria) at least 3470 mya. Some molecules involved in photosynthesis are among the most ancient, conservative structures known to biochemists, relicts, like membranes and RNA, of the first few hundred million years of life. Life itself originated 3900–3700 mya, when the temperature of the cooling planet fell to the point where macromolecules could be stable in hot water under pressure.

Land plants evolved at least 430 mya, multicellularity about 1000 mya, eucaryotes about 1700 mya, photosynthesis about 3500 mya and life about 3800 mya.

Like all organisms, the maple tree is a mosaic of parts of different ages; some had their evolutionary origins at vastly different times.

The history of its neighbors

Ferns and mosses dominated the Paleozoic (540–280 mya), gymnosperms the Mesozoic (280–65 mya), and angiosperms the Cenozoic (65 mya–present).

Now consider the other organisms in the landscape (Fig. 13.1 and Table 13.1). The oldest large plants are the mosses, ferns, and their relatives, which appeared in the fossil record about 360 mya and dominated the Carboniferous and Permian 355–250 mya. The tree of vascular plant life is dominated by mosses, horsetails, and ferns. At that scale, the seed plants are a minor branch of a big tree. Pines appeared in the fossil record 210 mya, about the same time as dinosaurs, but gymnosperms, the group to which pines belong, appeared at the beginning of the Carboniferous. They radiated in the Triassic and dominated from the mid-Triassic through the Cretaceous (225–65 mya). The orchids, like the maple, are flowering plants, angiosperms. Angiosperms originated in the early Cretaceous and have dominated plant communities in the Tertiary (65–0 mya). The phylogeny of the seed plants is dominated by the conifers and their relatives; at that scale, the angiosperms are a small branch within which the monocots (orchids, grasses, palms, lilies) are an insignificant twig. Thus the pteridophytes (ferns and their relatives), the gymnosperms, and the angiosperms are each characteristic dominants, in that order, of different major eras of Earth history (Niklas *et al.* 1983; Fig. 13.2). The greatest diversity of major plant types is to be found among the liverworts, mosses, horsetails, and ferns; compared to them, the seed plants form just one of many branches (Doyle 1998).

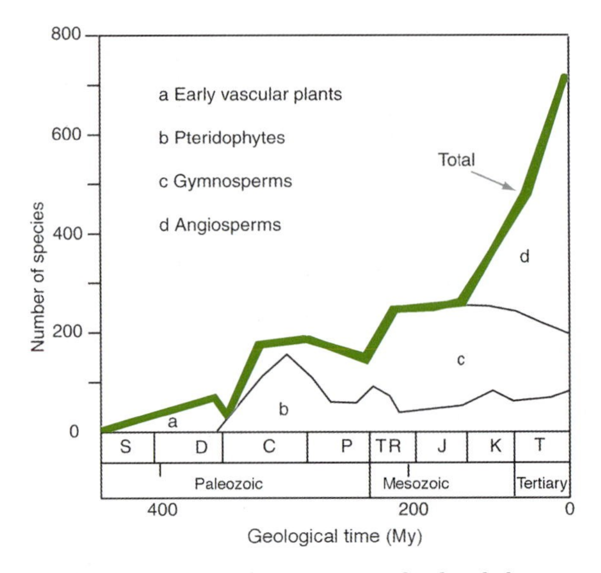

Fig. 13.2 The plants that are currently alive belong to groups and have forms that originated in the different major eras of Earth history. The ferns and mosses dominated the late Paleozoic, the gymnosperms were characteristic of the Mesozoic, and the angiosperms are characteristic of the Tertiary. S, Silurian; D, Devonian; C, Carboniferous; P, Permian; TR, Triassic; J, Jurassic; K, Cretaceous; T, Tertiary. (From Niklas *et al.* 1983, reproduced by kind permission of the authors and *Nature*.)

Among the terrestrial animals on this Alaska shoreline, the oldest forms are the dragonflies, which can be found as fossils 300 million years old. Frogs appear in the fossil record 190 mya, loons 70 mya, herons and rabbits 65 mya.

Sponges and coelenterates (650–550 mya) remain from the Cambrian; polychaetes, sea stars, crinoids, brachiopods, and ostracods (500–440 mya) remain from the Ordovician; clams, snails, sea urchins, crabs, lobsters, octopus, sharks, and bony fishes (230–65 mya) from the Mesozoic; and whales and teleost fish (65–40 mya) from the Tertiary.

Under water we see forms that are much older, for life evolved in the sea. These also make a series that dominated successive ages (Fig. 13.3). Not much is left of the fauna of the Cambrian, which was characterized by trilobites, monoplacophoran mollusks, inarticulate brachiopods, and several echinoderm classes, all of which are wholly or mostly extinct. From just before the Cambrian we can still see sponges and coelenterates (650–550 mya). The Paleozoic fauna originated primarily in the Ordovician, dominated to the end of the Permian, and has left many surviving forms: polychaete worms, sea stars, crinoids, articulate brachiopods, nautiloids, and ostracods (500–440 mya). The Mesozoic fauna contains more familiar forms: clams, snails, sea urchins, crabs, lobsters, octopus, sharks, and bony fishes (230–65 mya). Whales and teleost fish evolved 65–40 mya.

Modern communities are mosaics of forms of very different ages.

Many of the forms that dominated successive ages have disappeared, but there are survivors characteristic of a series of ancient communities. A modern community is a mosaic of forms of very different ages.

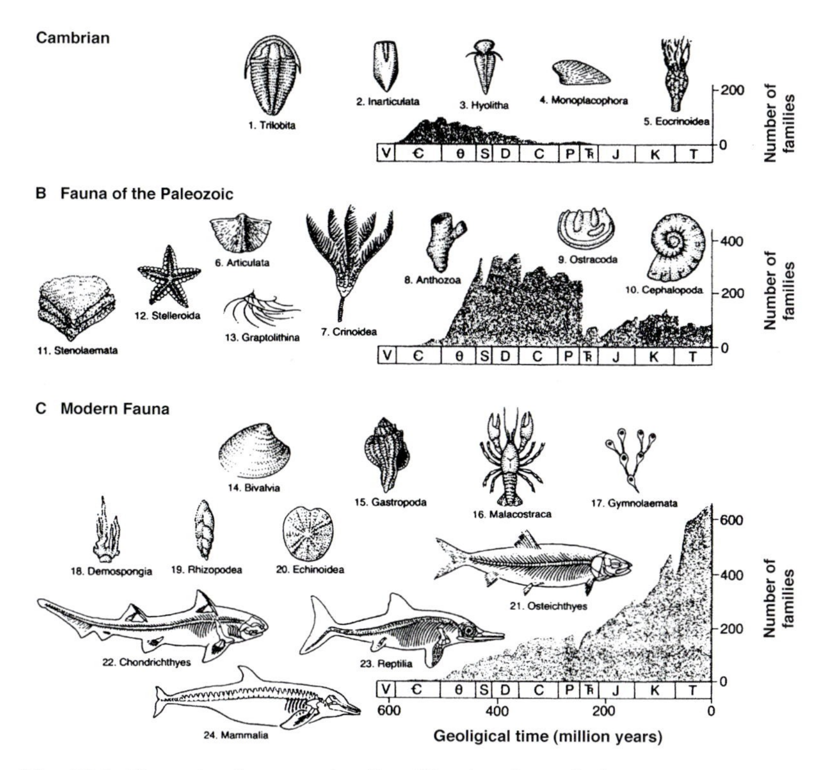

Fig. 13.3 The animals currently alive, like the plants, belong to groups and have forms that originated in the different major eras of Earth history. The sponges and coelenterates remain from the Cambrian; the polychaetes and sea stars from the later Paleozoic; the clams, sea urchins, octopus, and bony fishes from the Mesozoic; and whales and teleost fishes from the Tertiary. V, Vendian; C, Cambrian; Θ, Ordovician; S, Silurian; D, Devonian; C, Carboniferous; P, Permian; TR, Triassic; J, Jurassic; K, Cretaceous; T, Tertiary. (From Sepkoski 1984.)

The history of the landscape itself

Some terranes of Alaska came from far to the south.

The geological history of south-eastern Alaska is puzzling. The Pacific Coast of Alaska is built out of several **terranes**, pieces of continental crust that do not belong to the original core of the North American continent, whose edge lies to the east, near the western edge of the Rocky Mountains. Terranes originated as pieces of island arcs that lay offshore, or pieces of continents that were fragmented by active rifts, then drifted towards North American on subducting plates. As the Atlantic Ocean opened to the east, North America drifted west, and the islands and pieces of continents that it encountered accreted onto its western margin. Before accretion they looked like the Greater Antilles, Japan, or the Indonesian archipelago. Some pieces of Alaska seem to have come from as far away as present-day South America. Thus the rock on which the maple tree grows probably spent the Mesozoic and early Tertiary—

during which the flowering plants, birds, bees, wasps, and lizards originated, the dinosaurs and ammonites went extinct, and the mammals radiated—moving across the Pacific as an island.

The lessons of Messel

At certain rare places and times, conditions favored the fossilization of whole communities, preserving a snapshot of the past. One particularly well-analyzed community was found in the oil shales at Messel, between Frankfurt and Darmstadt in Germany (Schaal and Ziegler 1992). In the mid-Eocene, 49 million years ago, Messel was a lake or lagoon with an anoxic bottom layer. Organisms that fell into the anoxic layer were preserved virtually intact, including soft parts and stomach contents. Fossils found here include ferns, palms, water lilies, grapes, citrus, walnuts, giant ants, cicadas, cockroaches, bush crickets, stick insects, jewel beetles, stag beetles, longhorn beetles, salamanders, toads, frogs, freshwater turtles, crocodiles, lizards, owls, rollers, swifts, nightjars, opossums, insectivores, primates, bats, pangolins, rodents, a variety of carnivores, and early toed horses.

A European community 50 million years old contained some surprisingly familiar species and some totally unexpected ones.

Many of these fossils look like their modern relatives; those lineages have not changed much. Others, particularly the relatives of modern horses, carnivores, and rodents, hedgehog relatives that looked like kangaroo rats, and a flamingo relative with affinities to stilts and plovers, were intriguingly different. Those lineages have gone extinct or changed a lot.

Two finds at Messel are particularly surprising:

(1) many of the plants belong to tropical families, for example the screw pines (Pandanaceae) (which are currently confined to Central and West Africa, South-East Asia, Australia, New Zealand, and Polynesia); and

(2) a South American anteater (a tamandua) has been found.

This suggests that central Germany was then subtropical or tropical. Plants from present-day South-East Asia and animals from South America were then living together in western Europe.

The geographical distribution of climates and organisms in the past can have been different in ways that no one would now expect. Only the fossils can reveal such surprises. The shoreline of Alaska and the shale pits of Germany suggest what can be learned from the history of the planet and of life (have another look at Table 13.1).

The geological theater

The origin of the universe, solar system, and planet

The universe is about 15 000 million years old.

The universe formed about 15 000 mya; the leading hypothesis remains the Big Bang, which is plausible but not proven. According to that hypothesis, after a brief period of incredibly rapid expansion, a universe consisting of hydrogen and a little helium was formed. Gravitation then pulled concentrations of gas into primitive galaxies, within which stars formed, synthesizing heavier elements from hydrogen and helium nuclei, until they consisted mostly of iron, at which point the larger stars exploded as novas and supernovas, forming the elements heavier than iron.

Solar systems formed secondarily from the products of one or more cycles of star formation, ours about 4500 mya. There are two scenarios for the formation of the inner, Earth-like planets. In the first, they formed like the gas giants of the outer system, as accretion disks condensing into spheres consisting mostly of hydrogen and helium with a core of metal and rock debris. As the sun began to burn, the solar wind blew off the primitive atmosphere of light molecules, leaving the small, heavy cores that we now know as Mercury, Venus, Earth, and Mars. In the second scenario the inner planets were never gas giants; they formed by accretion of heavy debris, planetoids, and meteorites. In both scenarios, the inner planets underwent an early period of intensive bombardment by planetoids and meteors, a bombardment that raised temperatures high enough to melt the crust and make life impossible. The traces of that bombardment, which ended about 4000–3800 mya, can still be seen as giant impact craters on Mercury and the Moon.

Our solar system is about 4500 million years old.

As the inner planets cooled, an atmosphere formed by outgassing from the rocks. It was composed of compounds heavier than molecular hydrogen and helium, compounds that were more strongly held by gravity and less easily blown off in the solar wind. Water, carbon dioxide, methane, ammonia, and hydrogen sulfide were among them. Even this atmosphere was blown off Mercury, which is close to the Sun where the solar wind is strong, and could not be completely retained by Mars, whose gravity is weaker than Earth's. Venus and Earth, farther from the Sun than Mercury and larger and heavier than Mars, have retained thick atmospheres. The one on Venus, which has not been engineered by life, may resemble the early atmosphere of Earth. It consists mostly of carbon dioxide with some water and sulfuric acid, retains solar heat like a blanket, and raises temperatures on the surface of Venus well above the boiling point of water. The early atmosphere of the Earth was hot, toxic by our standards, reducing instead of oxidizing, and sulfurously smelly.

Our atmosphere may have formed by secondary outgassing. It began hot, toxic, reducing, sulfurous, and smelly.

In the molten mass of the cooling Earth, the heavier metals sank to the middle, forming a nickel–iron core, and the lighter rocks rose to the surface, forming a crust. Within this crust, which must have been driven by energy flows and wracked by movements more powerful and rapid than those in existence today, the cores of continents formed high ground. When the temperature of the surface fell below the boiling point of water, oceans formed. The continental cores and the first oceans formed about 4000 mya. The oceans have been remodeled several times since then, but traces of the original continental cores can be found in South Africa, the Canadian Shield, Greenland, Australia, and India. That is also where the oldest fossils occur.

Continents and oceans formed about 4000 mya.

Continental drift

Plates move slowly across the surface of the Earth.

Since then, the crust of the Earth has consisted of a set of continental plates floating on the upper mantle, moved by the flow of heat from the interior. The plates either slide by one another, forming transverse faults, such as the San Andreas Fault in California, or are torn down the middle, as is happening in the Red Sea and the great Rift Valleys of Africa, or collide with one another. Where continental plates collide, mountain ranges grow. The great mountain arc stretching from Iran to China formed in the Tertiary collision of India with Asia. The Alps formed when Africa shoved Italy into Europe.

Their collisions produce mountain ranges; their subduction produces volcanic chains and recycles the Earth's surface, wiping out history.

Because a continental plate is lighter than an oceanic plate, an oceanic plate is mostly subducted beneath a continental plate in a collision. The ocean-bottom sediment is scraped onto the continental margin to form a coastal mountain range. The oceanic plate melts at 300–700 km depth and sends up plumes of lava, which form chains of volcanoes 100–300 km back from the leading edge of the subduction zone. The volcanoes and raised montane arches around the entire Pacific Ring of Fire—Indonesia, the Philippines, Japan, Kamchatka, the Aleutians, southeastern Alaska, British Columbia, California, Mexico, Central America, and the Andes of South America—were so formed when the oceanic plates of the Pacific Basin were subducted beneath them.

The energy source reaches the surface at the mid-oceanic ridges . . .

The energy source for the movement of the plates reaches the surface at the mid-oceanic ridges, the largest mountain ranges on the planet, stretching more than 40 000 km around the globe. At the center of the ridges, which reach the surface in Iceland, lava wells up as the plates move apart. When the lava solidifies, it preserves the orientation of the Earth's magnetic field. The field changes polarity every few million years. As the plates move apart, the space between them is filled by ocean bottom with magnetic stripes whose polarity and age are symmetrical around the ridge crest, younger nearer the ridge, older farther away.

whose inflation and deflation has flooded and dried the continents.

The rate of sea-floor spreading has not been constant. At times the flow of heat from the Earth's interior speeds up, the mid-ocean ridges inflate with magma, the spreading from the ridges accelerates, the continents drift more rapidly, and the sea level rises, displaced by the increased volume of the mid-ocean ridges, flooding the continental plains. At such times shallow seas flooded large areas that are now dry land, providing increased habitat for marine organisms and creating some anoxic lagoons where fossils formed easily. At other times the flow of heat from the Earth's interior decreases, the mid-ocean ridges deflate, continental drift slows, and the seas retreat to the margins of the continental shelves. Habitat for marine organisms is limited, and there are not many places suitable for the deposit of fossils.

The history of the continents since the Paleozoic is roughly known and has left traces in the distributions of living organisms.

At the end of the Permian, the continents had accreted into a single mass, Pangaea, and there was a single world ocean. During the Mesozoic, that megacontinent began to break up (Fig. 13.4). Huge rift valleys formed as plates pulled apart. They broadened into arms of the sea, like the Red Sea, then into oceans. First two major blocks were formed: Laurasia, consisting of North America and Eurasia, and Gondwanaland, consisting of South America, Africa, India, Antarctica, and Australia. Fossils of the same terrestrial Triassic reptiles can be found on all four continents, and the distribution of ratite birds (ostriches, rheas, and their relatives), southern beeches (Genus *Nothofagus*, from Chile and New Zealand), and *Araucarias* (conifers from South America and the islands off Australia and New Zealand) still reflect a time when those continents were joined in a single land mass.

By the mid-Jurassic, 150 mya, Gondwanaland began to fragment. India was by then already an island continent, but for a time South America remained joined to Africa, and Antarctica to Australia. By the late Cretaceous, 70 mya, South America had completed its separation from Africa, India was still an island journeying towards Asia, and Australia was just separating from Antarctica. The Atlantic finished

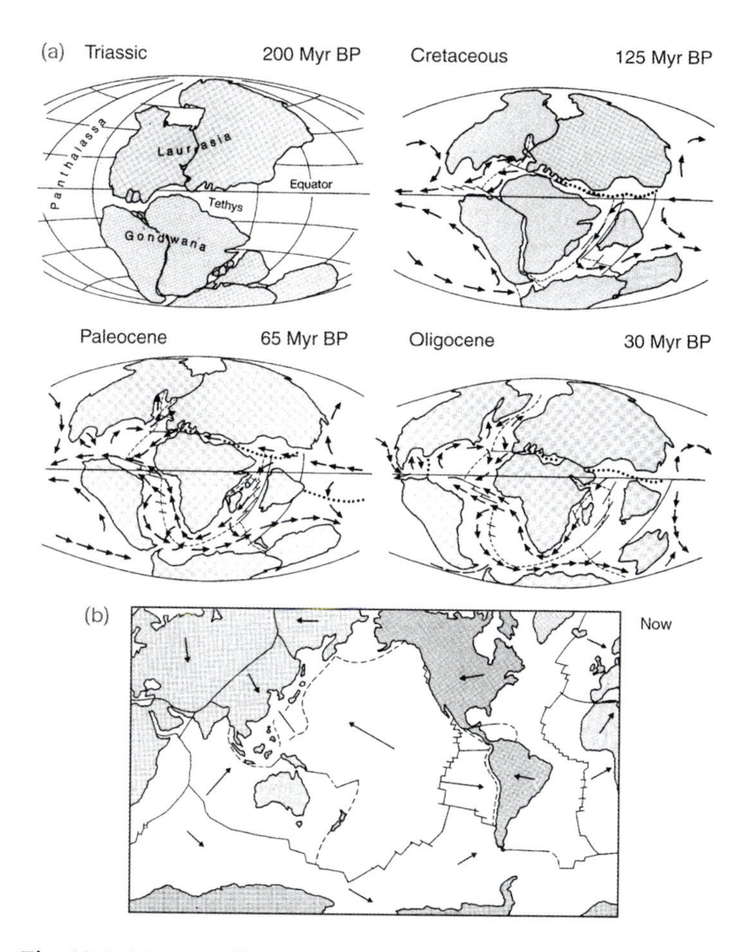

Fig. 13.4 (a) 200 million years of continental drift. (b) The current plates and their motions. (From Futuyma 1998.)

opening, from south to north, about 60 mya. Rocks found in Newfoundland match rocks in Scotland and Scandinavia, just as rocks in Brazil match rocks in West Africa. On a computer one can shift the eastern margins of North and South American back to match the western margins of Europe and Africa. North and South American continue to move west relative to Europe and Africa at 1–10 mm per year.

Fish underwent repeated radiations and extinctions in rift-valley lakes.

When North America began to separate from Europe, rift valleys opened, and in the bottoms of these valleys great rift lakes formed. As the climate changed, sometimes wetter, sometimes dryer, and as the landscape changed and with it the course of rivers, the lakes filled, dried up, and filled again. Each time they filled with water, a now-extinct clade of teleost fish, the semionotids, underwent a radiation, filling the lakes with species that displayed a diversity of feeding types. Each time the lakes dried up, the species flocks went extinct (McCune 1996). Thus the evolution of the great species flocks of cichlid fish in the African rift lakes was repeatedly foreshadowed in the Cretaceous lakes of eastern North America.

South America and India were island continents.

South America remained an island continent, like Australia and Antarctica today, until about 10 mya, when its northward drift into island arcs raised the Isthmus of Panama and parts of Central America, with major consequences. During its isolation a diverse endemic fauna and flora had evolved in South America. Some endemic species went extinct when invaders from North America arrived; others, including armadillos and opossums, moved north. Like South America, India spent several tens of millions of years as an island continent and also evolved an endemic fauna and flora, most of which vanished after it collided with Asia during the middle Tertiary. India continues to move north into Asia, and the Himalayas continue to rise. During the Tertiary, Africa moved north, closing the Tethys Sea at Suez and Gibraltar, which had been a continuous tropical ocean from Spain to Australia, and driving Italy into Europe, raising the Alps and the Jura.

While the continents drifted, major groups of living organisms originated. It is instructive to compare the broad history of the geological theater with some major events in the evolution of the biological players (Table 13.2).

Volcanic hot spots

Oceanic hot spots produce chains of isolated volcanic islands . . .

The floor of the Pacific Basin, over which moved the terranes that docked along the west coasts of Central and North America, has had a more complex history than the Atlantic. The middle of the Pacific Plate contains another important geological mechanism: a 'hot spot'. (Other hot spots can be found in French Polynesia, the Galápagos, and Yellowstone National Park in Wyoming.) As the Pacific Plate moved north-west, the hot spot now located south-east of the youngest emergent island, Hawaii, repeatedly broke through the plate and built volcanic islands. In less than a million years a volcano up to 10 km high and 300 km in diameter can form over that hot spot. The plate carried the

Table 13.2 Fossil and molecular evidence of origins

Group	First fossil evidence[1] (mya)	Molecular evidence (mya)	Comments
Procaryotes			
Cyanobacteria	3500		Warrawarra Deposit, Western Australia, and Onverwacht Deposit, South Africa
Split between procaryotes and eucaryotes[2]		2000	Controversial, approximate
Sex	1100–900		
Multicellular algae	1000–700		Little Dal Deposit, Canada. Many groups in Devonian
Coevolutionary associations			
Arthropod herbivory[3]	*c.* 400		Early Devonian, spore feeding, piercing and sucking
Gall formation[3]	*c.* 300		Mid to late Pennyslvanian
External foliage feeding[3]	*c.* 280		Early Permian
Insect pollination[4]	*c.* 180		Mid-Jurassic
Plants			
Mosses	400–280		Devonian, Carboniferous
Lycopodia, ferns	400–350		Devonian
Selagellinas (Equiseta)	280–330		Permian
Ancestor of seed plants[5]		300	Late Carboniferous
Gymnosperms	350		Early Carboniferous
Cycads	280		Early Permian
Podocarps	230		Early Triassic
Araucarias, pines	210		Mid-Triassic
Yews	190		Early Jurassic
Angiosperms	75–60		Late Cretaceous, Early Tertiary
Monocot–Dicot split[6,7]		200–300	Late Carboniferous to Mid-Triassic, controversial
Walnut family	75–65		Late Cretaceous
Rhodedendron family	65–60		Paleocene
Grass family	65–60		Paleocene
Maples, willows	60–55		Late Pleocene
Beeches, casuarinas, elms, tilias	50		Mid-Eocine
Orchids	10		Pliocene
Fungi	440–400		Silurian
Split between protists and other eucaryotes[2]		1230	Controversial, approximate
Split between plants, animals and fungi[2]		1000	Controversial, approximate
Sponges	570		Early Cambrian
Coelenterates	650		A few Precambrian, mostly Ordovician 500–440
Split between (sponges and coelenterates) and bilateral animals[7]			Much before the Cambrian—perhaps several hundred million years. Date imprecise
Mollusks			
Amphineura	570		Cambrian

Table 13.2 Continued

Group	First fossil evidence[1] (mya)	Molecular evidence (mya)	Comments
Mollusks (Cont.)			
Gastropods	570		Cambrian
Scaphopods	500–440		Ordovician
Bivalves	500–440		Ordovician
Cephalopods			
Ammonites	570		Cambrian, mostly Ordovician, repeated radiations, vanish in end-Cretaceous mass extinction 65 mya
Nautiloids	500–400		Ordovician–Silurian
Coleoids			
Squid	400–190		Devonian–Jurassic, depending on group
Octopus	150–135		Late Jurassic
Cuttlefish	65–60		Paleocene, early Tertiary
Annelids			
Polychaetes	570–440		Cambrian, but mostly Ordovician
Oligochaetes			
Tubifex	300		Late Carboniferous
Earthworms	15		Late Tertiary
Arthropods			
Trilobites	570		Cambrian, vanish in end-Permian extinction 230 mya
Chelicerates	570–440		Cambrian, mostly Ordovician, depending on group
Scorpions	440–400		Silurian
Harvestmen (Opilionids)	350		Early Carboniferous
Ticks	400		Early Devonian
Most spiders	40		Mid-Tertiary
Crustacea	570		Early Cambrian
'Shrimp'	400		Devonian, many groups Triassic–Cretaceous
'Crabs and lobsters'	190–130		Jurassic–early Cretaceous, depending on group
Daphnia	135		Early Cretaceous
Ostracods	570		Cambrian, but mostly Ordovician, Darwinulidae 350
Insects	400–300		Some Devonian, mostly Carboniferous
Odonates (dragonflies)	300–200		Late Carboniferous, but mostly Permian–Triassic
Termites	100		Mid-Cretaceous
Mantises	40		Mid-Tertiary
Orthopterans (grasshoppers)	300–100		Carboniferous, but mostly Triassic–Cretaceous
Diptera	150		Jurassic, but mostly Cretaceous–Tertiary
Mosquitoes	65		Earliest Tertiary, important disease vectors
Hymenoptera	150		Jurassic, but mainly Cretaceous–Tertiary

Table 13.2 Continued

Group	First fossil evidence[1] (mya)	Molecular evidence (mya)	Comments
Arthropods (Cont.)			
Insects (Cont.)			
Honeybees	70		Late Cretaceous
Ants	150		Mid-Jurassic
Wasps, parasitoid wasps	130		Early Cretaceous
Lepidoptera	150		Jurassic, but mostly mid-Tertiary (see plants)
Brachiopods	570		Cambrian, peak diversity in Ordovician, relicts survive
Bryozoans	500–440		Ordovician
Echinoderms	570		Many now-extinct classes in Cambrian and Ordovician
Sea lilies (crinoids)	570		Cambrian, mostly Ordovician
Most surviving classes	500–440		Ordovician (sea stars, cucumbers, urchins, brittle stars)
Vertebrates	500–440		Ordovician
Agnatha (jawless fishes)	500–400	564[9]	Ordovician, mostly Silurian
Placoderms	400–350		Devonian
Chondrichthyes	400–350	538[9]	Devonian
Sharks	190–135		Jurassic
Rays	190–65		Jurassic, mostly Cretaceous
Osteichthyes	440–280	450[9]	Silurian, mostly Devonian and Carboniferous
Teleosts	230–65		Triassic, mostly Jurassic and Cretaceous
Most modern fish	65–40		Early–mid-Tertiary (carp, cod, guppies, cichlids)
Sarcopterygians (lungfishes)	400–350		Devonian
Amphibians	360	360[9]	Late Devonian
Frogs, salamanders	190–135	197[9]	Jurassic
'Reptiles'	350–280		Carboniferous
Turtles	190–65		Jurassic, mostly Cretaceous
Lizards and snakes	135–65		Cretaceous, many first in Tertiary
Crocodiles	230–65		Triassic–Cretaceous, depending on group
Crocodile–alligator split[8]		75–55[9]	Immunological distances from albumin proteins
Dinosaurs	230–190		Triassic, vanish in end-Cretaceous mass extinction
Birds	155		Late Jurassic
Loons	135–65		Cretaceous
Herons, falcons, anhingas,	65–60		Early Tertiary
Avocets, parrots, rollers, many	50–30		Mid-Tertiary
Mammals	230–190		Triassic
Marsupials	90–60	173[9]	Triassic–early Tertiary
Placentals	65–60	129[9]	Cretaceous–early Tertiary
Rabbits	65–60	91[9]	Cretaceous–early Tertiary

Table 13.2 Continued

Group	First fossil evidence[1] (mya)	Molecular evidence (mya)	Comments
Vertebrates (Cont.)			
Mammals (Cont.)			
Placentals (Cont.)			
Whales	65–60	58[9]	Early Tertiary
Rodents	65–60	66[9]	End-Cretaceous–early Tertiary
Ungulates, carnivores	65–60	83–74[9]	End-Cretaceous
Bats, dogs, weasels, elephants	40–30	–	Mid-Tertiary

[1] Benton (1993), [2] Doolittle *et al.* (1996), [3] Labandeira (1998*b*), [4] Labandeira (1998*a*), [5] Savard *et al.* (1994), [6] Martin *et al.* (1993), [7] Laroche *et al.* (1995), [8] Hass *et al.* (1992), [9] Kumar and Hedges (1998). mya = million years ago

islands away, the conduits into the mantle were broken, the volcanoes stopped erupting, and their weight caused the crust to bend downwards, sinking the islands below the ocean surface. The Hawaiian shield volcanoes, larger in volume and higher from sea floor to summit than any other mountains on the planet, sink beneath the waves in 10–20 million years.

whose pattern constrains the history of neighboring continents.

The Hawaiian islands and their underwater continuation, the Emperor Seamounts, stretch 6500 km from the middle of the North Pacific into the subduction zone of the Kamchatka Trench. The oldest of the Emperor Seamounts, now about to be subducted, are about 300 million years old. Thus there have been islands breaking the surface near the present location of Hawaii at least since the Carboniferous. This constrains the potential motion of the terranes that have moved from the western and southern Pacific into Central and North America. About 150 million years ago there was a major change in the direction of motion of the plate, from north–north-west to west–north-west, recorded in a change in the direction of the Emperor Seamounts north-west of Midway Island, which coincided with an event at the plate's eastern edge: the docking onto California of the terrane containing the gold that formed the mother lode of the Sierra Nevada.

Comment

The dynamic geological theater has profoundly affected evolution.

During the course of evolution, the surface of this planet has been extensively remodeled by continental drift and volcanism. The continents have had a variety of configurations. The level of the sea relative to the mean continental elevation has varied by several hundred meters. Island archipelagos have been repeatedly formed, and some of them have vanished, either by accretion onto continental margins or by sinking beneath the waves. During much of the Earth's history, the climate was warmer than it is today, but there were also cold periods with continental glaciers in the late Proterozoic, the Carboniferous, the Permian, and the Quaternary, including the present. Because we live in one of the few colder periods that have been separated by much longer periods of warmer temperatures, our impression of the climate of the globe is not representative of most of its history. These global changes in the geological theater have profoundly influenced the course of evolution. In the next section we discuss some geological catastrophes that have had major impacts on local regions, if not on the planet as a whole.

Local geological catastrophes

When the Mediterranean was a desert

The Mediterranean was last dry about 18 mya.

The Mediterranean has probably been dry several times, the last time about 18 mya. This was discovered by ocean-bottom drilling that revealed salt deposits in the eastern Mediterranean which could only have formed if the entire sea had dried up. The mechanism requires that the Atlantic drop below the level of the rim of the basin at Gibraltar. Once it is cut off from the Atlantic, the Mediterranean steadily evaporates, for the flow of rivers into it is not enough to compensate the evaporative loss, and it dries up in about 15 000 years at a rate of 30–35 cm/year. When it was dry, the Rhone, the Nile, and the rivers flowing into the Black Sea (the Danube, Dnieper, and Don) cut deep canyons into its flanks, canyons that can still be traced under water. Animals and plants could then disperse directly onto what are now islands—Cyprus, Crete, Sicily, Sardinia, Corsica, the Balaerics. Then the Atlantic rose again, and the world's largest waterfall was formed as the sea surged in through the Straits of Hercules. It took about 750 years to fill the basin at a rate of about 1 cm/day.

Gigantic volcanic eruptions

Major volcanic outbreaks are regional catastrophes with impact on global climate.

Some past volcanic outbreaks dwarf the major eruptions of historic times, like Krakatoa, in Indonesia, which sent tsunamis into Japan and the west coast of North America. When the Phlegrean Fields, a suburb of Naples, erupted, they sent ash as far as the Ukraine. Oligocene and Miocene eruptions from volcanoes in the Cascades and Sierra Nevada of Oregon and California sent ash as far east as Nebraska, 2000 km away, burying herds of large mammals. About 1600 BC the eruption of Santorini in the Cyclades sent falling ash, and the accompanying earthquakes sent giant waves, into the Greek islands and Crete, destroying many Mycenaean settlements and probably weakening the Cretan civilization, which fell not long after to Greek invaders.

Not all giant eruptions have been explosive; there have also been episodes of massive volcanic flooding. During the middle Tertiary, 30–15 mya, lava poured from cracks kilometers long in the Columbia Plateau of Washington and Oregon and in the Ethiopean highlands, sterilizing hundreds of thousands of square kilometers with lava flows. Other flood basalts are found from north–central USA (Proterozoic), central Siberia (Triassic), the Karroo of South Africa (Jurassic), and the Paraná plateau of Brazil and Uruguay and the Deccan plateau of India (both Cretaceous). Individual flows were typically 10–30 m thick, the area flooded was up to 1 million km^2, and the volume of lava in such an area could exceed half a million km^3, or 50 times the volume of the largest terrestrial shield volcanoes (Mohr 1983).

Giant floods and waves

Evolution has not occurred in a stable, predictable, environment.

During the Pleistocene glaciations, floods repeatedly swept across the state of Washington. When the continental glaciers grew, as they did at least seven times, an arm of the Columbia Ice Field moved south to dam the Clark Fork of the Columbia River, near what is now Pend Oreille Lake in Idaho. The dam formed glacial Lake Missoula. As the water rose, it floated the ice that dammed it and swept out in a massive flood, carrying large icebergs with it, forming the Channeled Scablands of eastern Washington. Blocked by the Cascades, the flood could only drain through the Columbia River Gorge, where it was hundreds of meters deep. Blocked again between the Cascades and the coast, the water backed up in the Willamette Valley to a depth of 150 m over an area of several thousand km^2. Boulders lodged in the icebergs carried by the flood are scattered down the valley. During the last flood, there were probably native Americans living in the area who witnessed it. Those on high ground might have survived.

Very large waves can be generated by earthquakes, submarine landslides, and meteorite impacts. After the 1964 Anchorage earthquake a local wave confined to a coastal Alaskan fjord destroyed a structure more than 200 m above sea level. A much larger wave went over the top of the Island of Lanai, in Hawaii, about 18 000 years ago, leaving sea water perched at more than 1000 m elevation. The cause was probably a massive landslide with a volume of tens of cubic kilometers, such as those found on the ocean floor at the feet of the Hawaiian shield volcanoes. Huge undersea landslides also occur where unstable sediments are deposited at the mouths of rivers. Traces of such events have been found in the Mediterranean, off the Amazon, off the Carolinas, and off Norway; some moved 500 km^3 of material (Nisbet and Piper 1998). When a meteorite impacted in the eastern South Pacific between Chile and Antarctica 2.15 mya, it sent a tsunami up the western coast of South America that reached about 1 kilometer above sea level. The tsunami created by the end-Cretaceous meteorite impact, discussed below, was much larger.

This selection of dramatic local events in the history of the planet shows that evolution has not occurred in a stable, predictable environment. From time to time catastrophes occurred that could not be countered by adaptations produced by normal natural selection. They wiped out entire communities, some local, some larger. Next we discuss the largest, the mass extinctions.

The mass extinctions: when, who, and how

Overview

The Phanerozoic (570 mya–present) has been marked by a series of mass extinctions. It is no accident that they fall at the end of geological eras, for those eras are recognized as strata bearing characteristic fossils, and when species disappeared and were replaced by new forms in overlying strata, it was natural to invent a new name for the period with the new assemblage. There may have been 20 mass extinctions in the Phanerozoic, five of which are well documented—those at the end of the Ordovician, Devonian, Permian, Triassic, and Cretaceous. Three were truly massive.

There have been at least five mass extinctions in the last 570 million years.

The end-Ordovician extinction

At end of the Ordovician, about 440 mya, 22% of the families and nearly 60% of the genera of marine invertebrates vanished, including many trilobites, brachiopods, graptolites, echinoderms, and corals. The extinctions came in two waves, one about 10 million years before the end of the era, the other at the end. The mechanism is not clear. There is evidence of a glacial maximum associated with a drop in sea level that exposed the continental shelves, followed by deep flooding of the shelves with fresh water from melting ice caps and continental glaciers.

At the end of the Ordovician 60% of marine invertebrate genera vanished, mechanism unknown.

The end-Permian extinction

The extinction at the end of the Permian 280 mya was the largest ever. In it vanished about 50% of the families, more than 80% of the genera, and more than 90% of the species of marine invertebrates, including all the trilobites, all the tabulate and rugose corals, about 70% of the brachiopod families, 65% of the bryozoan families, and 47% of the cephalopod families, among them many ammonites. Its intensity—numbers of genera disappearing per unit time—was the greatest ever. The event was massive and brief, and the mechanism is not yet settled, but a consensus is starting to emerge (Knoll *et al.* 1996).

At the end of the Permian 80% of marine invertebrate genera vanished, including the trilobites and rugose corals.

At that time the continents had just accreted into a single mass, Pangaea, with northern China and Siberia finishing the process shortly before the end of the Permian. Continental ice sheets formed and retreated at least four times. When there were no continental glaciers, ocean circulation patterns probably could not supply enough oxygen to the deep basins and, except for a thin surface layer, the oceans became anoxic. Surface photosynthesis continued to draw carbon dioxide from the atmosphere and export it to the anoxic sediments. Deep water became enriched in carbon dioxide and hydrogen sulfide, and the greenhouse effect of carbon dioxide in the atmosphere was reduced, accelerating any cooling trend. When continental glaciers formed at high latitudes, their cooling induced vigorous circulation in the oceans, bringing deep water to the surface, poisoning the surface waters, and

The mechanism may have been global poisoning.

releasing vast quantities of carbon dioxide into the atmosphere, which raised the temperature, contributed to the melting of the glaciers, and started the cycle over again. It continued until Pangaea broke up and the drifting continents altered the ocean circulation patterns.

Speaking for this hypothesis are both the chemistry of the sediments and the selectivity of the extinctions. Terrestrial animals can compensate better for high carbon dioxide levels than marine animals, and marine animals with active circulation and gills compensate better than those without them. Animals forming carbonate skeletons should be particularly affected because of their sensitivity to disturbances of their internal acid–base balance. In fact, the groups most strongly affected were corals, articulate brachiopods, bryozoans, and echinoderms, all of which have carbonate skeletons, weak circulation, and low metabolic rates: they lost 65%, 67%, and 81% of extant genera in three waves of extinction. The mollusks, arthropods, and chordates, with gills, circulatory systems, and higher metabolic rates, lost 49%, 38%, and 38% of extant genera in each of those extinction waves.

Details are missing. Meteorite impacts, massive volcanism, and continental drift may all have contributed to the chemistry of the Late Permian oceans and atmosphere. That chemistry was certainly strange. The immediate cause of the most massive extinction in Earth history was probably poisoning.

The end-Cretaceous extinction

At the end of the Cretaceous 50% of marine invertebrate genera disappeared, including all the remaining ammonites and most of the dinosaurs.

In the mass extinction at the end of the Cretaceous, about 50% of extant genera disappeared. All marine invertebrates were affected. Prominent victims were foraminiferans, bivalves, bryozoans, all the remaining ammonites, gastropods, sponges, echinoderms, and ostracods. The dinosaurs appear to have been declining before the final extinction, and there are a few dinosaur fossils from the earliest Paleocene, so a few dinosaurs may have survived the mass extinction. However, many of the dinosaurs probably did die at the same time as the marine invertebrates.

It appears to have been caused by a meteorite impact that may also have triggered massive volcanism.

The end-Cretaceous extinction also took place under unusual conditions. Widely scattered soot deposits suggest fires on a hemispheric scale, as do spores from ferns that would have invaded habitats cleared by fire. Right at the Cretaceous–Paleocene boundary there is an enrichment of iridium, an element found at low concentrations on Earth and in higher concentrations in meteorites. The leading hypothesis is the impact of a meteorite 10 km in diameter that formed a crater 180 km in diameter. The ejected debris would have ignited hemispheric fires, and the impact would have injected so much dust into the atmosphere, both directly and through volcanic eruptions induced by the impact, that the Earth would have been dark and cold for several years. That most seed plants survived suggests that the period of deepest crisis did not exceed the time that seeds can survive in the soil, which is about a decade. The time scales for destruction and recovery are given in Table 13.3. Both took a long time.

Table 13.3 The end-Cretaceous impact: destruction and recovery (from Conway Morris 1998b)

Time	Effect
1 s	Annihilation around impact site (*c.* 30 000 km^2)
1 min	Earthquakes, Richter scale 10
10 min	Spontaneous ignition of North American forests
60 min	Impact ejecta cross North America
10 h	Tsunamis swamp the Tethyan coastal margins
1 week	First extinctions
9 months	Dust clouds begin to clear
10 years	Severe climatic disturbance (cooling) ends
1000 years	Continental vegetation recovers; end of 'Fern Spike'
1500 years	Deeper-water benthic ecosystems start to recover
7000	Full recovery of benthic ecosystems
70 000 years	Ocean anoxia diminishes
100 000 years	Final extinction of dinosaurs (?)
300 000 years	Final extinction of ammonites (?)
500 000 years	Ocean ecosystems start to stabilize
1 000 000 years	Open ocean ecosystems partly recovered
2 000 000 years	Marine mollusk faunas mostly recovered
2 500 000 years	Global ecosystems normal

The leading candidate for the impact crater is Chicxulub on the Yucatan Peninsula of Mexico; it dates precisely to the end-Cretaceous boundary. Shock quartz and tektite ejecta from that crater have been found both across the Caribbean and farther afield. More than 100 km inland in Texas are traces of waves more than a kilometer high that could have been caused by a meteorite strike in the Yucatan. Flood basalts in India may have been caused by the focusing of the energy of the collision through the spherical lens of the Earth to a point in the crust on the opposite side of the globe.

The meteorite hypothesis is not universally accepted. The major alternative is massive volcanic eruptions, which could also have been caused by such an impact. We know that a large meteorite did smash into the Yucatan at the right time, that its impact disturbed the entire globe, and that the disturbance lasted a long time. We also know that mass extinctions can occur without a meteorite impact, as at the end of the Permian, and that large impacts have occurred—the Montagnais impact structure, 45 km in diameter and 51 million years old, and a gigantic crater in the Kalahari, possibly 350 km in diameter and 145 million years old—without causing a mass extinction.

Comment

Mass extinctions repeatedly changed the course of evolution.

Mass extinctions repeatedly changed the course of evolution. It was probably not an accident that the reptiles radiated in the Triassic after the end-Permian mass extinction had eliminated many other groups, and that the mammals radiated in the Paleocene after the end-Cretaceous mass extinction had eliminated most of the reptiles. Molecular evidence indicates that the mammalian radiation started well before the end of the Cretaceous, but what form it might have taken if the dinosaurs had survived, we cannot say.

Patterns of stasis, speciation, and morphological change

The punctuated equilibrium hypothesis

Punctuated equilibrium: long periods without change broken by brief periods of rapid change.

In 1972 Eldredge and Gould published their hypothesis of **punctuated equilibrium**. They observed that in the history of many fossil lineages, long periods without change—called **stasis**—were broken by brief periods of change so rapid that it could not be observed in the fossils, and that these brief periods of rapid change were associated with apparent speciation events. They extrapolated these observations to infer that most morphological change occurred during speciation events, that during most of their lifetime (several million years) most species did not change very much, and that most evolutionary change was thus **cladogenetic** (occurring during speciation events) rather than **anagenetic** (occurring within a species). Their claims contradicted what they portrayed as the dominant neo-Darwinian orthodoxy, which viewed evolutionary change as gradual and continuous, and set off a controversy that is still being resolved. Its positive effect was to reinvigorate paleontology by showing that paleontology revealed patterns not predicted from microevolutionary processes, that paleontology had something unique to contribute. Its negative effect was to exaggerate the differences between neontologists and paleontologists and inhibit communication through the creation of opposing camps. This negative effect is now abating.

Stasis is real. Some lineages show punctuated change, others do not.

The intense scrutiny of the hypothesis of punctuated equilibrium since 1972 has led to the following progress report. Many characters, and some species, are static for long periods, some of them lasting tens to hundreds of millions of years. The fossil record of many lineages is also marked by brief periods of major change in which many new species appear and a great deal of morphological change occurs. Not all traits and lineages show this pattern of stasis and punctuation, and only in some lineages does most morphological change occur during or soon after speciation events. Some groups, such as Pleistocene corals in New Guinea and bryozoans in tropical America, display stasis over several million years broken by near simultaneous speciation events associated with major climatic change. Some groups, such as mollusks in Lake Victoria, display morphological change during apparent speciation events and stasis between speciation events. Jackson and Cheetham (1999) found that 29 of 31 species with well-documented fossil histories displayed patterns like the Lake Victoria mollusks—punctuated morphological change associated with cladogenesis. Other groups, such as rodents, display as much morphological change between as during speciation events.

Thus, while it does not describe a universal pattern, the punctuated equilibrium hypothesis is supported by enough evidence to require serious attention and has usefully highlighted two important patterns that often occur in the fossil record, stasis and brief periods of rapid change.

Stasis

Two classes of explanation for stasis in the fossil record are logical and not mutually exclusive: intrinsic explanations, which see stasis as the result of internal constraints on evolutionary change, and extrinsic explanations, which see stasis as the result of lack of change in the environment and therefore in selection pressures. Stasis cannot result from neutral evolution, which would produce drift in character states.

How internal constraints might function to restrict further evolution is not clear. Some traits change more easily than others, but it is not clear how one trait might change rapidly for a brief period, then not change at all for a long time. If an intrinsic constraint is to explain such a pattern, then the intrinsic constraint itself must rapidly evolve to stop further change.

Is stasis caused by intrinsic constraints or . . .

Extrinsic explanations are easier to imagine, which does not necessarily make them correct. For example, a marine invertebrate with a larval stage that disperses widely but only settles in a very restricted habitat can track the environment, finding places where adults can survive and reproduce, and greatly reduce the impact of the global changes that paleontologists infer from the climatic history recorded in the sediments. The larvae choose selection pressures for the adults. If they do so efficiently, and if the habitat that they choose occurs continuously over long periods of time within the geographic range of the dispersing larvae, then the selection pressures on the adult organisms, which is the stage usually fossilized, will remain constant over long periods. If that selection is stabilizing, for intermediate trait values, the result will be stasis. A shift in larval settling preference that exposed the population to new habitats and new selection pressures could then trigger rapid change. Such a shift might result from the disappearance of old and the appearance of new habitats, which could be correlated with climatic change.

by habitat selection and stabilizing selection?

Sudden bursts of change

Sudden change can also be explained by hypotheses that rely on intrinsic or extrinsic mechanisms. One intrinsic explanation involves the gradual evolution of a new state in a key trait which, once in place, sets of a sudden cascade of adjustments in other traits. Such key traits include air breathing (followed by the radiation of the amphibians and insects) and homeothermy (followed by the radiation of the mammals and birds).

Are sudden bursts of change caused by the emergence of key innovations or . . .

are they driven by external forces?

One extrinsic explanation involves sudden climatic change, exposing organisms to new habitats and new physical and biotic environments. Another extrinsic explanation involves removal of a group that had previously functioned as a competitor or predator and prevented speciation and morphological change. An example mentioned above is the hypothesis, unfortunately not testable, that Mesozoic reptiles inhibited the evolution of the already-extant mammals, and that the rapid mammalian radiation at the start of the Cenozoic was only possible after the disappearance of the dominant reptiles, including the dinosaurs, in the end-Cretaceous mass extinction.

An interaction between intrinsic and extrinsic causes of stasis and punctuation is also plausible.

The Cambrian explosion

The major multicellular animal groups emerged suddenly in the Cambrian.

One of the most dramatic sudden bursts of change in Earth history was the sudden appearance in the early Cambrian of the first fossils of many of the major animal phyla, including the mollusks, annelids, arthropods, brachiopods, and echinoderms (see Table 13.1). Prior to that fossils were mostly small and soft-bodied; after that fairly large organisms, up to a meter long, with hard body parts have been found in the communities that have been well preserved. The molecular evidence, particularly the deep molecular homologies of developmental genes controlling the formation of eyes, brains, and axial patterning (Chapter 15), suggests that at least some animals had acquired the complex body plans that typify the major phyla well before Cambrian fossils recorded that fact in the form of hard body parts. Thus in the Cambrian some extant lineages rapidly evolved armor and defensive structures. Other lineages probably did originate in the early Cambrian and rapidly evolved hard body parts. Both processes would account for the sudden appearance of fossils.

Thus the shallow marine communities of the early Cambrian were the scene of dramatic evolutionary developments. Predators became more efficient and selected for defensive structures in their prey, and both predators and prey evolved larger body sizes. The diversity of fossilized body plans increased impressively and rapidly. Once the vertebrates had appeared in the Ordovician, all the major body plans of animals that now dominate the planet had evolved.

A Cambrian community is preserved in the Burgess Shale.

The most famous fossil Cambrian community is the Burgess Shale in British Columbia, Canada; another deposit with superb preservation occurs at Chengjiang County in Yunnan Province, China. These deposits document a strange world. Some of its creatures would be familiar (particularly some arthropods that have changed little since then) and some looked like no living organism (including some echinoderms, some arthropods, and some strange creatures that have left no descendants). The early Cambrian abounds with puzzles and has generated some high-quality popular science; Conway Morris (1998*a*) and Gould (1990) offer strikingly different interpretations.

The Silurian–Devonian radiation of vascular land plants

The major groups of vascular land plants appeared later and somewhat more gradually in the Silurian (438–410 mya) and Devonian (410–355 mya). There is a gap in plant fossils in the Silurian, and clocklike molecules either change too slowly to capture key events, or change so rapidly that they have long since become oversaturated with mutations that erased the relevant historical traces. For those reasons we do not know whether the major groups of plants all emerged within a period of 100 million years or in 35–50 million years. In either case, their appearance was rapid compared to the rest of their history and a bit slower than the appearance of the major groups of multicellular animals (Bateman *et al.* 1998).

The vascular land plants radiated 438–355 mya.

Summary

This chapter reviews major events in the history of the planet and of life.

- The different parts of familiar organisms had evolutionary origins of vastly different ages, as did the landscapes on which they occur. Both organisms and landscapes are mosaics of modern and ancient forms.
- The chemical elements themselves formed at different times in the history of the universe. Some heavy metals involved in essential biochemical reactions were formed in supernovas; they are recycled starstuff.
- The geological theater in which life evolved formed about 4500 mya; continents formed about 4000 mya and have been drifting since then. Their motions create rifts, mountain ranges, and volcanoes, and are driven by upwelling at the mid-ocean ridges. Their configuration determines the pattern of ocean circulation and affects the global pattern of precipitation.
- Fossils from the few deposits in which nearly complete communities are preserved tell us that ancient biogeography could be surprisingly different from modern distributions.
- When continents drift rapidly, the inflation of the mid-ocean ridges can flood the continents with shallow seas; when they drift slowly, the continental seas recede and huge expanses of shallow marine habitat disappear. Modern biogeographical patterns bear the traces of post-Permian continental drift.
- Geological catastrophes have occasionally reshaped local environments and caused extinctions. These include the drying and refilling of the Mediterranean, gigantic volcanic explosions and floods, and giant floods and waves produced by glacial dams and meteorite impacts.
- The history of life has also been marked by at least five mass extinctions at the end of the Ordovician, Devonian, Permian, Triassic, and

Cretaceous. The end-Permian extinction was the largest; it appears to have been caused by poisoning of the world ocean; in it, the last trilobites disappeared. The end-Cretaceous extinction was the second largest; it appears to have been caused by a large meteorite impact; in it, the last ammonites and dinosaurs disappeared.

- Most of the major groups of multicellular animals appear as fossils in a relatively brief period of the Cambrian, apparently as a consequence of predator–prey coevolution in the sea. The major groups of land plants radiated in the Silurian and Devonian as a consequence of the invasion of land.
- The fossil record of many, but not all, groups is characterized by long periods of stasis interrupted by brief bursts of rapid change. For both stasis and bursts of change, there are intrinsic, constraint-oriented and extrinsic, environment-oriented explanations. The truth may lie in interactions between the two.

While the planet was changing and major groups were originating and going extinct, the basic nature of the evolving units was itself changing. Those key events are the topic of the next chapter.

Recommended reading

Carroll, R. L. (1997). *Patterns and processes of vertebrate evolution.* Cambridge University Press, Cambridge.

Conway Morris, S. (1998). *The crucible of creation: The Burgess Shale and the rise of animals.* Oxford University Press, Oxford.

Gould, S. J. (1990). *Wonderful life: The Burgess Shale and the nature of history.* W. W. Norton & Co., New York.

McPhee, J. (1998). *Annals of the former world.* Farrar Straus and Giroux, New York.

Raup, D. M. and Jablonski, D. (ed.) (1986). *Patterns and processes in the history of life.* Dahlem Konferenzen. Springer-Verlag, Berlin.

Raup, D. M. and Stanley, S. M. (1978). *Principles of paleontology,* (2nd edn). W. H. Freeman, New York.

Schaal, S. and Ziegler, W. (ed.) (1992). *Messel. An insight into the history of life and of the earth.* Oxford University Press, Oxford.

Questions

13.1 Compare the definition of 'major groups' from the point of view of an organism like us who sits at the living tip of a phylogenetic tree, and from the point of view of an organism that sits at its base. From the base, how important do the vertebrates, mammals, or primates seem? From the tip, how important does the diversity of procaryotes or non-seed plants seem?

13.2 Pick a favorite organism and use Tables 13.1 and 13.2 to reconstruct its history.

13.3 The volume of water stored in ice on the planet is roughly 10 million cubic kilometers. The surface of the oceans is roughly 330 million square kilometers. If global warming melted all the ice, how much would the oceans rise (ignoring continental flooding)? Discuss ways to make the calculation more realistic.

Chapter 14
The history of life II: key events in evolution

Introduction

Key events involve the origin of something fundamentally new with large impact on subsequent evolution.

Evolutionary biologists study both the history of life and the mechanisms that drive evolutionary change. Like geologists and astronomers they want to know what happened in the past, and they want to explain history with general principles. Therefore Chapter 13 described some highlights in the history of the planet and of life. Just as the history of the Earth was shaped by key events, key events in evolution shaped the life we observe today. In this chapter we consider the origins of fundamentally new things that had major impacts on subsequent evolution, drawing much of the material from Maynard Smith and Szathmáry (1995). First we make some cautionary remarks.

The importance of a key event can only be judged retrospectively

Evolutionary change does not occur because it produces benefits in the distant future.

According to Jacob (1977), evolution proceeds by tinkering: finding an *ad hoc* solution to an immediate problem. Evolutionary change occurs either because of the short-term advantage of genetic variants in a specific environment, or by chance. No mechanism can cause evolutionary change because of beneficial effects in the distant future. But we judge the importance of key events retrospectively by their long-term consequences. For example, the origin of sex had many long-term consequences, but it did not originate for those reasons, and many consequences could not have been predicted at the time sex originated.

Our perception of important events is biased

What we identify as a key event depends on where we sit in the phylogenetic tree. A bacterium would have a different view.

When thinking about evolution as history, it is tempting to start with our species and proceed backwards through a series of imagined ancestors to simple early forms. That script suggests that evolution has produced ever-increasing complexity and that our species is the culminating point in the evolution of life. When we view evolution from the bottom rather than the top of the tree of life (see Fig. 15.1), the progressive interpretation become questionable. All other existing species have an evolutionary ancestry as old as our own, and some originated later than *Homo sapiens*. We are biased by having much more information on the few lineages that produced plants and animals than on the many lineages that produced the enormous variety of microorganisms. In lesser known parts of the tree, key evolutionary events may have occurred that are unknown to us. This bias explains why most of the evolutionary events discussed in this chapter occurred in the lineages containing multicellular organisms: plants, fungi, and animals.

Our interpretation of key events is no more than constrained speculation

Because present-day traits are heavily modified, they are often not very informative about their origin.

Many key events in evolution occurred long ago. We can infer them from differences between existing organisms and from the fossil record, but what happened and why it occurred must remain speculations. The chance of reconstructing the evolution of novelties is small, for present-day structures have been extensively modified since their origin. Thus what we observe today is not directly informative about origins. Repeating key evolutionary events in experiments is difficult, for we do not know, or cannot create, the appropriate starting conditions. Speculations about the origins of key evolutionary novelties are, however, constrained by physics, chemistry, and biology.

There were a small number of key events.

Some key events in evolution were the following (Maynard Smith and Szathmáry 1995):

- the origin of replicating molecules
- the sequestration of replicating molecules in compartments
- the condensation of independent replicators into chromosomes
- the transition from RNA as gene and enzyme to DNA with its genetic code for proteins
- the symbiotic origin of eucaryotes from procaryotes
- the origin of asexual clones from sexual populations
- the origin of multicellularity in plants, animals, and fungi
- the origin of societies with reproductive castes
- the origin of language.

Rather than discussing all these key events, we describe cases that illustrate principles involved in the evolution of novelty, then discuss the principles.

The origin of life

Living things have metabolism and hereditary replication.

The most fundamental and perhaps the most intriguing key event was the origin of life. To be clear about its origin, we first have to define life. As for many other concepts in biology, a precise definition is difficult. A rough definition characterizes life by two of its essential features: a living thing should have metabolism (a coordinated system of chemical reactions contributing to its maintenance) and hereditary replication (a system of copying in which the new structure resembles the old).

Prebiotic chemical evolution produced a diverse array of organic compounds.

In the 1920s Oparin and Haldane independently suggested that in aqueous solutions under a reducing atmosphere with energy supplied by lightning or ultraviolet radiation (the 'primitive soup'), a variety of organic molecules, would be synthesized. In a famous experiment Miller (1953) mimicked these conditions. Using a gas mixture of methane (CH_4), ammonia (NH_3), and hydrogen (H_2), simulating lightning by electric discharges between two electrodes, and starting with a water solution containing some simple inorganic molecules, he found that several biologically important organic compounds had formed, notably amino acids. Such experiments have been repeated with similar results, including the production of nucleotide precursors. They demonstrate the possibility of chemical evolution of a diverse prebiotic chemical environment. However, many biologically significant compounds have never been obtained in such prebiotic synthesis experiments, notably nucleic acids, the chemical basis for heredity in all existing organisms.

But nucleic acid macromolecules have not yet been synthesized *de novo*.

More recently it has been suggested that prebiotic chemically diverse environments may have evolved not in aqueous solutions, but on the surfaces of pyrite crystals. Another alternative to the primitive soup is chemical evolution on the surface of droplets in clouds. Both possibilities can be defended on theoretical grounds.

Irrespective of the precise conditions of prebiotic chemical evolution, a collection of organic compounds does not imply life. In particular, we need to understand the origin of hereditary replicators in that chemical environment. Most experts think that RNA is a good candidate for the primitive hereditary replicator, for two important reasons. RNA can act as a template for replication, and it also can function to some extent as an enzyme, assisting the replication process. However, we do not yet know how RNA could be formed in the prebiotic environment.

And Eigen's paradox has not yet been solved: large genomes are only possible with replication enzymes, but replication enzymes require large genomes.

Even if we could find a plausible scenario for the evolution of simple replicators like RNA, the next problem is Eigen's paradox. Eigen (1971) noted a problem with the accuracy of replication. Suppose that the replicator is a polymeric molecule like RNA, a chain of nucleotides, and that a particular molecule is optimal. During replication, occasional errors will occur, generating a family of molecules. This family will contain some molecules identical to the original (replicated without errors) and a collection of molecules similar but not identical to the optimal molecule. Selection favoring the optimal type (perhaps because it permits faster replication) should counteract the accumulation of nonoptimal types. If errors in replication occur with a constant probability per subunit, then long molecules are replicated with more mistakes than short molecules. It can be shown that a critical size of a replicator molecule exists (the so called error threshold), above which larger molecules cannot be maintained because they are replicated with too many mistakes.

Eigen's paradox then follows. Nonenzymatic replication has low accuracy, so that only small molecules can be maintained. For accurate replication enzymes are needed, but the primitive genomes were too small to code for them. This is the catch-22 of prebiotic evolution: large genomes are only possible with replication enzymes, but replication enzymes require large genomes.

Theoretical solutions of Eigen's paradox have been proposed (see Maynard Smith and Szathmáry 1995). None are fully satisfactory.

The evolution of chromosomes

Chromosomal linkage makes possible synchronous replication and fair segregation.

In all known organisms genes are linked on chromosomes, but early in evolution there must have been a transition from unlinked to linked genes. Chromosomal linkage has two important consequences. Genes no longer replicate autonomously but in synchrony, and a type of cell division is possible that produces daughter cells with copies of all genes.

It is easier to control a few chromosomes than many unlinked genes.

Synchronous replication of genes reduces the scope for competition between genes within cells. If genes were not linked, synchronous replication would be much harder to achieve and control, particularly with thousands of genes. Although a similar argument applies to the chromosomes, which must also replicate in synchrony, the smaller number of chromosomes makes control of replication feasible. Interestingly, many species contain chromosomes called B chromosomes that escape replication control and accumulate through successive divisions. B chromosomes are not transcribed and do not contain information vital to the organism; they are genomic parasites. Excessive accumulation of B chromosomes reduces individual fitness and is countered by individual natural selection. B chromosomes show that replication competition within a cell can be a real danger.

Transposable elements escape from replication control on chromosomes.

The evolution of chromosomes does not, however, protect completely against mutant nuclear genes that escape strict replication control, for transposons ('jumping genes') inhabit the genomes of both procaryotes and eucaryotes. Transposons are either a gene, a small group of linked genes, or a stretch of DNA which is not transcribed. They can move to new positions on the same or on a different chromosome. Transposition often involves replication of the transposon, leaving a copy behind at the original site and increasing the number of transposons within the genome. Because transposition can reduce individual fitness—it induces mutations—several mechanisms have evolved to suppress it. Transposable elements seem to be present in all cells, making up about 10% of the *Drosophila* and more than 33% of the human genome. Transposons illustrate genomic conflict (Chapter 10) between selection favoring mutants that increase the replication rate of transposons and selection favoring the suppression of transposons through stronger replication control. Such conflicts arise as soon as a new level of replication originates, in this case linked genes on chromosomes.

The second consequence of linkage is that it contributes to orderly

Fair segregation inhibits genomic conflicts, and linkage of genes on chromosomes helps ensure fair segregation.

segregation of genes at cell division. Each daughter cell must get the full complement of genes through a mechanism ensuring chromosome duplication at cell division and proper segregation of the chromosomal copies into the daughter cells. Even a primitive chromosome would be better in this respect than a collection of unlinked genes. Fair segregation of genes at cell division is crucial in preventing the detrimental effects of genomic conflicts (Chapter 10). Linkage of genes on chromosomes helps to suppress such conflict but does not completely prevent it, for some genes can distort segregation.

The origin of multicellularity

Experiments suggest that multicellularity may originate to escape predators.

Boraas *et al.* (1998) studied the evolution of multicellularity in a model system. They inoculated chemostat cultures of the unicellular green alga *Chlorella vulgaris* with a predator. Within a few generations globular clusters of tens to hundreds of *Chlorella* cells appeared. After about 20 generations most clusters were eight-celled, and within 100 generations eight-celled colonies dominated the cultures. These colonies replicated their characteristic eight-celled form stably, both in continuous culture and when plated on solid medium. They were too large to be eaten by the predator but small enough for each cell to be in direct contact with the nutrient medium. In the absence of the predator, *Chlorella* did not evolve eight-celled colonies and remained unicellular.

This experiment shows that a simple form of multicellularity easily evolves in response to predation. It also suggests that large multicellular forms are only viable if structures exist to transport the substances needed by every cell, for simple diffusion from the medium into every cell is no longer sufficient. Because the eight-celled *Chlorella* form evolved so quickly, the transition from single celled to eight-celled probably required only a few mutations. There was no sign of cellular differentiation: all eight cells were equivalent.

Development in multicellular organisms starts from a single cell, which suppresses genetic conflict by guaranteeing maximal relatedness among the cells of an individual.

Even in these simple multicellular forms the cells have to cooperate as a functional unit. A cell that continued to divide would increase its representation in the colony but would lower colony performance: genomic conflict within such a colony is possible. However, conflict is unlikely because all eight cells descend from a single cell and are genetically identical except for mutations occurring during divisions from the single to the eight-celled stage. Genomic conflict between genetically identical cells is, by definition, impossible: any advantage or disadvantage to one of the cells is automatically an advantage or disadvantage to the others, because fitness gains or losses are translated into frequency changes of genes. If the colonies were formed by sticking together eight genetically different cells, there would be some scope for genomic conflict within colonies. Then selfish behavior of one cell could increase the frequency of genes causing such behavior because they are present in that cell line but not others. This may explain why development in multicellular organisms usually starts from a single cell, guaranteeing maximal relatedness between the cells of an individual.

A transition to multicellularity with different specialized cell types requires two things. First, for cells with identical genes to develop into types with different functions, genes expressed in some cell types must be shut off in others. This requires the evolution of mechanisms for gene regulation. In a simple form of gene regulation, a gene product directly influences the activity of its gene; other mechanisms also exist. Genes are regulated in single-celled organisms, which must express genes at different times. Therefore, the first requirement for differentiated multicellularity need not have been difficult.

Multicellularity with specialized cell types requires that different genes be active in each cell type . . .

The second requirement is some form of cellular memory or **epigenetic inheritance** that maintains the differentiated state through cell divisions (see Jablonka and Szathmáry 1995). In simple epigenetic inheritance, the state is transmitted by distributing a regulatory substance to daughter cells at cell division. In eucaryotes the functional state is often inherited through DNA methylation or proteins bound to DNA. If a stretch of DNA containing a gene is methylated or bound to a protein, the gene cannot be expressed. This epigenetic system may have evolved from defense systems found in bacteria. Specific DNA sequences on bacterial chromosomes are methylated to protect them from destruction by the hosts' restriction enzymes, which degrade foreign parasitic DNA, such as a viral genome, when it enters the host cell.

. . . and some form of cellular memory or epigenetic inheritance.

The evolution of reproductive and nonreproductive units: germ line and soma

Many multicellular organisms are organized into a reproductive and a nonreproductive part. In animals the separation between reproductive and nonreproductive cells is nearly absolute and almost always occurs early in development. In plants and fungi differentiation into reproductive and somatic cells also occurs, but it happens later during development and is less absolute, for somatic cells may dedifferentiate into stem cells that can form reproductive cells; the reverse is also possible.

The germ line–soma distinction is common in animals, less common and less absolute in plants and fungi.

Simple multicellular organisms with reproductive and nonreproductive cells occur in the order Volvocales. Among these green algae are several colony-forming genera, including *Volvox*. *Volvox* forms spherical colonies of 500–50 000 cells of two types. Most cells are somatic. A few cells, the gonidia, are much larger than the others and specialized for reproduction. In the absence of a sexual hormone the gonidia reproduce asexually, forming small daughter colonies which are released from the mother colony. In the presence of sexual hormone the gonidia differentiate into sexual structures (Fig. 14.1).

Volvox is a simple organism with a germ line–soma distinction.

The differentiation into germ line and soma is widespread and must have evolved independently in many lineages. What selected this differentiation? We assume that division of labor enhances individual fitness because it allows more efficient conversion of resources into reproductive capacity. Bell (1985) did find that *Volvox* colonies produce more

Volvox colonies produce more offspring than the equivalent number of unicells. Selection at the colony level outweighs selection at the cell level.

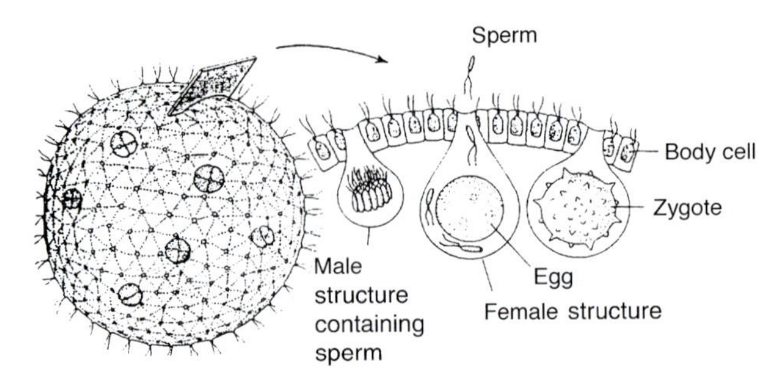

Fig. 14.1 Reproduction in *Volvox*. Left: a few gonidia have produced daughter colonies by asexual division. These daughters are still attached to the mother colony. Right: a cross-section through part of the maternal colony shows sexual reproductive structures formed by gonidia after induction by sexual hormone. Here a monecious (homothallic) species is shown, but other *Volvox* species are diecious (heterothallic), each colony producing only one kind of gamete. (From Gould and Keeton 1996.)

offspring than the equivalent number of unicells. But is their germ line–soma differentiation stable? After all, only gonidia are allowed to form copies of themselves, while somatic cells do not divide and have a cellular fitness of zero. If **altruism** is behavior that benefits others at a cost to oneself, somatic cells are altruistic. They contribute resources from which the reproductive cells profit by producing offspring. However, this is not altruistic behavior in the *genetic* sense because the somatic cells have the same genes as the gonidia and therefore the same genetic fitness. A mutation in a somatic cell that converted it into a reproductive cell would be selected against, for descendant colonies carrying this mutation would have more than the optimal number of gonidium cells. A mutation in a somatic cell causing it to continue cell division (a sort of cancer) would be selected against because it would produce colonies of suboptimal shape and size.

Social insect colonies appear to resemble *Volvox* but differ because their units are not genetically identical. This raises a problem.

In some lineages the differentiation into reproductive and nonreproductive units has occurred at a level higher than the cell. Beautiful examples are provided by the eusocial insects. Honeybee workers normally do not reproduce and perform nonreproductive tasks in the colony, contributing to the reproductive success of the queen. The potential for genomic conflict is greater here than in *Volvox,* where reproductive and nonreproductive units have the same genotype, for worker bees are not genetically identical to the queen. How can we explain this apparent reproductive altruism? We address this important question at the end of the next section.

Principles involved in key evolutionary events

Several general principles appear to be involved in many key evolutionary innovations.

Additional selection at a new level of replication

Units capable of independent replication become part of a larger whole, thereby giving up their independence (Fig. 14.2).

Previously independent units join a larger whole, losing independence.

Units A and B may be as similar as unlinked genes joining to form linked genes on a chromosome, or they may be as different as the independent organisms that merged to form the eucaryotic cell (Chapter 15). Key evolutionary events that resulted in a higher-level regulation of replication include the transition from unicellular to multicellular organisms and the origin of social groups.

The transition to higher replication levels requires higher-level replication structures and coordination and control of replication between levels. For example, genes replicate as parts of chromosomes, chromosomes replicate as part of a cell, cells replicate as part of a multicellular organism, and sexual organisms replicate as part of a sexual population or social group. Such a hierarchical structure allows multilevel selection (see Chapter 10), which can generate genomic conflicts. Thus the problem of controlling genomic conflict often arises in evolutionary transitions to higher replication levels.

Mitochondria and chloroplasts in eucaryotic cells, genes on chromosomes, cells in multicellular organisms, and individuals in societies are hierarchical replication structures with multilevel selection and genomic conflicts.

'Progress' in evolution

The addition of replication levels creates a hierarchical structure. Evolutionary 'progress' might be defined as a tendency to expand replication hierarchies, at least in the lineage that produced multicellular organisms. We should, however, be careful about proposing general evolutionary 'progress'. In lineages that produced unicellular micro-organisms an evolutionary trend to add replication levels is less clear, although different levels of replication do occur. For example, in bacteria replication occurs at the levels of transposable elements, plasmids, the bacterial chromosome, and the cell. Whether developments that increase the potential for conflict can be defined as progress is open to discussion.

Can developments that increase the potential for conflict be defined as progress?

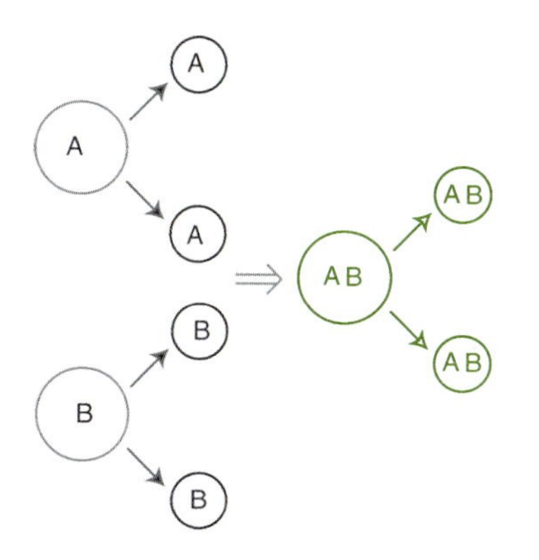

Fig. 14.2 Two independent replicators, A and B, merge into a unit, AB, in which their replication is no longer independent.

Functional specialization (division of labor)

Tasks that were carried out by one unit become distributed over different units (Fig. 14.3).

Tasks previously carried out by one unit become distributed over different units, e.g. cell types in multicellular organisms, anisogamy and male–female differentiation, the germ line–soma distinction, and castes in social insects.

In some sense this is the reverse of the above process. Units A and B have arisen that are independent replicators undergoing separate adaptive evolution. However, A and B are not fully independent because they continue to share a common gene pool and interact, for they have to function within the same individual or population of individuals. Some key origins that involved this process include the differentiation of cell types in multicellular organisms, the origin of anisogamy and male–female differentiation, the evolution of the germ line–soma distinction, and the evolution of castes in social insects (see Chapters 9 and 10). Functional differentiation also creates scope for conflict, because genes good for one differentiated type may be bad for another and because uncontrolled replication of one type will jeopardize cooperation between types.

Change to another system of information transmission

Some key events are marked by a change to a new system of information transmission.

Hereditary information has to be transmitted in some form. Information storage and transmission varies from simple systems with few possibilities, like chemical autocatalysis in which a compound helps to form copies of itself, to complex systems such as genetic coding in nucleic acids and human language, which can transmit an almost infinite number of different messages. Transitions to new systems of information transmission have greatly affected subsequent evolution. Examples include the origin of the genetic code, the origin of epigenetic inheritance, and the origin of human language.

The evolution and stability of cooperation

The evolutionary origins of differentiated multicellular organisms and of groups of social individuals show the importance of cooperation be-

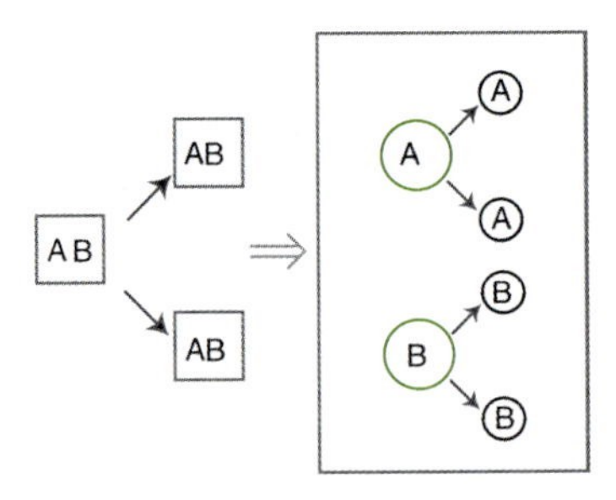

Fig. 14.3 A unit, AB, that carries out two functions differentiates into two units, A and B, each performing one of the functions. The box surrounding A and B indicates that they remain contained within an individual organism or population of organisms.

tween parts of organisms and types of individual. In such systems selfish mutants can 'cheat' by no longer providing resources for the reproductive parts or individuals but reproducing themselves instead. Thus the stabilization of cooperation is a necessary feature of several key events in evolution. The basic idea explaining the evolution of cooperation is that being cooperative results in greater reproductive success than being noncooperative. It can, however, be difficult to see how cooperation yields more fitness than noncooperation. Does a sterile worker bee have higher fitness than she would if she stopped working and became reproductive herself? The key to such questions is the concept of kin selection.

The basic idea: cooperation increases fitness. Cooperation among relatives involves kin selection.

Kin selected altruism

Adaptive evolution occurs when organisms, through differences in their survival and reproduction, contribute different numbers of copies of their genes to future generations. Kin selection occurs when these competing individuals are related to one another. Then the genes that are transmitted are identical with a probability that depends on the degree of relatedness of the individuals that carry them. This insight led Hamilton (1964) to formulate the theory of kin selection. He saw that relatedness has to be taken into account in the evolution of behavioral traits. If you help a relative to increase its reproductive success, you also enhance your own fitness because some of the genes in your relative are identical to your own genes. The closer the relatedness, the more likely that this will be the case. Suppose you carry an allele *A* that induces you to help relatives. This allele will spread in the population if the fitness gained by relatives that also carry the *A* allele more than compensates the fitness you lose as a consequence of your behavior.

Increasing the fitness of a relative enhances personal fitness because some genes in the relative are identical.

The condition under which cooperative behavior is expected to evolve is a simple inequality, known as Hamilton's rule:

Hamilton's rule: altruism pays when $br > c$.

$$b \cdot r > c$$

where b is the fitness benefit to the recipient, r is the relatedness of the recipient to the helper, and c is the fitness cost of the behavior to the helper.

Let us apply Hamilton's rule to germ line–soma differentiation in *Volvox*. The somatic cells are helpers and the gonidia are recipients. The genetic relatedness between the two types of cells is 1.0 (they are identical), and cooperation will evolve if $b > c$, if the fitness benefit is greater than the cost. This is the case if a colony consisting of N cells produces more offspring than a collection of N unicells, as was found.

Hamilton's rule in *Volvox* implies that a colony must produce more offspring than the same number of unicells—as was found.

We can also apply Hamilton's theory to the evolution of sterile workers in social insects. By forgoing reproduction and helping the queen to raise more siblings, sterile workers display reproductive altruism. It is striking that **eusociality** (a social system with nonreproductive workers) sometimes evolved in haplo-diploid organisms. In a haplo-diploid genetic system, females are diploid and develop from fertilized eggs, while males are haploid and develop from unfertilized eggs, and genetic

Kin selection can explain reproductive altruism in social insects.

relationships between relatives are unusual. Because males are haploid, they transmit the same genes to all offspring, which are only daughters. Therefore these daughters all share the half of their genes that came from their father as well as half of the other half that came from their mother. Thus sisters have a relatedness of 0.75. Because the relatedness between a mother and her daughters is 0.50, females are more closely related to their sisters than to their daughters.

Being able to distinguish the sex of siblings makes the evolution of eusociality easier.

At first sight, kin selection theory helps to explain the evolution of sterile workers, for they gain more from helping to raise sisters than from producing daughters themselves. However, the relatedness of a female worker to her brothers is only 0.25, for none of her paternally and only half of her maternally derived genes are in her brothers, so that her average relatedness to her sibs of both sexes is 0.50, as in a normal diploid. Thus, the argument only holds if workers raise more sisters than brothers. If workers could not distinguish the sexes of their siblings, then their relatedness to their sibs would be $r = 0.5$, which they would have with their own offspring if they reproduced. Then raising own offspring would be genetically equivalent to raising siblings, and a worker should forego reproduction to help the queen raise sibs only if by doing so the queen could produce more additional offspring than the worker could produce by herself. When workers can distinguish the sex of siblings and preferentially raise sisters, the evolution of eusociality is easier.

Cooperation and altruism are more likely among relatives.

Thus the theory of kin selection predicts that cooperation between different levels of replicators in a system will be stabilized by high relatedness between replicators. This is probably why natural selection has produced reproductive systems in which offspring start developing from one or a very few cells. Kin selection also predicts that cooperation and altruism is more likely to evolve between related than between unrelated individuals.

Summary

This chapter discusses a small number of key changes that have occurred in evolutionary history with long-lasting effects on subsequent evolution.

- Key events included the origin of a genetic system, the origin of eucaryotic cells, the emergence of sexual reproduction, and the appearance of differentiated multicellular organisms.
- Principles often involved in these events are the addition of a new level of replication, functional specialization, and ways to ensure cooperation between different parts.
- Three important events within the lineage that produced multicellular organisms are the evolution of chromosomes, the evolution of multicellularity, and differentiation into reproductive and nonreproductive parts.
- These events show a tendency to add new replication levels, which

creates the need for stable cooperation and the danger of genomic conflict between levels.

- Genetic relatedness and kin selection are key concepts in the understanding of the evolutionary stabilization of cooperation.

Thus key events can be deduced from the existing organization of life. The next chapter discusses insights into major historical events that can be inferred from molecular data.

Recommended reading

Maynard Smith, J. and Szathmáry, E. (1995). *The major transitions in evolution*. Oxford University Press, Oxford.

Maynard Smith, J. and Szathmáry, E. (1999). *The origins of life: from the birth of life to the origin of language*. Oxford University Press, Oxford.

Questions

14.1 Explain why genomic conflict is relevant for understanding the origin of many evolutionary novelties.

14.2 Chemical communication often occurs between the cells in bacterial colonies, resulting in coordinated behaviors and cellular division of labor and differentiation into distinct cell types. Would you therefore conclude that bacteria can be viewed as multicellular organisms? Is the concept of kin selection relevant here?

14.3 Evolution depends on storage and transmission of information. It has been argued that human language, a relatively recent novel information transmission system, has greatly influenced our evolution and will continue to do so. Can you think of examples that show the role of language in evolution?

Chapter 15
Molecular insights into history

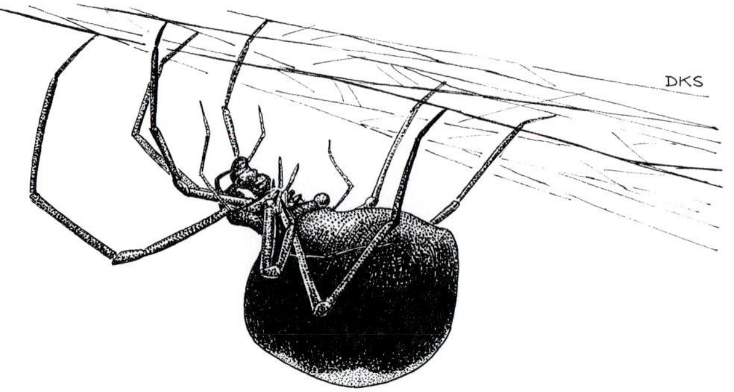

Introduction

New methods are revolutionizing fields, creating new ones, and flooding us with data.

Chapter 13 described some high points in the history of the universe, this planet, and life. Chapter 14 looked at the same history from a different point of view, concentrating on key events marking major transitions in evolution. This chapter describes insights into the history of life made possible by new methods: by large-scale DNA sequencing, by the creation of data banks of DNA sequences, by the invention of computer algorithms that efficiently find similar sequences, and by gene technology that allows gene libraries to be built and the expression of genes to be visualized *in situ*. These methods are revolutionizing fields, creating new ones, and flooding us with data. No other chapter in this book is so likely to be out of date before you read it as this one.

Their impact is illustrated in this chapter, with four classic case studies.

From this flood of results, many of which await confirmation, and of new interpretations, many of which are still controversial, we have selected four case studies that are already classics. They show that the new methods illuminate almost all of evolutionary history, including

(1) the evolution of basic cell structure more than 3400 million years ago;

(2) the evolution of developmental mechanisms that appear to account for the body plans shared by large groups of organisms 1000–500 million years ago;

(3) the diversity of the genes involved in our immune defenses against pathogens, which have a history of about 35 million years;

(4) the genetic diversity of human mitochondria, which has a history of less than 1 million years; and

(5) the human migrations and colonizations that followed the invention of agriculture 10 000–8000 years ago.

The genome stores information about events that occurred at many different times in history, both ancient and recent. The period of history that one can analyze depends on the genes and organisms selected and the kinds of comparisons one makes with them. Each of the four examples listed was addressed with data chosen to allow inferences into that part of the past.

They make the point that the genome stores information about events that occurred at many different times in history, both ancient and recent.

Not all past events are visible to such methods. Molecules that change very slowly can be used to look far into the past, but they cannot record changes that occurred quickly but long ago. Molecules that change rapidly can record recent changes precisely, but they become saturated with mutations too quickly to record changes that happened long ago. Thus we cannot expect the molecules to tell us how fast the Cambrian radiation of the animals or the Silurian–Devonian radiation of the vascular plants happened.

But the genome does not record events that happened rapidly a long time ago.

Deep time: from the first bacteria to the first eucaryotes

The oldest fossils

The earliest known fossils come from rocks in Western Australia about 3465 million years old. Eleven species have been described, seven of which appear to be cyanobacteria (Schopf 1993). Because the temperature of the Earth's surface only fell below the boiling point of water about 3800 million years ago (Chapter 13), the discovery of fossils this old suggests that life originated and complex organisms such as cyanobacteria, capable of photosynthesis, evolved within 400 million years. Considering the complexity of cells and of photosynthesis, that is impressively rapid evolution.

The oldest fossils are cyanobacteria 3465 million years old.

The discovery of such ancient fossils puts the rest of evolution into an interesting perspective. If there were bacteria 3500 million years ago, and multicellular organisms with fossilizable parts only since the start of the Cambrian 550 million years ago, then 85% of the history of life, or about 3000 million years, was dominated by micro-organisms. During the first 400 million years of that period biochemical pathways, membranes, translation, transcription, replication, and recombination evolved and were improved.

85% of the history of life—3000 million years—belonged to the micro-organisms.

The three domains of life: Archaea, Eubacteria, and Eukarya

Now let us compare this fossil evidence to the traces of history contained in the DNA sequences of microbial genomes. In the 1970s, Woese discovered that the DNA sequences coding for ribosomal RNA can be used to determine the structure of extremely ancient relationships. Those relationships included three branches in the deep structure of the tree of life, one of them quite unexpected. These branches do not correspond to

Life appears to have three ancient branches: Eubacteria, Archaea, and Eukarya.

the procaryote–eucaryote dichotomy, nor do they support the notion that there are five kingdoms of life (procaryotes, protists, fungi, plants, and animals: Margulis 1970). Woese and his colleagues found that some organisms that had been thought to be bacteria were so different in their transcriptional and translational machinery, which resembles that of eucaryotes, and in their membranes, which contain tetraether lipids, that they formed a third major group. He therefore proposed that the three basic branches be named Eubacteria, Archaea, and Eukarya (Fig. 15.1), with Archaea designating the new group of bacteria-like organisms (Woese *et al.* 1990). Some Archaea inhabit extreme environments, including thermophiles that can grow at 110 °C and pH 0.0, halophiles with internal salt concentrations of up to 4 M K^+ and 1 M Na^+ (4–5 times the concentration of sea water), and methanogens that can only grow in anoxic, reducing environments, such as deep-sea hydrothermal vents. Other Archaea occur widely in more normal environments, including agricultural and forest soils and well-oxygenated, cold, ocean waters.

The structure of Woese's tree of life implies that the split between Eubacteria, Archaea, and the eucaryotic nuclear genome is extremely ancient. Because cyanobacterial fossils are 3465 million years old, the Archaea and the lineage containing the eucaryote nuclear genome had branched off well before then. Thus eucaryote nuclear genes are not directly descended from any of the Eubacteria or Archaea, but from the

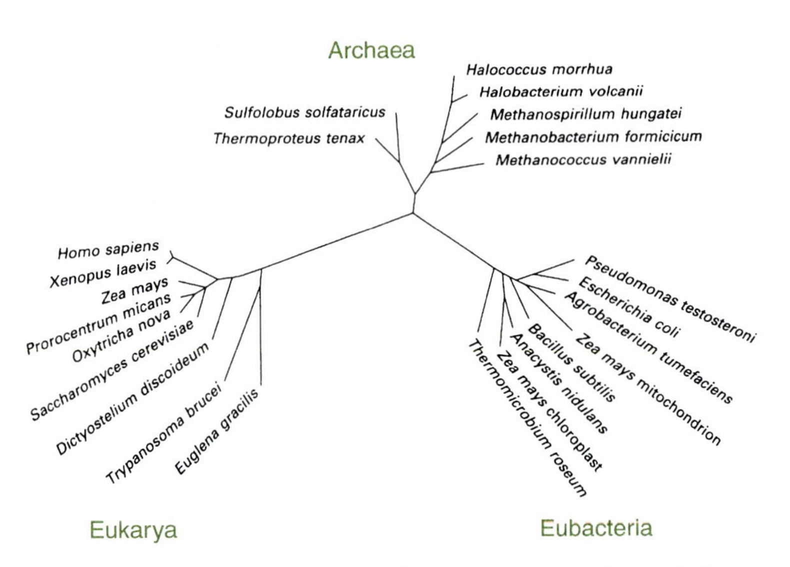

Fig. 15.1 This tree of life, based on the DNA sequences that code for ribosomes, has three main branches: the Eubacteria, the Archaea, and the Eukarya. The branch point at the center is known from fossils to be at least 3465 million years old. From this perspective, animals, plants and fungi are minor twigs on the left-hand branch. Mitochondria and chloroplasts group with the Eubacteria. Because the universal ancestor is unknown, the tree is unrooted. (From Darnell *et al.* 1990.)

more ancient common ancestor of all three groups. That ancestor existed more than 3500 million years ago.

What kind of organism was it? It probably had a circular genome without introns (introns are a typically eucaryotic structure consisting of DNA sequences that are transcribed into messenger RNA but then spliced out and not translated into proteins); with genes linked into operons, often in the same order that they are linked in extant organisms (operons occur in Eubacteria and Archaea, where they control gene expression); ribosomes with most of the modern ribosome proteins; the 'universal' genetic code; modern replication and recombination mechanisms, similar to those found in eucaryotes; and many of the enzymes of modern metabolic pathways (Doolittle 1998).

Their common ancestor was complex and in some sense 'modern'.

Note how complex that organism was. It had already evolved considerably since the origin of life, and it contained many basic structures and processes that have hardly changed since and are found in every living cell. Life originated much earlier than the first fossils, and there may have been a long period of sorting among lineages, only one of which survived.

Woese's tree further implies that the basic divisions of life are to be found among the unicellular micro-organisms. The fungi, plants, and animals are minor twigs on one of its three main branches. Because micro-organisms dominated 3000 million years of evolutionary history, whereas the animals, plants, and fungi have existed for less than 1000 million years, this result should not surprise us. The reason why it does surprise us at first sight, and the reason why it encountered resistance, is that all previous classifications had started with animals and plants and worked down, instead of starting with the simplest living things and working up. The ability to start near the beginning and to work up from there is one of the many benefits of molecular systematics.

Most of the tree of life is found among the unicellular micro-organisms. The fungi, plants, and animals are twigs on one of three main branches.

The tree of life, with its three main branches of Eubacteria, Archaea, and Eukarya, is based on a phylogeny constructed with a single molecule. Recently the entire genomes of more than a dozen micro-organisms have been sequenced, and some of the story they tell is not consistent with the hypothesis of these three main branches. For example, eucaryotic proteins have more affinities to Eubacteria than to Archaea, to which Eukarya are supposed to be more closely related. Some inconsistencies in the conclusions drawn from different genes can be resolved by postulating horizontal gene transfers between lineages. If such events happened often enough, we might not be able to decide whether there are three domains of life or some more complex network. See the recent literature to learn how opinions have changed as more data have come in.

Some data from entire genomes is not consistent with three domains.

The origin of the Eukarya

Margulis (1970) suggested that the eucaryotic cell originated through the symbiotic fusion of two (for animals) or three (for plants) previously

The eucaryotic cell originated in the symbiotic fusion of previously independent genomes. Mitochondria appear to be derived from α-proteobacteria, chloroplasts from cyanobacteria, and the eucaryotic nucleus from an ancestor more like the Archaea than the Eubacteria.

independent genomes. Her view was that mitochondria and chloroplasts (the **plastids**), which have ring-shaped, bacteria-like genomes, had previously been independent organisms. Molecular systematics supports this hypothesis, which is now generally accepted. Mitochondria appear to be derived from an ancestor similar to α-proteobacteria like the present-day *Paracoccus,* and chloroplasts appear to be derived from cyanobacteria.

Some plastid genes were transferred to the nuclear genome, stabilizing intragenomic conflicts and allowing nuclear diploidy and recombination to support the repair of mutations.

After the symbiosis was established, some genes in the plastid genomes were transferred to the nuclear genome. This transfer stabilized conflicts that arose with sexual reproduction, when the transmission patterns of nuclear and cytoplasmic genes became different (Chapter 10). Putting some essential plastid genes into the nuclear genome meant that the nucleus and the plastids partially shared the same transmission pattern, reducing the potential for conflict. Transfer of plastid genes to the nucleus also allowed nuclear diploidy and recombination to repair mutations (Chapter 7). The two explanations—DNA repair and conflict resolution—are not exclusive.

The eucaryotic nuclear genome also contains bacterial genes that are not related to the plastid genes, suggesting that the history of gene exchange has been more complex than one would expect from the surviving plastids.

The ancestral eucaryotic genome existed as a procaryote without plastids for 1500–2000 million years; the fusion with the ancestors of mitochondria happened about 2000–1500 million years ago and with the ancestors of chloroplasts about 1000 million years ago.

More mitochondrial than chloroplast genes have been transferred into the nuclear genome. Together with the greater number of neutral changes in mitochondrial than in chloroplast genes, this suggests that the symbiotic event involving mitochondria occurred earlier than that involving chloroplasts. The ancestral eucaryotic genome appears to have existed as a procaryote without plastids for about 1500 million years, for the symbiotic fusion with the ancestors of mitochondria is thought to have occurred about 2000–1500 million years ago. The fusion with the ancestors of chloroplasts probably occurred about 1000 million years ago. These estimates are rough and will be refined. That they can be made at all illustrates how much information about the sequence of historical events more than 1000 million years ago can be extracted from the genes.

Cell structure suggests that the origin of the symbiosis was complex.

The ancestral eucaryotic nucleus appears to have been adapted to anaerobic conditions. It may have become a nucleus inside a cell when a gram-negative bacterium with a cytoskeleton (necessary for the engulfing maneuver) ate the ancestral eucaryote, which then expelled the bacterial genome from the cytoplasm. This would explain both the membrane-within-a-membrane structure of the eucaryotic nucleus and the phylogenetic patterns of some proteins. Later the ancestral eucaryote, which then had a nucleus but no mitochondria, is thought to have engulfed other procaryotes. At times the engulfed procaryote was not digested and continued to metabolize, which opened the door to the evolution of symbiosis based on the mutual exchange of metabolic products.

Two reasons for the original symbiotic fusion:

What was the reason for the original symbiotic fusion, which may have occurred several times? There are two views on the metabolic exchange that stabilized the symbiosis by providing benefits for both parties (Doolittle 1998).

The original view was that the proto-mitochondria, thought to have had aerobic metabolism, exchanged respiration-derived ATP for metabolizable substrates and physical protection from the nuclear partner.

(1) the proto-mitochondria exchanged ATP for substrates and protection from the nuclear partner;

A more recent hypothesis is that instead of ATP the proto-mitochondrion excreted hydrogen and carbon dioxide, the waste products of anaerobic fermentation of reduced organic compounds, which the nuclear partner, like a modern methane-producing Archaean, used as its sole sources of energy and carbon. When external hydrogen was scarce, the proto-nuclear host became dependent on the symbiont, leading to tighter physical association and surface contact, transfer of genes for membrane proteins and the enzymes for glycolysis, and eventual fusion (Martin and Müller 1998). In this view, the fusion occurred after a mutualistic relationship had been established in a reducing environment in which the future partners remained in stable contact. This hypothesis explains the energy metabolism of recently discovered eucaryotes lacking mitochondria better than the view based on ATP exchange.

(2) the proto-mitochondria excreted hydrogen and carbon dioxide, which the nuclear partner used as its only sources of energy and carbon.

While it is not clear which view will win out, it is exciting that Woese's discovery of the Archaea and Margulis's symbiosis hypothesis now combine with molecular systematics and evolutionary biochemistry to generate new, plausible, and detailed ideas on the reasons for the evolutionary events that established the basic structure of eucaryotic cells.

The evolution of developmental mechanisms

Nineteenth-century biology asked grand questions for which there were, at the time, no adequate answers. Are the basic developmental mechanisms that generate the diversity of animals variations on a single theme, or has each group evolved its own idiosyncratic mechanism for generating form? Are arthropods, as Geoffrey St.-Hilaire suggested in 1822, upside-down chordates? The position of their nervous systems in relation to their digestive tracts suggests that they are. Are the hearts, eyes, muscles, and brains of vertebrates in some sense homologous to the hearts, eyes, muscles, and brains of arthropods, or did they arise independently? Comparative morphology could pose these questions, but because it lacked direct access to the relevant mechanisms and long remained a descriptive science, it fell out of favor and was replaced by experimental methods in genetics, development, and molecular biology. These methods led to discoveries that have now made it possible, at the end of the twentieth and start of the twenty-first century, to answer those grand nineteenth-century questions and many new ones.

Some grand questions of nineteenth-century comparative morphology can now be answered with molecular developmental genetics.

Discoveries in the comparative molecular genetics of development—known as 'Evo-Devo'—are a third approach to understanding the history of life, as valuable as systematics or paleontology, and especially powerful when combined with them. They illuminate events that

This field, 'Evo-Devo', studies mechanisms responsible for major morphological change.

happened between 1000 million and 500 million years ago, as multicellular organisms originated, radiated, and evolved different body plans. They also suggest mechanisms responsible for major morphological change. Figure 15.2 gives a classical view of the major events in the phylogeny of multicellular animals; changes in developmental regulation were involved in all the transitions depicted. Some of the relationships depicted are now controversial.

Some genes that regulate development are conserved in evolution. They include genes coding for transcription factors, cell membrane receptors, and extracellular matrix molecules.

The breakthrough discovery showed that the genes that regulate development—genes that code for transcription factors, receptors, and extracellular matrix molecules—are deeply conserved. Transcription factors are proteins that bind to DNA, interact with the general transcription machinery, and thus control the expression of specific genes. Receptors are structures in cell membranes that receive signals from outside the cell and transform them into messages inside the cell. Extracellular matrix molecules, particularly the cell adhesion molecules, determine the outcome of cell–cell contacts, guiding cell growth and movement. For the past 550–1000 million years, large-scale animal evolution has been shaped by changes in these developmental regulatory

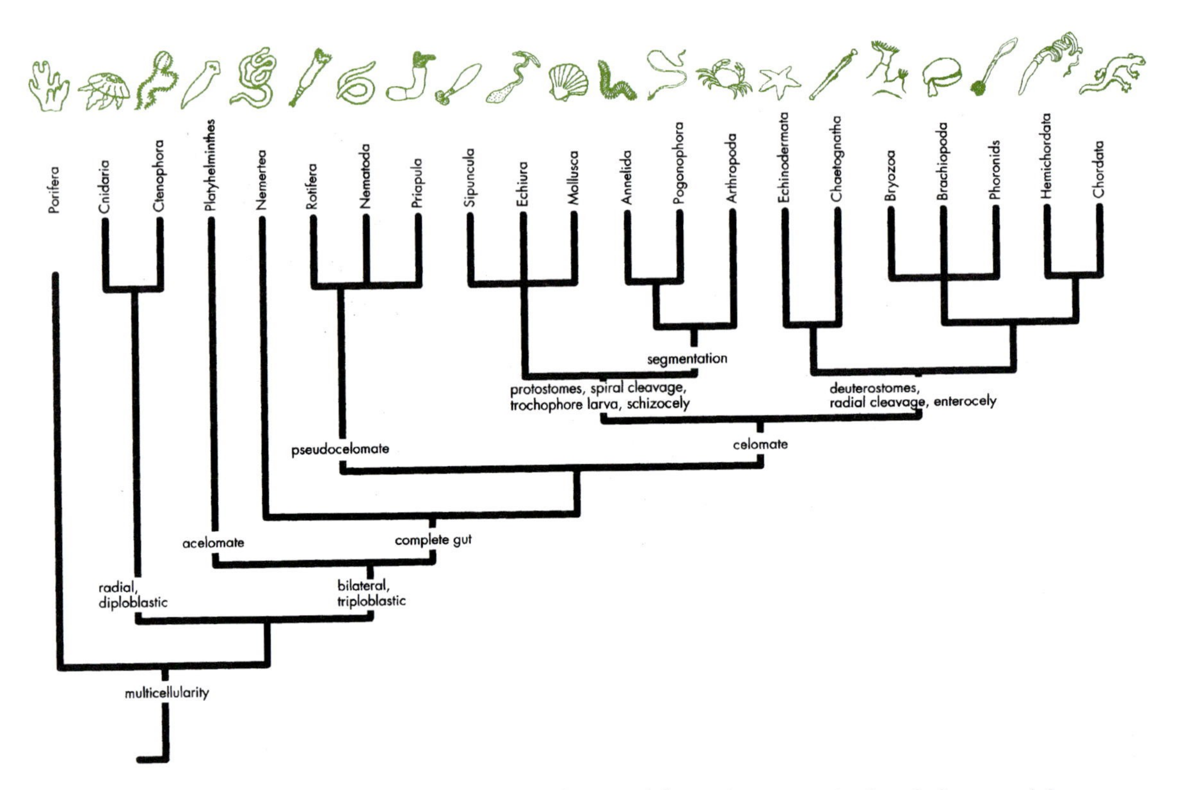

Fig. 15.2 A classical view of the major events in the phylogeny of the multicellular animals. The basic events were in the number of layers of tissues and in the number and the developmental origin of body cavities. (From Gerhard and Kirschner 1997.)

mechanisms based on a set of genes shared by all metazoans. The evolution of animal diversity has also utilized a very small sample of the possible protein families. Both at the level of developmental control genes, where evolution has varied a few basic patterns to produce diverse structures, and at the level of protein domains, where evolution has combined a small number of domains to produce a huge array of proteins, the surprise has been the discovery of simplicity underlying complexity.

How the basic developmental regulatory mechanisms evolved is not yet clear, but we do know about when it must have happened. The transcription factors that specify the anteroposterior axis and when and where hearts, eyes, and brains should form share DNA homologies in vertebrates and invertebrates. Thus sequence and function have been preserved since they originated in a common Precambrian ancestor. *In situ* methods, which allow us to visualize where these regulatory genes are expressed, connect the genes to the developmental patterns they control.

The hearts, eyes, brains, and body axes of vertebrates and invertebrates share deep homologies.

The use of *in situ* expressions in comparative studies suggests homologies between specific body regions, even between insects and vertebrates—flies and mice. Perhaps the most exciting thing about these methods is that the expression patterns used to establish homology are not just passive labels but directly track 'the tools that sculpt morphology' (Akam *et al.* 1994*a*), the key developmental mechanisms. We no longer have to be content with comparing morphologies, whose extensive evolutionary modification can hide widely shared underlying mechanisms. We can now visualize parts of the mechanisms themselves.

The expression patterns used to establish homology track 'the tools that sculpt morphology'.

That comparative morphology can now be enlightened by molecular genetics emphasizes the need to retain the former while developing the latter.

The homeobox system

The new molecular data bear on basic questions about body plans. Here the homeobox genes are the best example. **Homeobox** genes contain a special DNA sequence, the homeobox, that codes for a 60-amino-acid sequence called the homeodomain. The homeodomain forms part of the gene product, a transcription factor. Transcription factors regulate differentiation by binding to specific sites on DNA and turning particular genes on or off. Some homeobox genes specify a region of the body where a structure will form, such as a head, thorax, abdomen, limb, or eye. The *Hox* genes are a subset of the homeobox genes that keep the segments along the anterioposterior axis from being the same. They were originally discovered in *Drosophila* and given special names, but now they are usually more simply referred to by numbers (*Hox1*, *Hox2*, etc.) to make clear their relationship to similar genes in mice, the nematode *Caenorhabdites elegans*, and the cephalochordate *Amphioxus*.

Homeobox genes contain a DNA sequence, the homeobox, that codes for part of a protein transcription factor. Transcription factors regulate differentiation by turning genes on or off.

The *Hox* genes have two remarkable characteristics: **colinearity** and **conserved function**. Colinearity means that the genes lie next to each other along the chromosome and that their order along the

Homeobox genes are colinear and have conserved function. Colinearity means that the order of the genes along the chromosome to the region of the body over which they exert their control chromosomes corresponds to the region of the body over which they exert their control (Fig. 15.3), which in most regions depends on the expression of several genes in combination. These genes control the order in which things get built along the anteroposterior axis of the organism.

Conserved function means that genes with similar DNA sequences have similar developmental roles in distantly related organisms.

Conserved function means that genes that are DNA sequence homologs play the same or similar developmental roles in organisms that are phylogenetically widely separated. (Sequence homology is now called orthology to distinguish it from the concept of homology as applied to morphological structures.) The *Hox* genes were the first gene family shown to have homologous functions in insect and mammal

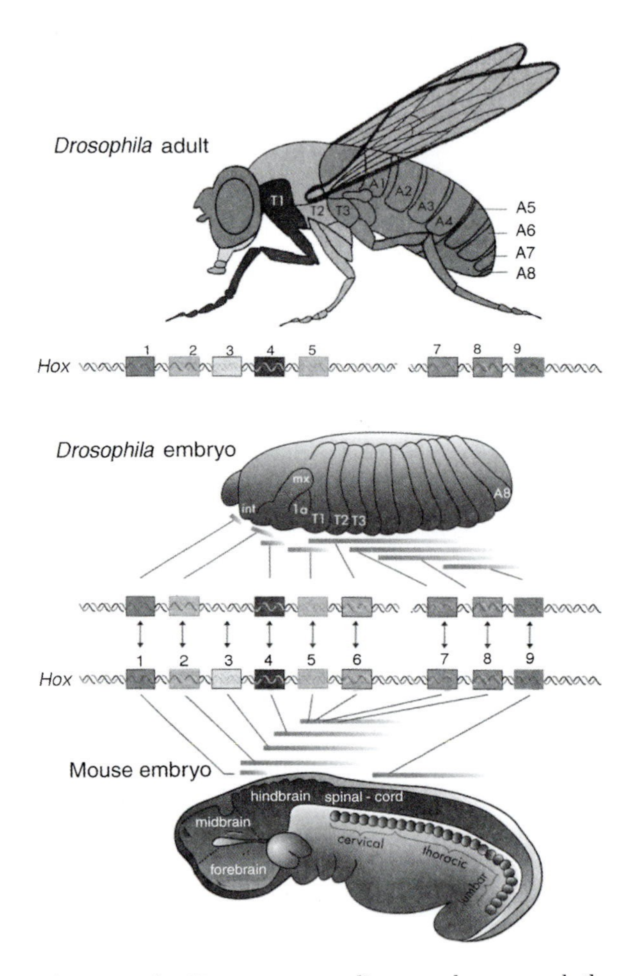

Fig. 15.3 The *Hox* genes are colinear and conserved: they control the development of parts of the body that occur along the anteroposterior axis in the same order that the genes occur on the chromosomes; and genes with a high degree of sequence homology control similar positions of the body in flies and vertebrates. T1–T3, thoracic segments 1–3. A1–A8, abdominal segments 1–8. *Hox* 1–9, the nine *Hox* genes. (From Gerhart and Kirschner 1997.)

development. Since then other gene families involved in developmental patterning have been discovered, with similarly conserved sequences and functions. While the *Hox* genes specify segment identity as far forward as the hindbrain, other genes specify segment identity in the anterior brain and the front part of the head.

The *Hox* genes occur in a single cluster in nematodes, arthropods, and cephalochordates (*Amphioxus*). There are two *Hox* clusters in hagfish, signaling a gene duplication event between cephalochordates and hagfish. It is associated with, but not necessarily causally related to, the origin of paired sense organs. There are four *Hox* clusters in all jawed vertebrates, signaling another gene duplication event between the Agnatha and the Gnathostomes, this time associated with, but not necessarily causally related to, the origin of jaws and possibly of paired fins.

In the vertebrate line the homeobox genes were duplicated twice.

The *Hox* genes in arthropods that are expressed in posterior head, thorax, or abdomen are **paralogous** to genes in chordates that are expressed in posterior head, trunk, or tail (Fig. 15.3). The conserved function of paralogous genes is identified by a procedure called knockout and rescue. Knockout refers to mutations that destroy the function of the native gene, for example one of the *Hox* genes in a fly. Rescue refers to rescuing the original function of the native gene, which is no longer functioning because it has been knocked out, by the gene or gene product of the paralog, such as a gene from a mouse. If the gene product of a mouse gene can successfully rescue the missing function of a fly gene, then one can assert that they have conserved function.

Conserved function is identified by knockout and rescue.

Conserved colinearity and conserved function of paralogous genes is strong evidence that they had similar order and similar function in a common ancestor. Because they are shared by chordates, arthropods, and in part by nematodes, and because similar genes (the order along the chromosome and function of which is not as precisely known) occur in mollusks, flatworms, and cnidarians, the common ancestor must have lived before the Cambrian radiation, perhaps 1000–800 mya. Current evidence suggests that some developmental control genes have been conserved in order and function since they first occurred in an organism with the morphology of a segmented worm.

Homeobox genes stem from a common ancestor that lived 1000–800 mya and looked like a segmented worm.

The conserved function of developmental control genes suggests that evolution has varied a few basic elements to produce incredibly diverse structures. The morphological diversity of at least all animals with three tissue layers, and possibly of all multicellular animals, consists of variations within a framework provided by conserved genes.

The morphological diversity of most animals consists of variations on a theme provided by conserved *Hox* genes.

Dorsoventral patterning in amphibians, mammals, and flies

The dorsoventral patterning of the embryos of amphibians, mammals, and flies is controlled by orthologous genes. The expression of some of these genes occurs in a striking pattern. Those that are expressed dorsally in the fly embryo are paralogous to genes expressed ventrally in

Arthropods appear to be upside-down chordates.

vertebrate embryos. This supports the hypothesis that somewhere in the evolutionary divergence of arthropods and chordates there was a reversal of the dorsoventral axis in one of the two lines (De Robertis and Sasai 1996). Whether the genes that control the development of the gut are also paralogous and have expression patterns that support this hypothesis should soon be known.

Cautionary remarks

Downstream genes also evolve.

The fact that a sequence in a developmental control gene has been conserved for a long period of time does not mean that the structure whose development that gene controls has likewise been conserved. Genes coding for transcription factors control the spatial and temporal expression of **downstream genes**. The downstream genes can evolve. This makes it difficult to infer the morphological phenotype of the last common ancestor from the fact that certain organisms currently living share developmental control genes with impressive DNA sequence homologies. For example, the gene for the transcription factor that initiates the development of eyes in mice is paralogous to the one that initiates the development of eyes in flies (Halder *et al.* 1995), but the mouse eye differs so dramatically from the fly eye that all we can infer about the eye of the ancestor is that it was a light-sensitive organ connected to the central nervous system in a multicellular organism.

Many more genes are needed to build an organ than to initiate its development.

Nor should we let the discoveries of deeply conserved developmental control genes cause us to forget the number of genes involved in building a complex organ. There are hundreds to thousands of genes involved in the development of an eye, a limb, or a brain. Many more genes are involved in building a structure than in controlling when and where it will appear in development, and the degree to which they are conserved must vary considerably. Otherwise the diversity of life would not be possible.

The echinoderms: new functions for old genes

No group better displays new uses for *Hox* genes than the echinoderms, which change symmetry patterns during development. In echinoderms genes orthologous to *Hox* genes in arthropods and chordates are used to control the patterning of structures with a geometry unlike anything found in other phyla, including fivefold radial symmetry in adults (Fig. 15.4).

The role of transcription factors changed in echinoderms to organize the development of new structures.

The way these genes are used, which differs from class to class among the echinoderms, suggests that the role of some transcription factors was changed in evolution to organize new developmental functions. Transcription factors changed roles both in the divergence of radial adult echinoderms from bilateral ancestors and in the divergence of the extant echinoderm classes from each other. This required transcription of the genes at new times and places to change the determination of basic symmetry patterns and the developmental control over very different larval and adult morphologies (Lowe and Wray 1997; Fig. 15.5).

Egg Embryo Early larva Late larva Juvenile
Bilaterally symmetrical
Left–right asymmetrical
Radially symmetrical

Fig. 15.4 Although *Hox* genes are associated with the patterning of the bilateral body axis in arthropods and chordates, the echinoderms have put them to new uses in different ways in several classes. The radially symmetrical juvenile sea urchin develops from a disk of cells that are sequestered in the bilaterally symmetrical larva. (From Lowe and Wray 1997, reproduced by kind permission of the authors and *Nature*.)

Molecular genetics sheds light on the origin of the developmental control genes used to build the diverse morphology found in much of the animal kingdom. Many of them originated between 1000 and 500 million years ago and have been conserved in sequence since then.

African Eve and polymorphisms in genes for immune response

Methods of molecular systematics can estimate the age of the most recent common ancestor of genes. Mitochondrial and MHC data only apparently disagree on the size of ancestral human populations.

Molecular sequences also contain information about much more recent events, including estimates of the ages of the most recent common ancestors of genes. Two examples provide a particularly instructive contrast. One is based on DNA sequences from human mitochondria from many populations and most continents. These data suggest that all human mitochondria are derived from a single female who lived less than 1 million years ago. The other example concerns genes in the major histocompatibility complex (MHC), which are involved in the vertebrate immune response. These data suggest that some MHC polymorphisms—the different alleles at a single MHC locus—have persisted in the primate lineage for at least 35 million years. The minimum size of the ancestral populations necessary to maintain this much genetic diversity was not less than 10 000 and probably closer to 100 000 individuals through several speciation events.

After reviewing gene genealogies and coalescence times, which pertain to both examples, we will see why the mitochondrial and the MHC data are not in conflict over the age of human genetic polymorphisms and the size of ancestral populations.

The concepts of time to coalescence and gene geneology

Consider the DNA sequences of two copies of the same gene. They could be two alleles from a single population or from two related species. Assume that there has been no recombination, and that mutations have

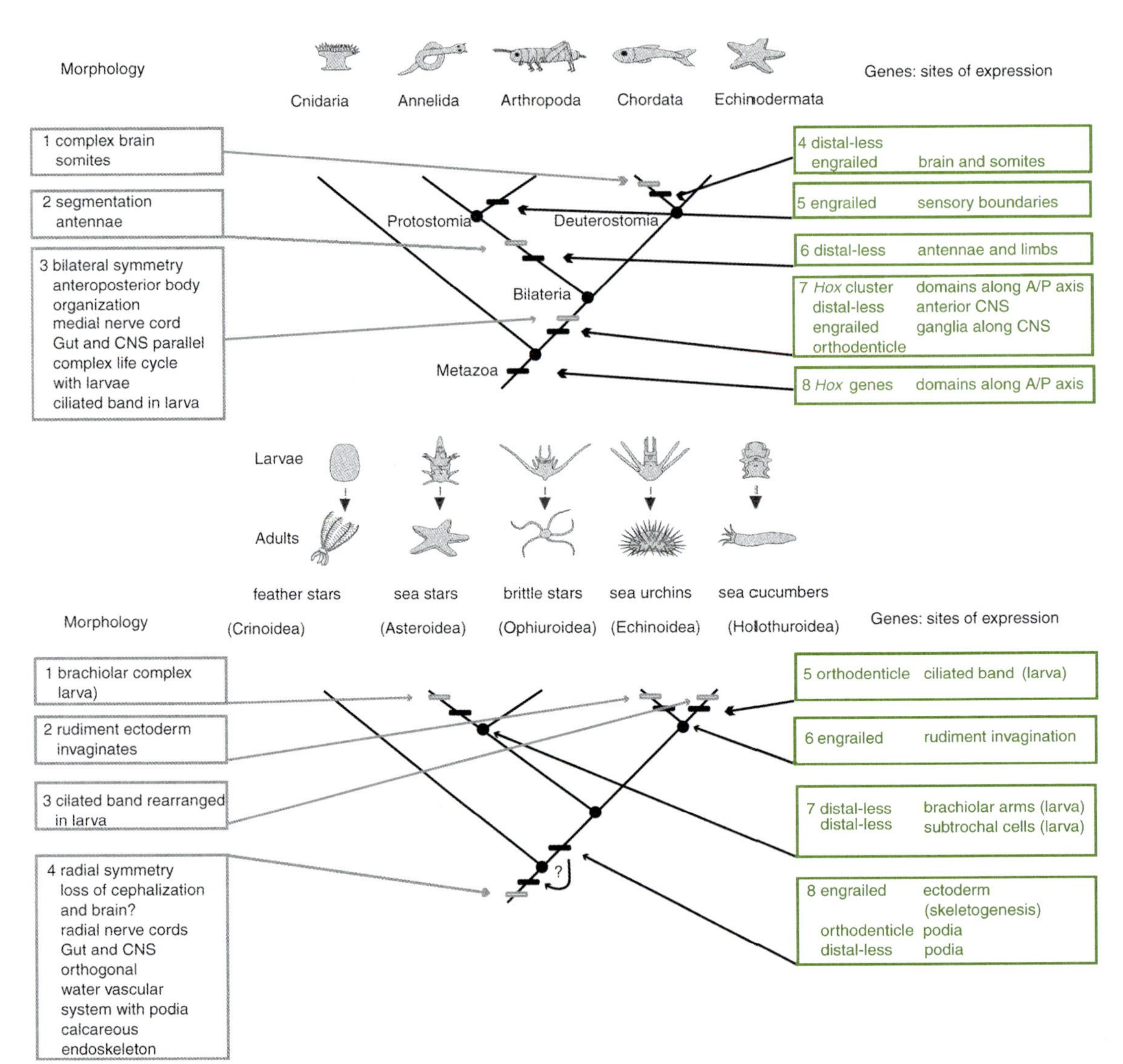

Fig. 15.5 Developmental control genes in echinoderms have acquired control over the development of larval and adult morphologies that are very different from the structures over which they had control in the ancestors that echinoderms share with chordates. (From Lowe and Wray 1997, reproduced by kind permission of the authors and *Nature*.)

been neutral. That would be the case if the DNA were sampled from mitochondria or other haploid asexuals and if mutations produced no change in protein function. Such changes should be neutral or nearly so.

The two copies of the gene differ at several neutral sites. At some time in the past, when both copies of the gene derived from a common ancestor, there were no differences between them. How long did it take for this many differences to accumulate? To make that calculation, we must assume a constant mutation rate, but we do not have to make any assumptions about population size or selection at nearby loci, for neutral mutations accumulate within genes at rates that do not depend on those factors. Other important genetic properties of the population do depend on population size and selection; these include the number of mutations that will be fixed in the entire population and the amount of polymorphism that exists in the population at any time. But the number of mutations that have been fixed along an individual lineage since the last common ancestor depends only on the mutation rate and the time elapsed (Hudson 1990).

A time to coalescence estimates the time since the most recent common ancestor from the difference in two DNA sequences under the assumption of neutral evolution.

Mutation rates for single nucleotides are about 10^{-8}–10^{-9} per organism per generation. With that information, we can use the neutral theory of evolution (Chapter 3) to estimate, with some error, how long ago the common ancestor existed. The size of the error depends on the lengths of the DNA sequences and the number of mutations detected in them. We build the tree starting from the tips of the branches in the present, then work back in time, calculating where the branches **coalesce** into a common ancestor. The trees that result are not phylogenies of species but genealogies of genes, and the process is called a coalescent process, because the calculations yield the age at which the differences coalesce into the same ancestral sequence (Fig. 15.6).

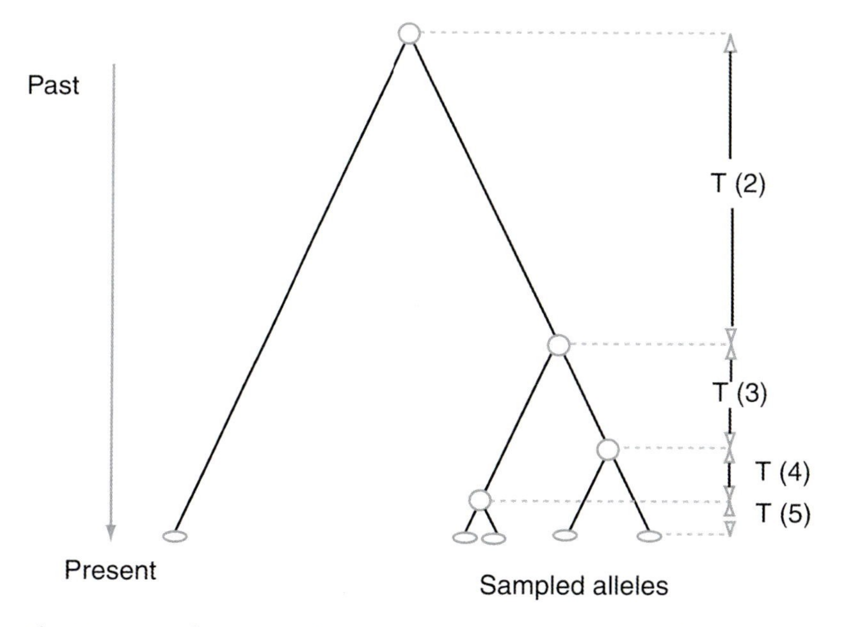

Fig. 15.6 In coalescence analysis one uses a mathematical model of evolution that assumes a similar rate of fixation of mutations in all branches of the tree, and calculates backwards from an array of existing species until all branches of the tree coalesce in a single root. T(5) is the estimated age of the youngest allelic relatives, T(4) the next youngest, and so forth. (From Hudson 1990.)

Once a reliable genealogy is obtained, the number of mutations separating the molecular ancestor from its tips, and the assumption of a molecular clock, can be used to date when this most common recent ancestor lived. Reliable genealogies allow us to compare old and young polymorphisms. The age of gene polymorphisms can be either greater or less than the age of the common ancestor of two species.

We have here described only the simplest possibility—no recombination, no geographical structure, and no selection—but the methods have been extended and can, to some extent, deal with more complex scenarios.

Human mitochondrial DNA and 'African Eve'

Because mitochondrial DNA (mtDNA) mutates 10 times faster than nuclear DNA, it can trace the divergence of populations that separated recently. The molecular ancestor of all human mtDNAs appears to have lived about 200 000 years ago.

Recently many studies have reported on the molecular diversity of human mitochondrial DNA (mtDNA). It represents only 1/200 000th of our genome, is haploid, is transmitted through females, and does not appear to undergo recombination. Because it also mutates on average at least 10 times faster than nuclear DNA, it can be used to track the divergence of populations that separated recently. An analysis of 144 mtDNAs sampled in different human groups (Cann *et al.* 1987), on the assumption that humans separated from chimpanzees 4–5 million years ago, suggested that the molecular ancestor of all human mtDNA molecules lived about 200 000 years ago. This result was confirmed by Vigilant *et al.* (1991), who estimated the coalescence time at 166 000–249 000 years (Fig. 15.7).

The molecular ancestor of all the human Y chromosomes lived about 270 000 years ago

Similar studies done on the human Y chromosome, which is inherited through males as a haploid that does not recombine, yielded a similar estimate of coalescence time: 270 000 years ago (Dorit *et al.* 1995).

There are two sources of error in such estimates. The first error results from the sampling process itself, which determines how many se-

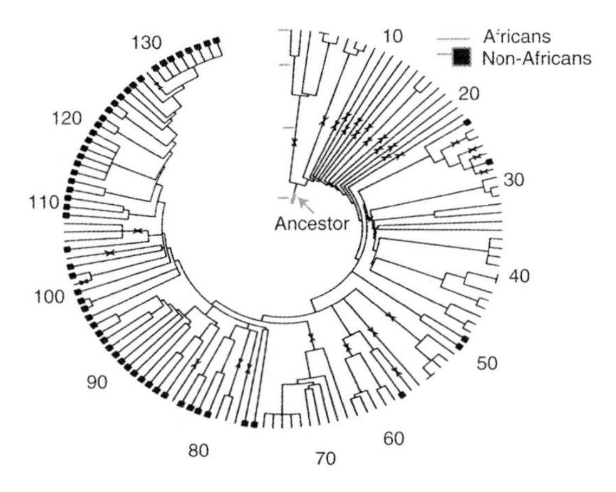

Fig. 15.7 Coalesence analysis of human mitochondria suggests that the common ancestor of all extant mitochondria existed in a woman who lived about 200 000 years ago. (From Vigilant *et al.* 1991.)

quences of what length were used. The second error arises in the calibration of the molecular clock (see Chapter 3), for the date of the human–chimpanzee split is not known accurately. When both errors are taken into account, then the coalescence times for both the Y chromosome and the mitochondrial data cannot be estimated precisely (Hillis *et al.* 1996). Thus the estimate of 200 000 years must be interpreted with caution. The branching pattern in the geneology is estimated more reliably than the date of the last common ancestor, and despite the inaccuracy of the estimate of coalescence time, results like those shown in Fig. 15.7 do contain information about relationships.

The errors in these estimates are considerable.

The authors also claimed that the ancestral mtDNA molecule was present in an African woman (the African Eve theory), but equally plausible mtDNA phylogenies suggest a European or Asian origin (Hillis *et al.* 1996). An African origin is the best way to reconcile paleontological, archeological, and genetic data, but the best support does not come from molecules—it comes from fossils (Barbujani and Excoffier 1998). Nor can we conclude from the mitochondrial data that we all share a single common ancestor, for these data only concern mitochondrial DNA and do not tell us where our nuclear DNA came from or where its many ancestors were located. The human population at that time was not necessarily small; the data are consistent with large populations.

Our ancestors probably came from Africa, but it is the fossils, not the molecules, that tell us that.

It is true that more mtDNA diversity is found within sub-Saharan African populations than in European, Asian, or Amerindian populations. The greater mtDNA differentiation in Africa could result from African populations having been larger than non-African populations for much of our evolution. It may also have been caused by earlier separation of populations from each other within Africa than on other continents.

The age of human MHC polymorphisms

What do other human genes, nuclear genes on chromosomes 6 and 15, tell us about the same period of human history? The major histocompatibility complex consists of 30–35 genes that code for polypeptide chains. These chains combine to form a huge variety of antigens. The genes form two classes (I and II) that each have two subclasses, A and B, a classification based on the structure of the antigens produced by the combination of the polypeptide chains. The products of the A gene subclass are labeled α-polypeptides, those of the B gene subclass are labeled β-polypeptides. The human leukocyte antigen (HLA) proteins consist of a combination of any α- with any β-polypeptide from any of the genes within an MHC class. The genes within one class all produce polypeptides that combine with those from the other class to form antigen molecules.

The genes of the MHC complex produce polypeptides that combine to form a huge diversity of antigens.

Some of these genes have many alleles; up to 59 have been identified so far for the highly polymorphic loci and it is likely that many more exist. These are the most highly polymorphic genes known. Other MHC

Some MHC genes are the most polymorphic genes known.

genes are represented by only one or a few alleles (Fig. 15.8). The reason for this diversity is thought to be the role these molecules play in immune defense and in the discrimination of self from nonself. Every individual can be effectively discriminated on the basis of combinations of MHC polypeptides, and some of the MHC genes are associated with particular infectious diseases, including tuberculosis, leprosy, dengue fever, HIV, hepatitis, malaria, leishmaniasis, and schistosomiasis (Singh *et al.* 1997).

Some MHC polymorphisms are older than the speciation event that separated humans from chimpanzees.

Even more remarkable, some of the alleles at the same locus differ by up to 90 nucleotide substitutions and their products by up to 30 amino acid substitutions. It takes a long time for that many differences to accumulate. They must be old. Comparing the DNA sequences for the alleles at one locus between humans and chimpanzees yielded a striking result: some human alleles are more closely related to chimpanzee alleles than they are to other human alleles (Fig. 15.9). Remember, this is a comparison only of alleles within a single locus. At least some of the MHC polymorphism must be older than the speciation event that separated humans from chimpanzees.

The molecular clock for primates has been estimated at $1.2–1.56 \times 10^{-9}$ substitutions per site per year. If so, the youngest alleles in Fig. 15.9 separated 2.7–3.5 million years ago, most of the branches separated 5–15 million years ago, probably before the human–chimpanzee separation, and the main branches separated 15–35 million years ago, perhaps before the separation of Old World and New World monkeys.

This places restrictions on minimum population sizes over that period of time. To pass 100 alleles from one generation to the next, one needs a minimum of 50 individuals, each heterozygous for that MHC gene and each heterozygote different from the other 49. The minimum number, however, is serious underestimate for several reasons. First, some individuals share alleles. Secondly, the minimum estimate assumes just one locus, but at least four MHC loci are highly polymorphic, and all of them

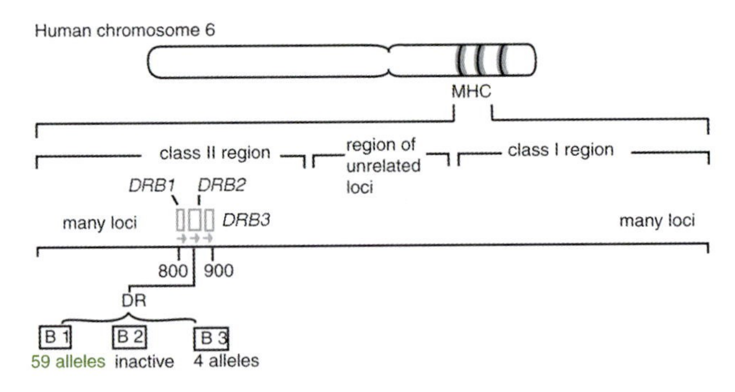

Fig. 15.8 The human MHC complex is located in two regions of chromosome 6. The class II region is shown in some detail. Within that region three MHC genes—the *DRB* genes—have been selected to illustrate the polymorphism of MHC loci. For example the *DRB1* gene is present in human populations in at least 59 different allelic forms. (From Klein *et al.* 1993.)

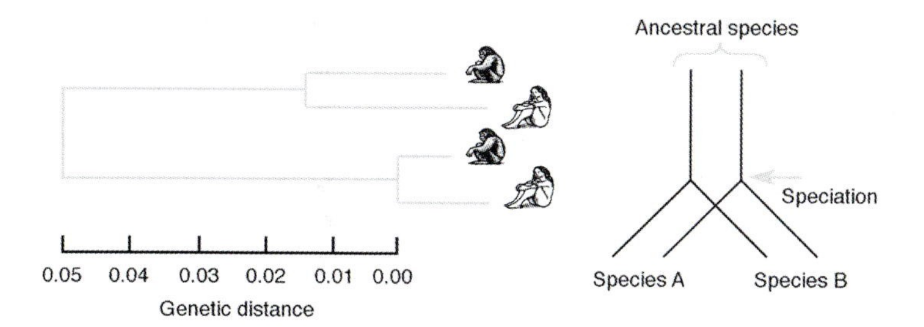

Fig. 15.9 Trans-species polymorphism is indicated when two alleles at the same locus from two different species are more similar than any two alleles from one of the species, suggesting that the polymorphism is older than the speciation event that separated the two species. This is the case for certain MHC polymorphisms in chimpanzees and humans. (From Klein *et al.* 1993.)

contain polymorphisms older than the species. Thirdly, even if the minimum number of individuals did, at one point in time, actually contain the maximum possible number of alleles, random drift would have led to the loss of most of the polymorphisms in a small population. To avoid such loss, the polymorphisms must have been maintained by selection pressure.

Computer simulations show that even with a population of 10 000 individuals, most neutral alleles initially present would be lost within 10 000 generations, which is roughly the time elapsed since *Homo sapiens* speciated from *Homo erectus*. With selection for heterozygotes with an advantage of 0.3 in a population of 1000 individuals, most alleles are lost within 1000 generations (Klein *et al.* 1990). But MHC loci are highly polymorphic for many alleles that are millions of years old. This suggests that since we shared common ancestors with chimpanzees, the effective human population size has not dropped much below 10 000 individuals.

To avoid the loss of such polymorphisms, ancestral populations must have included at least 10 000 individuals.

Polymorphisms shared among species because of descent from common ancestors are called trans-species polymorphisms (see Fig.12.10). Another well-studied example concerns polymorphisms in the self-incompatibility loci in flowering plants. These evolved to prevent self-fertilization and promote outcrossing and, like the MHC loci, some have large numbers of alleles, more than 30 in tobacco and more than 100 in cabbage and its relatives. Some alleles in tobacco are more closely related to alleles in petunias than they are to other alleles in tobacco, confirming the trans-species nature of the polymorphism (Klein *et al.* 1998).

Comparing mitochondrial and MHC evolution

The MHC results suggest that human populations for the last several million years were moderately large, certainly not very small. There were at least 10 000 people alive then, probably many more, and each of us probably has a nuclear gene from many of them. In contrast, the mitochondrial results suggest that all human mitochondria descend from one female who lived less than 1 million years ago; the other mitochondria

We now contain a tiny sample of the germ-line mitochondria from a large ancestral population.

in the population at the time that individual lived have since disappeared. This is expected with random drift of mitochondria. Since the ancestral populations were moderately large, the drift was probably caused by variation in female reproductive success rather than by sampling effects in small populations. The surviving mitochondria may have spread in part because they contributed to reproductive success, but we do not have to posit an adaptive advantage to explain the result. We might have the best of the mitochondria present long ago, but we might also have just some random sample of the mitochondria then present.

Gene trees do not track the branching patterns of historical populations until populations have been separated for very long times.

The common ancestor of the alleles of a polymorphic gene can be much older than the population in which it is found. Therefore gene trees do not reflect the branching history patterns of populations until those populations have been separated for very long times, about six times their effective population size in number of generations. For human populations with an effective size of 10 000 individuals and a generation time of 20 years, this means separation times of about 1.2 million years, longer than the estimated age of our species. It is not surprising that we find related alleles in very divergent populations, as is the case with the MHC alleles, and their presence does not necessarily indicate recent episodes of gene flow.

Recent human migrations and colonizations

Cavalli-Sforza and his colleagues (e.g. Cavalli-Sforza *et al.* 1994) have been collecting data on genetic variation systematically within and between human populations around the world for several decades. Their most extensive global analysis is based on genetic variation detected in blood samples. Many of these polymorphisms involve genes in the immune system, including those in the human MHC region. Their analyses are based typically on more than 100 polymorphic genes.

Geographic patterns of human genetic variation can be summarized in a few measures plotted on a map.

Phylogenetic trees are one way to summarize such complex data. Another method, especially suited to visualizing geographic patterns, is to summarize as much as possible of the variation in all the genes in a few artificial variables, then plot the values for each of those artificial variables onto a map. The method used to summarize the variation is called principle components analysis (PCA). The resulting artificial variables are statistically independent from one another; that is, the first principle component (PC1) summarizes one pattern of variation, and the second principle component (PC2) summarizes a second, independent pattern of variation. Because the principle components are based on samples from specific places, the value of each principle component can be calculated for each locality and plotted on a map.

Not all the variation in the original data can be represented this way. Typically PC1 represents 20–40% of the overall variation and describes the dominant pattern, PC2 represents perhaps 15–30% of the variation and describes the next most important pattern, and so forth. For details on how to calculate and to interpret principle components, consult a text on multivariate statistics and look up the details in Cavalli-Sforza *et al.* (1994).

The expansion of agriculture into Europe

The expansion of farming in Europe can be traced with archeological methods. Agriculture originated in what is now Iraq and Turkey about 10 000 years ago and spread in a wave across Europe from south-east to north-west, reaching Spain, France, and Germany 6000–6500 years ago, and England and Scandinavia less than 6000 years ago (Fig. 15.10).

Let us compare the archeological evidence with the traces that history has left in European genes (Fig. 15.11). The geographical pattern of the first principle component of human genetic variation in Europe matches the archeological data on the geographical expansion of farming quite well, implying that the spread of agriculture involved more than neighboring groups peacefully learning new methods from each other. The farmers dispersed into new territory, introducing both their new technology and their genes, and displacing or eliminating the local populations. The disappearance of the hunter–gatherers of Europe was both a cultural and a genetic event.

The geographical pattern of human genetic variation in Europe matches the geographical expansion of farming almost precisely. The expanding farmers introduced both a new technology and their genes.

The Greek colonies in Italy

In the seventh and sixth centuries BC the Greek city-states sent colonists to southern Italy, Spain, southern France, the coast of Albania and Croatia, and the Black Sea. The map of the first principle component of genetic variation within Italy, which is particularly well studied, shows a clinal pattern that radiates from the toe of Italy at the Straits of Messina northward towards Rome and westward across Sicily towards Palermo

The genes of Greek colonists introduced to Italy 2600 years ago left a signal strong enough to be detected in present-day populations.

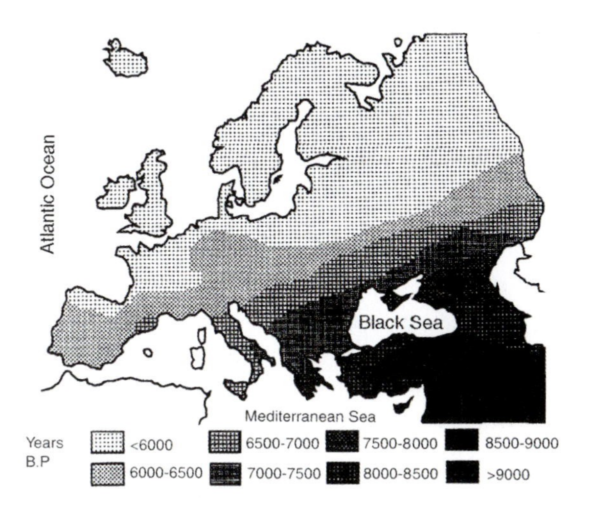

Fig. 15.10 Agriculture expanded from the Near East towards the north-west through Europe from about 9000 years ago until less than 6000 years ago. (From Cavalli-Sforza *et al.* 1994.)

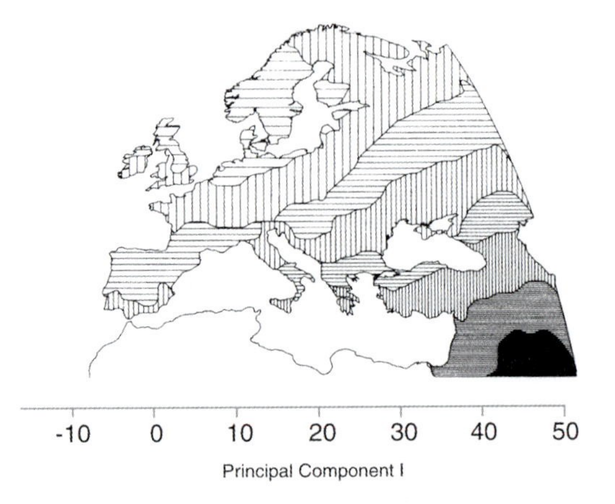

Fig. 15.11 Gene frequencies in Europe still carry the trace of the expansion of agriculturalists from the Near East towards the north-west. The graph displays the pattern in the first principal component of the sample of gene frequencies. Note the similarity to Fig. 15.10. (From Cavalli-Sforza *et al.* 1994.)

(Fig. 15.12a). It corresponds to a well-known gradient in body size and skin pigmentation (smaller and darker in the south).

The distribution of Greek surnames in southern Italy has much the same pattern. The highest frequency of Greek surnames is near the Straits of Messina, both on the mainland and on Sicily, and declines from there (Fig. 15.12b). Greek was spoken in southern Italy up to about 800 years ago. Introduction into Italy of Greek genes by colonists about 2600 years ago left a strong enough trace in the population that subsequent history, including the Roman expansion, the wars with Carthage 2200 years ago, the fall of the Roman empire, the invasion of Italy by the Goths 1600 years ago, the invasion and colonization of Sicily by the Saracens and Normans 1300–1000 years ago, and all the subsequent movements of populations, have not been able to wipe it out. Patterns of gene frequencies can thus be used to confirm and complement the interpretations of written human history.

Neutral genetic variation does not support the concept of human races

Human population geneticists know that patterns of genetic variation do not support the concept of human races. The prevalent stereotypes based on skin color, on hair color and form, and on facial traits are not well correlated with neutral genetic variation. Most human genetic variation occurs within populations among individuals, not between continents, which means that two individuals drawn from the same population may well differ as much genetically as two individuals drawn from populations originating on two different continents. Moreover, human genetic variation is clinal, which means that it changes gradually

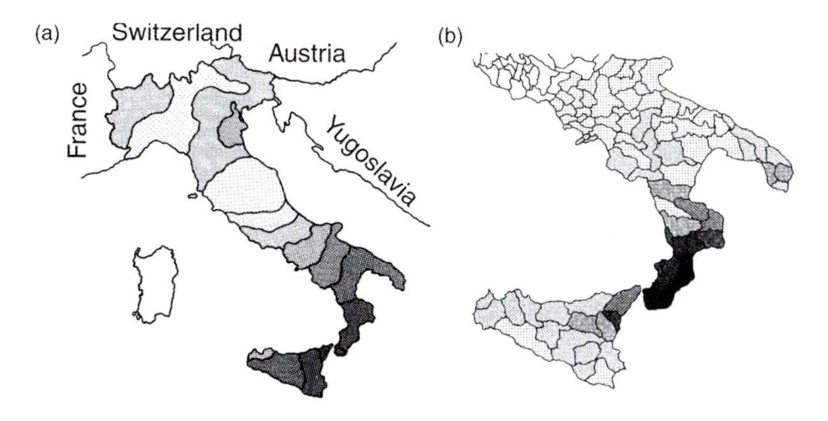

Fig. 15.12 (a) The Greek colonies of southern Italy, established about 2600 years ago, have left their traces in modern gene frequencies. The darker areas indicate Greek alleles. (b) The distribution of Greek surnames in southern Italy. The darker areas indicate Greek surnames. (From Cavalli-Sforza *et al.* 1994.)

along geographical transects. There are no sharp boundaries between groups.

If we look at the maps of worldwide genetic variation for the principle components derived from blood samples (Fig. 15.13), we can see that PC1 places Europeans and Africans in one group and East Asians and North American tribes in another. PC2 separates European from North Africans and North Africans from sub-Saharan Africans, but the differences found across the vast Asian continent in PC1 largely disappear in PC2.

The message is simple and important: the racial categories commonly used in everyday life do not reflect the multidimensional diversity and geographical complexity of human genetic variation. They are misleading. Nevertheless, the members of traditional racial groups do resemble

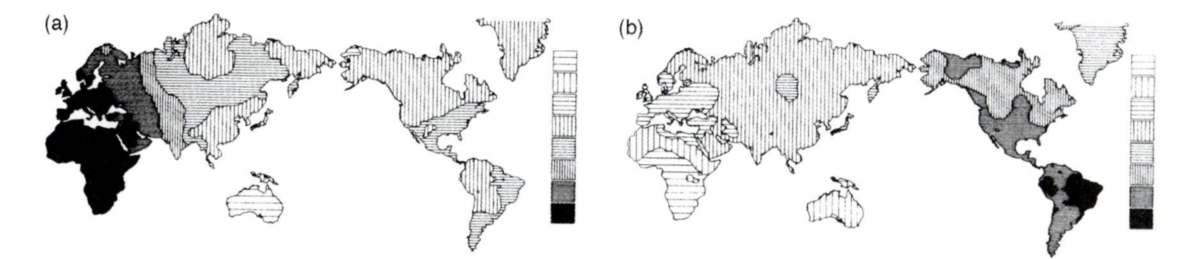

Fig. 15.13 Similar shading on the map represents similar gene frequencies. (a) Blood group analysis places Europeans and Africans in one group and East Asians and North American Indians in another. The graph displays the pattern in the first principal component of the sample of gene frequencies. (b) New patterns appear in the second principal component. Conclusion: the racial categories used in everyday life are misleading. (From Cavalli-Sforza *et al.* 1994.)

The racial categories used in everyday life do not reflect the diversity and geography of human genetic variation. They mislead.

each other more than they resemble the members of other groups in skin color, in skull shape, and in body proportions. The traits upon which the traditional racial classification is founded are based on a relatively few genes. Thus there are some genetically based differences among traditional racial groups, but they are minor compared to the similarities.

Summary

This chapter shows how the history of life is recorded in patterns of genetic variation among and within species and major groups at many different levels. New methods of detecting variation in DNA and protein sequences have produced a flood of new information.

- The DNA coding for ribosomes contains signatures of events that happened very early in the history of life; the pattern reveals three major groups of procaryotes, the bacteria, the Archaea, and a group from which the eucaryote nuclear genome was derived.
- Deeply conserved developmental control genes, including the homeobox genes, chronicle the events that led to the morphological diversification of the multicellular animals between 1000 and 500 million years ago. Many of these genes have continued to play similar general roles in arthropods and chordates over that entire span of time.
- Polymorphisms in human MHC genes have existed for tens of millions of years, suggesting that at no time was the ancestral population smaller than about 10 000 individuals and constraining the possible ways that our ancestors might have speciated. Founder events appear to be ruled out.
- Since the agricultural revolution, human migrations have left genetic patterns that are still detectable in living populations. These include the displacement of hunter–gatherers by farmers in Europe 6000–9000 years ago and the colonization of Italy by Greeks about 2600 years ago.
- Geographical patterns of neutral human genetic variation do not support traditional concepts of human races.

Thus the genomes of extant organisms can be compared to yield information on almost the entire history of life. At the moment the only period which is not open to some inspection is the period between the origin of life and the common ancestor of the Eubacteria, the Archaea, and the eucaryotes. The next chapter shows how comparative methods can be used to make reliable inferences from broad patterns of trait variation within phylogenies.

Recommended reading

Akam, M., Holland, P., Ingram, P., and Wray, G. (ed.) (1994). *The evolution of developmental mechanisms.* Development Supplement. The Company of Biologists, Cambridge

Cavalli-Sforza, L. L., Menozzi, P., and Piazza, A. (1994). *The history and geography of human genes.* Princeton University Press, Princeton.

Gerhart, J. and Kirschner, M. (1997). *Cells, embryos and evolution.* Blackwell Science, Oxford.

Hall, B. K. (1998). *Evolutionary developmental biology.* Kluwer Academic, Leiden.

Klein, J. and Klein, D. (1992). *Molecular evolution of the major histocompatibility complex.* Springer Verlag, Berlin.

Raff, R. A. (1996). *The shape of life. Genes, development, and the evolution of animal form.* University of Chicago Press, Chicago.

Questions

15.1 What was the more impressive accomplishment of natural selection, the evolution of the bacterial cell from abiotic matter in about 400 million years or the resolution of conflict in the symbiotic origin of the eucaryotic cell?

15.2 What does MHC polymorphism imply about coevolution with pathogens? What does it lead you to expect in the genetics of pathogen populations?

15.3 If maps of gene frequencies constructed from human mitochondrial genes and genes on the Y-chromosome differ strikingly, what does that imply for the history of the populations involved? (Such patterns are found around Basque populations near the Pyrenees.)

Chapter 16
Comparative methods

Introduction

Chapter 15 showed how molecular systematics provides insights ranging from deep to shallow time. Some comparative methods were used without being discussed explicitly. Now we look at the methods themselves and show that they can be applied to any kind of variation among species and higher taxa, not just in the context of molecular systematics.

Comparative methods can be applied to any kind of variation among species and higher taxa and answer questions that other approaches cannot.

Did the freshwater stingrays of South America invade from the Atlantic, after the Andes rose, or from the Pacific, before the Andes rose? Do mammals with longer gestation times also provide more maternal investment after birth? How much variation within a clade should we attribute to events that happened before the most recent speciation events, and how much to more recent changes? Are the roles of developmental control genes always conserved, as they seem to be in arthropods and chordates, or may they acquire new roles in groups with entirely new morphological structures, such as echinoderms (Chapter 15)? Such questions concern traits that vary mostly among, rather than within, species, and deal with historical events that may not be detected with fossils. Here comparative methods can provide answers where other approaches cannot.

When experiments are possible, they yield more reliable insights than comparisons, but a well-controlled comparison is more informative than a nonexistent experiment.

When traits vary within species and the causes of variation can be analyzed experimentally, experiments offer more reliable insights than comparisons. We cannot, however, perform experiments on many species at once, nor can we carry out experimental evolution on traits and patterns that only change over thousands or millions of years. In such cases, a well-controlled comparison is certainly more informative than a nonexistent experiment.

Comparative methods extract information from comparisons of two or more species in large-scale patterns not accessible to experimentation. They can tell us how much of a pattern of variation can be attributed to history, to inheritance from ancestors, rather than to adaptation to current environments. They can tell us in what sequence a set of character states probably evolved. They can give us a view of patterns in large clades, allowing us to generalize at a much broader level than is possible with experiments. The scope of comparative methods has been expanding rapidly.

Comparative methods tell us what can be attributed to history, in what sequence characters evolved, and how general a pattern is.

There are many different comparative methods (Brooks and McLennan 1991; Harvey and Pagel 1991; Miles and Dunham 1993). Here we introduce two of the simpler and more important. In the first, which aims to reconstruct history, one constructs a phylogeny (Chapter 12), plots the traits being compared on the branches of the phylogeny, infers the origins of the traits from the principles of cladistics, and argues the history of events from the geometry of the tree and the placement of the traits on its branches. If the traits analyzed include the geographic region in which the species live, the phylogeny can be drawn onto a map. We call this first method **phylogenetic trait analysis**.

In phylogenetic trait analysis we infer the history of change from the structure of a tree and the placement of traits on its branches.

In the second method, one compares changes in two or more traits across species or higher groups to reveal correlated changes. Such trends can be interpreted either as revealing intrinsic constraints associated with particular body plans or as documenting adaptation to extrinsic selection forces. Comparative methods often cannot determine whether constraint or adaptation caused the pattern. We call this second method **comparative trend analysis**.

In comparative trend analysis we compare changes in two or more traits across species within a clade and examine the correlated changes for evidence of constraints and adaptations.

Examples of phylogenetic trait analysis

Before the Andes rose

South America is famous for the diversity of its freshwater fishes, some of them quite unusual. Among the unusual fish are the stingrays of an endemic family, the Potamotygonidae. Most stingrays can enter fresh water but retain the ability to live in salt water. Because those in this family have lost the ability to adapt to salt water, they have probably been living in fresh water for a long time. Because most sting rays live in the sea and the large rivers of South America drain into the Atlantic, one might think that the ancestors of the freshwater stingrays in the South American rivers invaded from the Atlantic. That is not the case. Their ancestors came from the Pacific.

As the continent drifted west during the Cretaceous, the Pacific plate was subducted under its leading margin, causing the Andes mountains to rise from south to north. The Amazon, which probably then emptied into the Pacific–Caribbean (the isthmus of Panama had not yet formed) near what is now Lake Maracaibo, changed course, formed an inland sea, then later drained into the Atlantic. In the process, stingrays and other marine fishes, including trigger fishes, were cut off from the sea and isolated in fresh water. This surprising conclusion was achieved with phylogenetic trait analysis. The traits involved were the parasites that live in the stingrays and the geographic locations in which the stingrays are found.

The freshwater stingrays of South American rivers invaded from the Pacific before the Andes rose, as inferred from their helminth parasites, whose closest relatives live in Pacific stingrays, and as checked by the parasites of freshwater dolphins, which invaded after the Andes rose and whose closest relatives live in Atlantic dolphins.

The helminth parasites that live in the stingrays trace stingray history. They form monophyletic groups within which the closest relatives of the helminths of stingrays in one river system live in stingrays in another river system. They are not derived from parasites of other freshwater fish but entered South America with their host stingrays. The closest relatives of the whole group of parasites live in Pacific stingrays, and the simplest interpretation is that the closest relatives of South American freshwater stingrays now live in the Pacific, not in the Atlantic. To check this conclusion, the same methods were used on the parasites of the dolphins that live in South American rivers. The dolphins are a much younger group than the stingrays. By the time they evolved, the Andes had risen and rivers drained to the east, no longer to the north. Thus the dolphins must have invaded from the Atlantic, and their parasites indicate an Atlantic origin, for the closest relatives of the parasites that live in South American freshwater dolphins live in Atlantic dolphins. This comparison checks the reliability of the stingray conclusions by applying the same methods to a group that must have invaded from the Atlantic. Since the method led to the correct conclusion of an Atlantic origin for the dolphins and their parasites, it was probably correct in suggesting a Pacific origin for the stingrays and their parasites (Brooks *et al.* 1981).

Endoparasitic ancestors of ectoparasites?

In one family of hymenopteran parasites, endoparasitism is ancestral and ectoparasitism is derived . . .

There are about 115 000 described species of Hymenoptera (ants, bees, and wasps). They include most of the social insects (the exception being the termites), and the parasitic Hymenoptera are among the important sources of mortality for many insect species, including crop pests. Hymenopteran parasites can be either ectoparasitic or endoparasitic. Ectoparasite larvae live on the surface of their host and feed with their mouthparts buried into its body; endoparasitoid larvae live and feed within the body of their host. Traditional scenarios all suggested that endoparasites evolved from ectoparasites (Godfray 1994). However, if one plots onto the phylogeny of one important family, the Braconidae, the ectoparasitic and endoparasitic states of the species (Fig. 16.1), it appears that endoparasitism is the ancestral state and ectoparasitism is the derived state (Dowton *et al.* 1998). This suggests that, at least within this family, ectoparasites evolved from endoparasites.

implying that the ectoparasitic trait has undergone evolutionary reversal.

The endoparasitic braconid wasps almost certainly had ectoparasitic ancestors (Dowton *et al.* 1997). That would preserve the traditional interpretation of how endoparasitism evolved, but it would mean that these ectoparasites were secondarily derived from endoparasites. If so, the ectoparasitic trait underwent an evolutionary reversal. This analysis is based on an incomplete sample of species in a large family. It needs to be confirmed by studies involving more species, more molecules, and more traits.

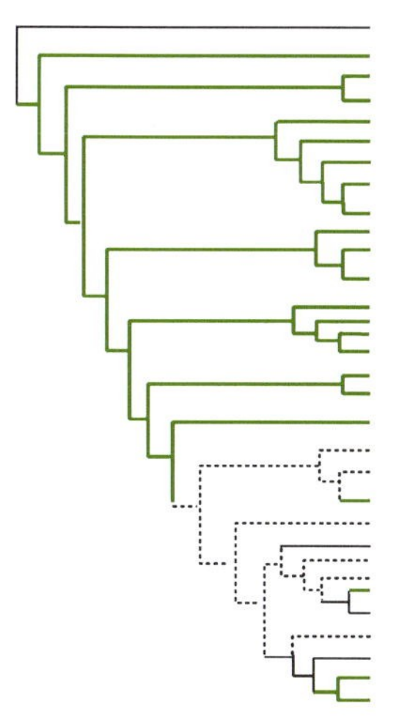

Fig. 16.1 The phylogeny of parasitic wasps with endoparasitism plotted as dark lines and ectoparasitism plotted as broken lines. It appears that ectoparasitism is a derived state. (From Dowton *et al.* 1998.)

Conclusions are only as reliable as the data and methods used

In recent years many phylogenies have been revised because of new data and better methods (Chapter 12). This could happen to the South American stingrays and their parasites and to the braconid wasps. While the conclusions in both cases are plausible given the evidence, they could change with new data. If the phylogenetic conclusions stop changing when significant data are added, we can put more trust in them.

Carnivorous plants

Carnivorous plants present us with radical modifications of leaf and stem morphology, including pitcher and flypaper traps for catching insects and other small animals. Are they all descended from a common ancestor, or did they arise more than once, independently? Are pitcher and flypaper traps evolutionarily independent, or is one type more likely to arise in a lineage in which the other is already present? Darwin thought that all flypaper traps were descended from a single common ancestor. Was he correct?

Albert *et al.* (1992) answered these questions with a molecular phylogenetic analysis of taxa from 72 plant families, using a chloroplast gene that codes for a photosynthetic enzyme. The rates of nucleotide substitution in this gene make it appropriate for analysis of phylogenetic events from species to families within the seed plants. For each species analyzed, they sequenced at least 1325 base pairs. They then performed a maximum parsimony analysis, which resulted in 396 'equally good' trees. In displaying their results, we have indicated with colored lines the relationships that were well supported in all 396 'equally good' trees.

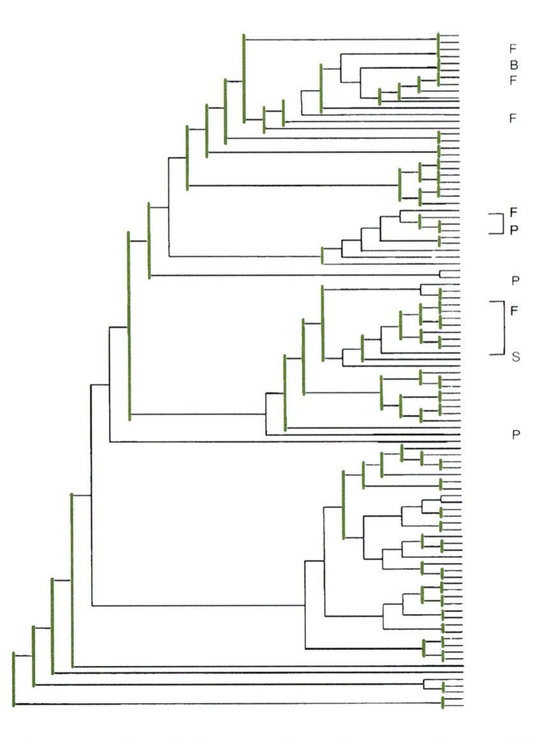

Fig. 16.2 The phylogeny of carnivorous plants. All species not labeled with a letter are noncarnivorous. F indicates flypaper traps, P indicates pitchers, B indicates bladder traps, S indicates snap traps. The vertical green lines indicate strict consensus—all of the most parsimonious trees support them. Flypaper traps have evolved independently at least five times, pitchers at least three times, and two pitcher-plant clades are sister groups to two flypaper-trap clades. (From Albert *et al.* 1992.)

Flypaper traps evolved independently at least five times, and pitcher plants at least three times. Both are polyphyletic, and some pitchers are homoplasies.

The results were surprising (Fig. 16.2). Flypaper traps appear to have evolved independently at least five times, and pitcher plants at least three times. Both are polyphyletic. Two of the pitcher-plant clades appear to be sister groups to two of the flypaper-trap clades. For example, the Old World pitcher plants, *Nepenthes,* seem to be the sister group to the sundews and the venus flytrap, the Droseraceae. Pitcher-plant anatomy has probably developed from the simpler, flytrap anatomy at least twice. Thus the structural similarity of pitcher plants misleads us about their phylogenetic relationships. Some pitchers are homoplasies.

A parasite myth revealed for what it is

Parasites are not unusually simple and degenerate.

Brooks and McLennan (1993) used phylogenetic trait analysis to test generalizations about parasite evolution and to refute many of them. For example, it was claimed that once an organism becomes a parasite, it cannot reverse its way of life and become free living again. Analysis of a group of protists reveals that it is simpler to assume that that mode of life

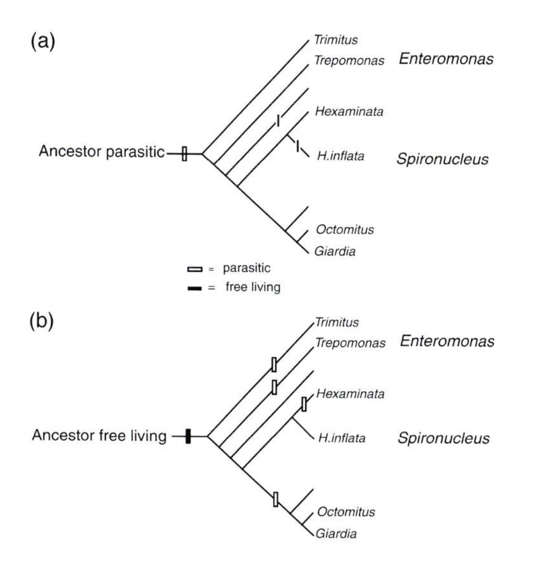

Fig. 16.3 It is simpler to assume in this case that some free-living organisms have evolved from parasitic ancestors (a) than it is to assume that once a lineage becomes parasitic it cannot become free living again (b). Thus phylogenetic analysis can destroy long-standing myths. The organisms are flagellate protists. (From Brooks and McLennan 1993.)

is reversible than to assume that commitment to a parasitic way of life is irreversible (Fig. 16.3)—another case of evolutionary reversals.

With similar methods, they showed that parasites are not especially simple and degenerate. That false claim resulted from comparing parasites with their hosts rather than with their free-living sister groups. Thus parasite evolution is more like the evolution of free-living organisms than the traditions of parasitology would have us believe.

An example of comparative trend analysis

Life-history evolution in the mammalian radiation

Life-history traits vary much more dramatically among than within species. Among the mammals, for example, adult body size ranges from 5 g (a shrew) to 100 000 000 g (a blue whale), and life span ranges from a few months to about 100 years. Do traits vary together when body size and life span change across the scale of the entire mammalian radiation? How much of that variation can be attributed to weight alone, and how much to phylogenetic affinity or body plan? Such questions can be answered with comparative trend analysis.

Comparisons of life history traits across species must control for body size and phylogeny.

Because body weight and phylogenetic affinity have important effects on most life-history traits, the first attempts to answer these questions took body weight into account, expressing traits as deviations of the values that would be expected for a mammal of that size, and used the means of families or orders as the units of variation analyzed (Stearns 1983; Read and Harvey 1989).

To see what difference it makes if we do or do not take the effects of body size into account, consider the relationship between gestation period and age at maturity among the major extant orders of mammals. When the absolute values of the two traits are plotted against each other as mean values for each of the orders, there is a strong positive relationship (Fig. 16.4a, $r^2 = 0.81$). When each trait is expressed as the positive or negative deviation of the value expected for a mammal of that adult body weight, the relationship is still positive but significantly weaker (Fig. 16.4b, $r^2 = 0.36$). The reduction in the squared correlation coefficients ($0.81–0.36 = 0.45$) when the effects of weight are removed tells us that nearly half (45%) of the covariation between these two traits among mammal orders is associated with variation in body weight.

Bats have the longest and rabbits the shortest gestation times among nonmarsupial mammals, primates have unusually late maturity and tree shrews unusually early maturity for their gestation time.

From the position of the orders on Fig. 16.4b, we can also infer that, for their body weights, bats have the longest and rabbits the shortest

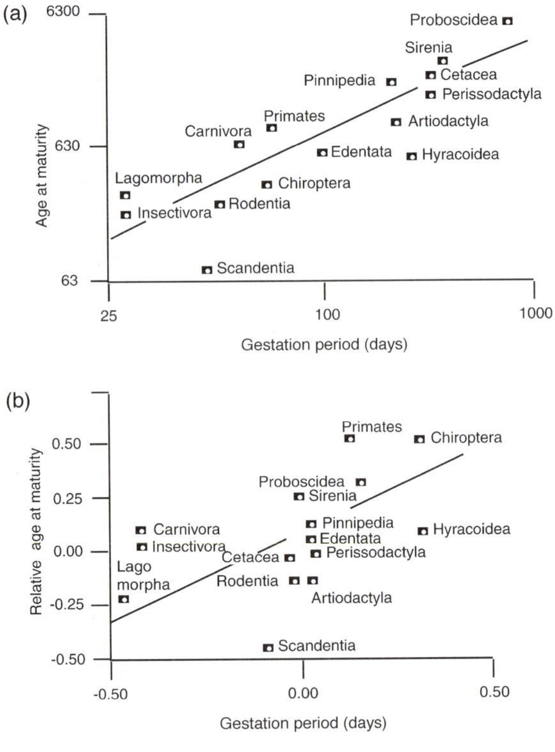

Fig. 16.4 Before the effects of weight are removed (a), there is a strong relationship between gestation period and age at maturity among the mammal orders. Removing the effects of weight (b) changes the ranking of some orders, such as Proboscidea, Primates and Cetacea, but not that of others, such as Lagomorpha and Insectivora. Thus the correlations of these traits with body weight depends strongly on taxonomic group. (From Read and Harvey 1989.)

gestation times of all the nonmarsupial mammals, that primates have unusually late maturity for a mammal with that gestation time, and that tree shrews have unusually early maturity.

The analysis was extended to other traits. When the effects of weight are removed, there are strong negative relations among the mammal orders between annual fecundity and period of maternal investment (Fig. 16.5, $r^2 = 0.79$). In other words, mammals that have few offspring per year, for their body weight, have relatively long periods of maternal care. Again, the bats and rabbits define the ends of the spectrum.

Mammals with few offspring per year have long periods of maternal care.

Such patterns of large-scale variation within the mammalian radiation suggest many questions to ask about what is special in the biology of each order. That is one strength of phylogenetic trend analysis: it suggests questions that would not occur if one did not view patterns so broadly.

Species are not independent samples

The statistical methods used in comparative trend analysis assume that each point is an independent sample. However, the species within a clade share a common history and a common set of biological characteristics. They are part of a nested hierarchy of relationships and cannot be regarded as statistically independent samples. This criticism applies to the trends depicted in Figs 16.4 and 16.5, which control for effects of adult body weight but do not control for effects of shared history. For example, there are more than 68 species of lizards in the genus *Sceloporus*, and at least 28 of them are viviparous. However, when Shine (1985) plotted the trait viviparity onto a phylogeny of the genus, he saw that the trait had arisen at most 4–6 times within the genus. Thus the sample size of independent evolutionary events for many traits can be much smaller than the number of species being compared.

The number independent evolutionary events is usually much smaller than the number of species being compared.

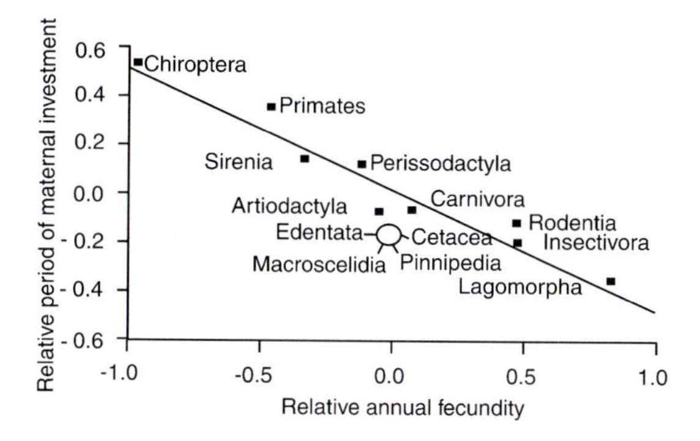

Fig. 16.5 After the effects of weight are removed, the bats have the lowest annual fecundity and the longest period of maternal investment of all the mammal orders. The rabbits and hares, the Lagomorpha, have the greatest annual fecundity and the shortest period of investment. (From Read and Harvey 1989.)

The solution is to compare independent phylogenetic contrasts.

The solution, suggested by Felsenstein (1985), is depicted in Fig. 16.6. The key insight in this figure is that the change that occurs after a speciation event in one daughter species is independent of the change that occurs after that event in the other daughter species. The figure depicts eight related species: A, B, C, D, E, F, G and H. A and B, C and D, E and F, and G and H form pairs of closest relatives. The changes that occurred between A and B after they speciated are independent of the changes that occurred between C and D after they speciated, although all four species share a common ancestor. Thus if we measure two traits, x and y, on each species, then the *difference* between A and B in trait x is independent of the *difference* between C and D in x. The same holds for trait y. One can then plot differences in x against differences in y, or group x and y according to some third trait, and analyze the trend with statistical methods for which the assumption of independence has now been satisfied. Differences so calculated are called independent contrasts (x_1–x_2 is independent of x_3–x_4), and this is called the **method of independent contrasts**. As presented here, it assumes that all branch lengths are equal; if they are not, appropriate corrections must be included.

The contrasts are most reliable and easiest to interpret when made at the tips of the branches of a phylogenetic tree.

The method can be applied to contrasts formed all the way down the phylogenetic tree. One can infer the value of a trait in a presumed ancestor by calculating the mean value of the trait among all the surviving descendants. That associates a value of the trait with every node on the phylogenetic tree. The contrasts—or differences—are then calculated just as they are for the species at the tips of the tree. The method is most reliable, and the contrasts are easiest to interpret, when one only calculates the contrasts as high on the phylogenetic tree, as close to the tips of the branches, as possible, where the species being compared share much of their ecology, and where ancestral states can be inferred with greater reliability.

An imaginary example shows that the inference of ancestral states can be questionable and should be checked . . .

To see how this works, we work out an imaginary example for the species in Fig. 16.6. Assume that they are mammals, that we want to

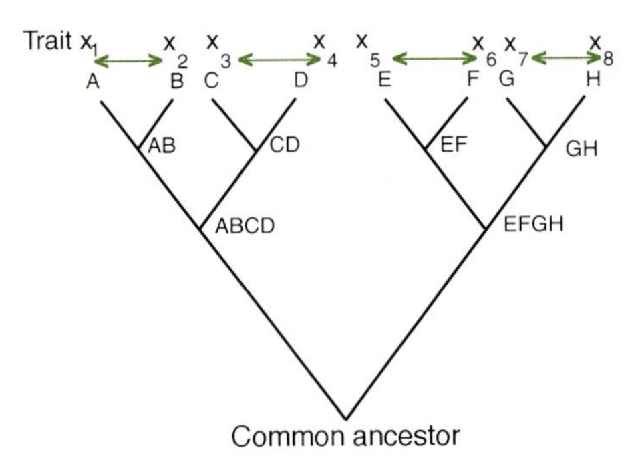

Fig. 16.6 The method of independent contrasts. The key idea is that changes that occur after one speciation event, such as (x_1–x_2), are independent of changes that occur after another speciation event, such as (x_3–x_4). (From Felsenstein 1985.)

know whether there is a relationship between gestation period and age at maturity in this clade, and that the mean values for the traits are given in Table 16.1.

The values were chosen so that gestation times would be 1–2 months, ages at maturity would be 1, 2, or 3 years, and species with longer gestation times would tend to have greater ages at maturity. If we plot the untransformed species means (Fig. 16.7a) to ask whether gestation time is correlated with age at maturity, it appears that there is a significant trend. When we calculate the values for the inferred ancestors, a questionable point emerges. Some of the inferred ancestors are maturing at 1.5 years, some at 2.5 years, and so forth. If the original set of eight species had evolved in a seasonal environment, where maturation was timed to the season, then the inference about ancestors violates the constraint of seasonality. If inferred ancestors gave birth in mid-winter, then the inference might be wrong.

When we plot the independent contrasts to ask whether an evolutionary change in one trait is associated with an evolutionary change in another trait, quite a different picture emerges (Fig. 16.7b): the trend disappears, and the only data point that hints at a trend is a contrast in which the inferred ancestors had ages at maturity of 2.25 and 1.25 years.

Table 16.1 An imaginary example of independent contrasts

Species	Gestation time (Days)	Age at maturity (Days)
A	30	365
B	40	365
C	40	365
D	50	730
E	50	730
F	60	730
G	60	730
H	70	1095
AB ancestor	35	365
CD ancestor	45	547.5
DF ancestor	55	730
GH ancestor	65	912.5
ABCD ancestor	40	456.25
EFGH ancestor	60	821.25
Contrasts		
B–A	10	0
D–C	10	365
F–E	10	0
H–G	10	365
CD–AB	10	182.5
GH–EF	10	182.5
EFGH–ABCD	20	365

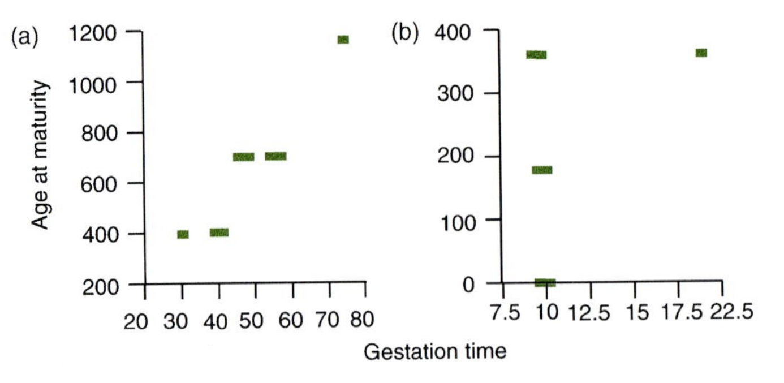

Fig. 16.7 (a) The untransformed species means for gestation time and age at maturity from the example in Table 16.1. There appears to be a strong positive relationship. (b) The relationship using the independent contrasts. The relationship disappears completely.

There is no trend at all in the contrasts formed from the tips of the tree, all of which involve a 10-day difference in gestation time.

and that when trait values are strongly associated with phylogeny, their pattern of change can be attributed to history rather than current adaptation.

There are three lessons to be drawn from this example. The first is that the inference of ancestral states can be questionable and should be checked. The second is that when trait values are strongly associated with phylogeny, the method of independent contrasts, which eliminates phylogenetic effects, may reveal that there is no evolutionary trend at all in the relationship of two traits. In this case, the relationship could be entirely attributed to phylogeny. In other cases, hidden trends can be revealed (see the next example). The third lesson is that comparing the related tips of a phylogenetic tree is more reliable than using deep contrasts involving inferred ancestral states.

The method of independent contrasts: patterns of seed size

Plotting contrasts for species, then for genera, can reveal suggestive patterns.

The seeds of trees span many orders of magnitude in weight, from milligrams to kilograms, and botanists have long sought to understand the causes of this enormous variation in offspring size, a trait directly related to fitness. One hypothesis is that the seeds of shade-tolerant trees, which must germinate and establish themselves on leaf litter beneath the forest canopy, should be larger than the seeds of light-demanding trees, which germinate in the open and have access to more light energy as soon as they produce their first leaves. Grubb and Metcalfe (1996) tested this idea with an insightful application of the method of independent contrasts. They first contrasted the seed size of shade-tolerant and light-demanding species within a genus, then they contrasted the mean seed size of genera all of whose species were either shade tolerant or light demanding. They compared species with their closest relatives within a genus and genera with their closest relatives within a family. In other words, they worked downwards from the reliable, directly observed tips of the phylogenetic tree to the less reliable, inferred, ancestral states.

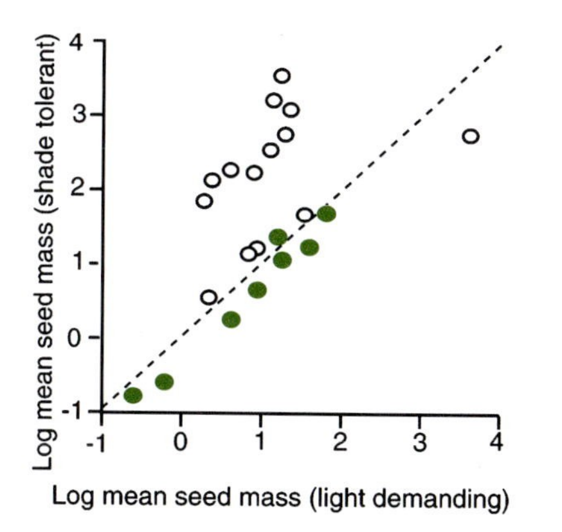

Fig. 16.8 Phylogenetically controlled comparisons of the dry seed mass of shade-tolerant and light-demanding trees. The open circles compare the mean values of related genera within families; the closed circles compare the mean values of related species within genera. (From Grubb and Metcalfe 1996.)

Their results (Fig. 16.8) were intriguing. The comparison of species within genera revealed no tendency for the shade-tolerant trees to have larger seeds than the light demanding trees. Closest relatives living in different habitats had seeds of nearly the same size. However, the comparison of genera within families supported the hypothesis. Most of the shade-tolerant genera had larger seeds than their closest light-demanding relatives. Only one shade-tolerant genus formed an exception.

The authors interpreted this as follows. The interspecific, intrageneric differences evolved relatively recently. If the shade-light contrast imposed very strong selection pressure on seed size, we would see bigger differences in comparisons at this level. That the comparisons of genera within families reveal an apparent response to the shade-light contrast may simply reflect the longer time that the difference has had to evolve. Those genera might also have evolved their seed sizes in other habitats before they invaded the environments in which they are currently found, and their seeds may have those sizes because their ancestors lived in other habitats and evolved those seed sizes for reasons that have nothing to do with the shade–light contrast. This case exemplifies both the power and the limits of the method of independent contrasts.

An exaggerated cure can be worse than the disease

The inference of ancestors may suggest states that are biologically implausible, and when deep and shallow contrasts are plotted as points on the same graph, information of variable reliability is mixed.

The method of independent contrasts aims to cure the disease of lack of independence in comparisons. When it is applied to the tips of the branches on a reliable phylogeny and to traits that have been measured reliably, it does its job well. What happens, however, when one uses all the independent contrasts that are possible in a large phylogenetic tree, where some contrasts are formed from the comparison of putative ancestors? If the actual history of the lineage has been complex and has included many trait reversals, then the more deeply into the phylogeny one infers history, the more one may mislead oneself. If, as in the artificial example above, the averaging used to infer ancestors results in traits that are biologically implausible, then one must question the method of inference. Another factor making it hard to interpret the reliability of deep contrasts, which are calculated only from species currently living, is that they ignore values from species that have gone extinct. Thus when deep and shallow contrasts are plotted as points on the same graph, information of different degrees of reliability is mixed. We recommend using the shallowest contrasts possible, those that require the least extrapolation.

General comments on comparative methods

History cannot be ignored; all comparisons should start with a reliable phylogeny.

There are many more comparative methods than we have described here. Like those described here, they fall into two broad categories: those designed to control or remove the effects of phylogeny, and those designed to infer correlated change or the origin of functional trait complexes (Miles and Dunham 1993). Most rely on methods of phylogenetic reconstruction or statistical analysis that can be traced through the recommended reading at the end of the chapter. Several general points about all comparative methods are important.

1. Life has a history that is ignored at the peril of anyone comparing two or more species. A proper comparison is based on a reliable phylogeny constructed with modern systematic methods (Chapter 12). A good start is to plot the traits compared onto such a phylogeny. That immediately reveals how many times the traits evolved in the clade and suggests hypotheses about primitive and derived states. This simple procedure is ignored by anyone who compares just a nematode, a fly, and a mouse—three species sampled from a gigantic tree, all popular model systems for molecular genetics and developmental biology—and dares to talk about ancestral and derived states.

Species are not independent samples.

2. Species are not independent samples, for all the descendants of a single ancestral species share some of that species' traits. Only by plotting traits onto reliable phylogenies can we infer how many times a trait originated in a given clade, and thus the number of independent events. If that means that something very interesting only happened a few times, then we have to accept that we cannot decide whether trends associated with such events are significant.

Even the best comparisons can only suggest, not reveal, causes.

3. Even the best controlled and most reliable comparative inferences cannot reveal causes. At their best they reveal strong correlations. For example, all species of macaques forage in groups with a similar social structure, but they occupy a very wide range of habitats. Does that mean that their social structure was inherited from their common ancestor and has not yet adapted to each habitat inhabited by macaques, or does it mean that the social structure is a superbly flexible adaptation capable of dealing with all the problems posed by a wide range of

habitats? To answer that question we need more than comparative information.

4. Comparative methods cannot, by themselves, tell us how much of a pattern to ascribe to adaptation and how much to constraint, for the historical component of a pattern is usually a mixture of the two. However, they can greatly increase the enlightenment gained from mechanistic, experimental approaches to those issues by placing the experimental results for single species into a broad phylogenetic context. Comparisons test the generality of both the questions asked and the conclusions reached in experimental research programs.

The historical component of a pattern is a mixture of adaptation and constraint. Comparisons test the generality of the questions asked and the conclusions reached in experimental research programs.

Summary

This chapter describes some important comparative methods with examples.

- Comparative methods aim to ascribe to history the component of variation that belongs to the historical part of an evolutionary explanation and to identify broad trends at a phylogenetic scale that cannot be approached experimentally.
- Two important types of methods are phylogenetic trait analysis, in which traits are plotted on trees, and comparative trend analysis, which analyzes how traits change together in evolutionary radiations.
- A technical problem in comparative trend analysis is to get data points that are independent in the statistical sense, i.e. where two measurements are not both influenced by a shared ancestor. The method of independent contrasts solves that problem.
- Comparative methods deliver answers to big questions that experiments cannot address, but they cannot determine causation. Their strength lies in reliable description of large-scale patterns and their ability to suggest questions for further analysis.

The next chapter concludes the book with comments on the status of evolutionary biology as a whole.

Recommended reading

Brooks, D. R. and McLennan, D. A. (1991). *Phylogeny, ecology, and behavior. A research program in comparative biology.* University of Chicago Press, Chicago.

Harvey, P. H. and Pagel, M. D. (1991). *The comparative method in evolutionary biology.* Oxford University Press, Oxford.

Miles, D. B. and Dunham, A. E. (1993). Historical perspectives in ecology and evolutionary biology: The use of phylogenetic comparative analyses. *Annual Review of Ecology and Systematics,* **24**, 587–619.

Questions

16.1 You have the opportunity to do two research projects, but you only have time and money for one of them. The first would be a comparative study of sexual selection in lizards in which you would apply the method of independent contrasts to look at evolutionary changes in the skin color of males associated with changes in mating systems. For this project you would use data gathered by other people and already available in the literature. The second project would be an experimental study of one lizard species in which you would manipulate parasite loads in males and study their changes in skin color and mating success. Which project would you choose and why?

16.2 In Fig. 16.5, after the effects of weight were removed, bats had the lowest annual fecundity and the longest period of maternal investment of all the mammal orders. The rabbits and hares had the greatest annual fecundity and the shortest period of investment. Assume those differences are evolved adaptations (why is that an assumption?). What does that tell us about the difference in adult mortality rates between bats and rabbits (see Chapter 8)?

Chapter 17
Conclusion

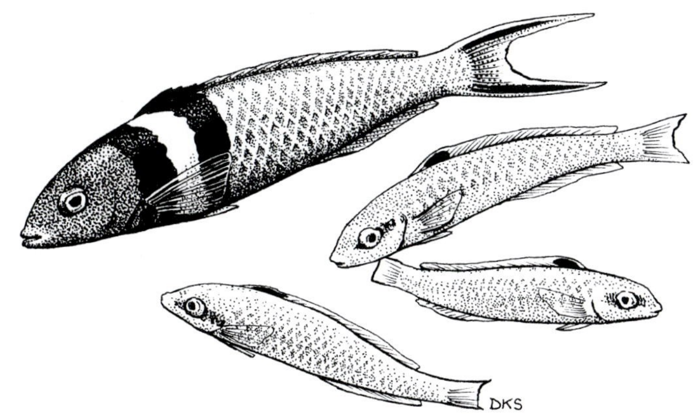

In this concluding chapter we comment on the status, nature, and preoccupations of evolutionary biology and describe some unsolved problems.

The reality and reliability of evolution

Evolution did happen and continues.

Some people think of evolution as a 'theory' that is not well supported. Nothing could be farther from the truth. Evolution is as well tested as any major idea in science. Evolution happened and is continuing. Natural selection is always at work; gene frequencies do drift; both have been observed repeatedly in the laboratory, in organisms with short generation times, and in the field. These mechanisms are well documented. The fossil record, molecular systematics, and the mechanisms shared by all living cells clearly demonstrate the ancient origin of life and the continuity of descent with modification from common ancestors. Evolution is a reality that cannot be ignored.

Evolutionary biology has a complex causal structure

Everything in biology has both proximate and evolutionary causes; the evolutionary causes mix historical accidents with adaptation.

Evolutionary theory resembles theoretical physics in its attempt to derive empirically sufficient predictions from simple first principles. However, evolution has an historical element, not easily incorporated into the theory, shared with astronomy and geology, where unique events have also had important, long-term consequences. Thus evolutionary causation is complex. Everything in biology has both proximate (or mechanistic) and ultimate (or evolutionary) causes, and the ultimate causes are themselves a mixture of history, accidents, and adaptation.

Because nothing is completely understood until all its causes have been established, biologists need to understand enough of the different causes of biological phenomena, while perhaps emphasizing one approach, to communicate with those who use the other approaches.

Evolution is happening all around us—and to us

Evolution can be fast.

Evolution is not something that takes millions of years. It produces results in a hurry. Bacteria evolve resistance to antibiotics within a few months, and insects evolve resistance to insecticides within a few years, of the introduction of new substances. Evolution has certainly not stopped in humans: it occurs in all populations in which individuals vary in their reproductive success and in which some heritable traits are correlated with reproductive success. Both conditions are true for humans, at least for some heritable traits. The key question is, which ones?

The scope of evolutionary explanation

Evolution greatly extends the reach of materialism, an extension with which people still struggle.

Science assumes that everything has a material explanation. This assumption has been so successful that it is now often regarded as a result. Evolution and molecular biology extend the scope of materialistic explanation to all things biological. As long as the things explained do not include humans, the implications of materialism have not been too controversial, but at least since the publication of Darwin's *Origin of species* in 1859, people inside and outside science have struggled with materialistic explanations of humans.

The scope of application of evolutionary explanation is controversial, both among lay people . . .

How far does evolutionary explanation apply to humans? The lack of a generally accepted answer continues to be an important reason why evolution is not more widely understood and accepted. Most scientists accept evolution as a fact, and virtually all biologists use it in their thinking, but in society as a whole opinions are much more diverse. They range from complete rejection by fundamentalist Christians and Moslems, through the position of the Catholic Church, which is that evolution did occur and is a good scientific theory but that the Holy Ghost descended uniquely into humans to give them an immortal soul, and the variable positions of other religions, to the situation of many informed lay people, who think that evolution probably did occur but do not think through its implications. Thus, in the general public, evolution encounters resistance because it contradicts some aspects of some religions.

and within the academic community.

Within the academic community, evolutionary explanations are controversial in disciplines that focus on humans—anthropology, psychology, medicine, economics, and sociology—for quite different reasons. Here it is not materialism that causes the resistance, but the competition of evolution with explanatory paradigms pitched at other levels. Those raised in other disciplines defend their intellectual territory and are not happy to see it taken over. Nevertheless, evolutionary explanations have gained some ground, and there are now subdisciplines of evolutionary anthropology, evolutionary psychology, and evolutionary epistemology. In the analysis of optimality, trade-offs, game theory, risks, and conflicts there is considerable overlap between evolution and economics. Even so, the spread of evolutionary explanations into related disciplines continues to meet with resistance, some of it justified.

At least two important and interrelated questions about the scope of evolutionary explanation remain unanswered. First, how does biological evolution interact with cultural change? How should we analyze phenomena with both biological and cultural causes? Secondly, are humans different from other animals in some essential respect, or are the boundaries between humans and other animals diffuse? Do other animals have culture, consciousness, and language? By some criteria they do, by others they do not. Because several borders between humans and chimpanzees that had previously seemed impervious have recently proven to be unclear, we think it unwise to draw hard borders between humans and animals.

Two important, unanswered questions concern the interaction of biological evolution with cultural change and the nature of the boundaries between humans and other animals.

Thus the scope of evolutionary explanation is not yet determined. It has taken society 140 years to partially digest the implications of evolution, and it will take a lot longer before they are fully digested. The scope of evolutionary explanation will continue to expand, will continue to meet resistance, and will continue to inject new ideas and insights into cognate disciplines, but we would not be surprised if a book written at the end of the twenty-first century also concluded that the implications of evolution had not yet been fully assimilated, for they go to the heart of what we are. People fight very hard over self-definitions.

The major preoccupations of evolutionary biology

That organisms might be related by descent was recognized by Aristotle, and evolutionary ideas were widespread before Darwin, but Darwin changed everything by proposing a mechanism that could produce adaptation: natural selection on heritable traits. He did not, however, have a plausible mechanism for inheritance that could maintain the genetic variation needed to sustain a response to selection. That problem was so troublesome in the late nineteenth century that when Mendel's laws were rediscovered in 1900, and it became accepted that genes were material particles located on chromosomes whose behavior followed Mendel's laws, genetics became the main preoccupation of evolutionary biology. The whole first half of the twentieth century was devoted to the assimilation of Mendelian genetics into evolutionary theory. This assimilation consisted primarily of the demonstration that phenomena from the population through the species to higher taxonomic groups and the fossil record were *consistent* with the new-found genetic mechanisms.

During the twentieth century genetics was the main preoccupation of evolutionary biology,

which established the *consistency* of all evolutionary phenomena with genetics, not the *sufficiency* of genetics to explain all evolutionary phenomena.

There are three criteria used in logic to judge how well we understand the causation of any phenomenon. The weakest of these is *consistency;* it is the weakest because it does not rule out alternatives. The next strongest is *necessity;* it is not the strongest because a cause can be necessary without being sufficient. The strongest criterion is *sufficiency*. If one can show that a cause is both necessary and sufficient, then the cause must be present for the phenomenon to exist and nothing else is necessary. It was the consistency of all evolutionary phenomena with genetics that was established, not the sufficiency of genetics to explain all evolutionary phenomena. At the time that consistency was established, some exaggerated the logical state of affairs and claimed sufficiency when only consistency had been demonstrated.

The second half of the twentieth century has seen the emergence of molecular biology, the assimilation into evolutionary biology of molecular genetics, the rapprochement of evolutionary and developmental biology, and the flowering of evolutionary and behavioral ecology. The influx of these alternative approaches to explanation—molecular, developmental, and ecological—and the recognition of the complexity of the determination of phenotypes—through the interaction of developmental mechanisms with the environment—have brought evolutionary biology closer to the goal of logical sufficiency.

But it is not there yet.

The next main preoccupation of evolutionary biology, and many other parts of biology, will be to clarify the genotype–phenotype relationship.

At the turn of the twenty-first century, it is clear that we will see rapid advances made in the use of developmental molecular genetics in a comparative context to understand the mechanistic basis of body plans. One consequence will be the clarification of the genotype–phenotype relationship, whose obscurity has blocked progress in twentieth-century evolutionary biology just as surely as the lack of a genetic mechanism blocked progress in nineteenth century-evolutionary biology. In the process, two major puzzles may be solved.

Two major puzzles: the fixed and the variable

Some traits are fixed within large lineages whereas others remain variable within populations. For example, all tetrapod vertebrates have four limbs, two eyes, and one backbone, but quite variable adult body weights, numbers of offspring, and life spans, and all arthropods have chitinous exoskeletons and grow discontinuously by molting, but the morphology of their appendages varies dramatically among lineages and species. Because all genes are exposed to mutation, and because environments change considerably over long periods of time, it is unlikely that extrinsic stabilizing selection has maintained so many traits in so many lineages in a fixed state for hundreds of millions of years. It seems reasonable to look for additional reasons for long-term maintenance of traits in fixed states in the genotype–phenotype relation, in development, and in the constraints inherent in multi-trait evolution (Stearns 1994).

We do not yet know how varying traits evolve into fixed traits, or how fixed traits become constraints on the further evolution of the still-varying traits.

Microevolution does not yet explain how varying traits evolve into fixed traits, nor does it explain how fixed traits become constraints on the further evolution of the still-varying traits. Microevolutionary theories of phenotypic evolution are nonhistorical, as is physics. To make successful predictions about phenotypes, they must assume, explicitly or implicitly, that certain traits are fixed and that other traits vary. The traits that happen to be fixed in the lineage are historical particulars which appear as boundary conditions or empirically fitted parameters in the models. This is not satisfactory. It should be possible to understand the transformation of variable traits into fixed ones. Doing so would help to connect macro- to microevolution.

Two major puzzles concern the relation of the traits that are fixed within clades to the traits that still vary among individuals. The first is, how did the fixed traits—which are not just fixed genes—become invariant? How did trait fixation evolve? It must have had something to do with the developmental control of gene expression, for all DNA sequences mutate, all genes become variable. The second puzzle is, what are the effects of the fixed traits on the further evolution of the variable ones? Is the expression of the genetic variation of the variable traits affected by which other traits happen to be fixed in that clade? Do the fixed traits affect the further evolution of the traits that remain genetically variable, thus producing clade-specific patterns of response to selection? Clade-specific patterns of variation were documented by Vavilov (1922) in crop plants, and theoretical and experimental analyses of the causes of clade-specificity in patterns of variation have been made by Alberch (1989) for amphibian limbs, by Nijhout (1991) for butterfly wings, and by Ebert (1994) for crustacean life histories. Their work (see Chapter 6) provides starting points for the solution of these two major puzzles.

Other unsolved problems

Stasis: why is evolution sometimes so slow when it can go so fast?

If evolution can be fast, why is it often so slow?

Two remarkable and apparently contradictory facts have emerged in the past 30 years. On the one hand, there is stasis documented in the fossil record, where a normal evolutionary pattern is for morphology not to change over periods of several million years. On the other hand, field studies and laboratory experiments have shown that evolutionary change can happen with surprising speed when selection is strong and genetic variation is available. If evolution can be fast, why is it often so slow? Getting an answer will not be easy, for we do not live long enough to see such changes in our lifetimes, and the fossils are beyond the reach of experiment.

What is the nature of species?

Does ecology cause clumps to evolve in trait space?

Variation in life forms is not continuous. Organisms form discrete clusters that we call species. The variation within species is generally small compared to variation between species; although there is a continuum of variation, there are clusters within the continuum. We do not understand why this is so. It seems plausible that sexual reproduction plays an important role in maintaining relative uniformity within species, but it cannot be the whole explanation, because discrete species also exist among organisms where sexual recombination is absent or very rare, as in many imperfect fungi. Do ecological conditions only allow a finite number of discrete forms, with intermediate forms less well adapted? We are still far from good answers.

What are the limits to evolutionary prediction?

How much evolution can predict is not yet clear.

Evolution may share a theoretical element with physics and historical elements with astronomy and geology, but it also shares an element of unpredictability with meteorology. Evolutionary theory makes some successful predictions—so prediction is not impossible—but evolution is a complex, nonlinear, dynamic process, and such processes are often characterized by unpredictability: tiny differences in initial conditions can lead to huge differences in outcomes. Just how much can be predicted in principle is not yet clear. To illustrate the scope and limits of evolutionary predictions, we describe two case studies.

Natural selection on RNA molecules

RNA is a system that is simple enough to let us understand the relationship of genotype to phenotype, and of phenotype to fitness. The results are intriguing, but we do not yet know if they can be generalized to organisms.

RNA molecules are linear chains of four nucleotides: adenine, uracil, guanine, and cytosine, abbreviated A, U, G, and C. RNA molecules are among the simplest entities that can evolve by natural selection, as is apparent from the test-tube evolution experiments described in Chapter 1. The nucleotide sequence of an RNA molecule is its genotype, and the spatial structure of the molecule is its phenotype. Schuster and colleagues have made a detailed study of the relationships between genotype and phenotype of RNA molecules (Schuster 1993; Schuster *et al.* 1994). The result is the first complete characterization of the genotype–phenotype relationships of any system.

Their findings can be summarized as follows:

1. There are far fewer possible shapes (phenotypes) than there are sequences (genotypes).
2. A few of the shapes produced by a large collection of RNA sequences are common, many are rare.
3. The same phenotype (shape) may result from quite different genotypes.
4. All the main phenotypes can be realized from a relatively small set of similar genotypes.

In this system one can, in principle, compute which phenotypes (molecular shapes) allow the highest replication efficiency by a certain enzyme. This means that we can rank all genotypes with respect to replication efficiency in an environment where this enzyme is available, and thus predict the evolution of a population of RNA molecules, given a particular starting composition.

It would be interesting to know to what extent conclusions from this study apply to the evolution of traits in organisms, but for no trait in a real organism do we have such complete information on mutations and development as in the case of the RNA molecules. Perhaps a few genetic diseases in humans, where we know that specific mutations in a single gene can produce a specific disease phenotype, come closest (see Chapter 5). But for most traits studied by evolutionary biologists there is very little knowledge of the genes affecting the trait, of the phenotypic consequences of the possible mutations in these genes, and of the fitness differences between the possible phenotypes.

A more limited evolutionary prediction would concern just one aspect of the evolution of a population. For example, one can establish a population of known genetic composition for some trait under controlled conditions, and ask the theory to predict how fast and how far the trait will change.

The disappearance of a lethal trait

The dynamics of a simple gene, a lethal, agree with quantitative prediction,

Wallace (1963) measured the rate at which a recessive lethal trait disappears from a population (Fig. 17.1). A *Drosophila melanogaster* population was started from flies that were all heterozygous for a recessive lethal allele. Therefore the initial allele frequency was $q = 0.5$. In each generation a random sample of the surviving flies was used as parents. The frequency of heterozygotes was determined each generation by test matings, from which the allele frequency was estimated. The expected change in the frequency of the lethal allele can be predicted by the model in Table 4.2. Substituting $s = -1$ and $h = 0$ yields the case of a recessive lethal allele and gives us the formula for the expected change in allele frequency $q = 1/(1 + q')$, which predicts the change in allele frequency given by the smooth curve in Fig. 17.1. The data fit the prediction reasonably well.

but we do not need theory to predict that insects will evolve resistance to pesticides, or bacteria to antibiotics.

Both these cases are examples of experimental evolution, and in both evolutionary change can be predicted quantitatively. The quantitative precision could only be achieved with detailed knowledge of genetics. However, genetic details often do not matter for qualitative predictions. For example, genetic details are not required to predict that an initially nonresistant population of insects fed food containing insecticide will evolve resistance to the pesticide. Predictions of changes in evolutionary research must be qualitative if the relevant genetics are unknown. This is often the case.

Looking ahead

How long will it take to follow Darwin's insights to their conclusions?

Neither science nor society has found it easy to assimilate evolution. The implications of evolution are often thought to be in conflict with

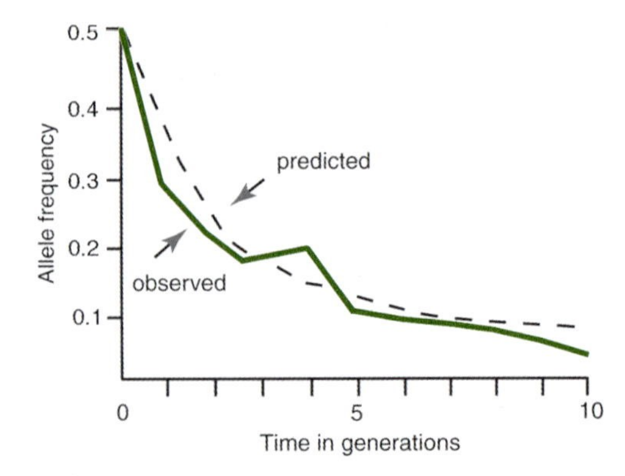

Fig. 17.1 Change of the frequency of a recessive lethal allele in a laboratory population of *Drosophila melanogaster.* The dashed curve shows the predicted change; the points are the outcome of the experiment. (From Wallace 1963.)

It is taking a long time to assimilate Darwin's insights.

traditional religions and philosophies. Evolution has also been falsely interpreted to serve unpopular, even evil, political ends, in Social Darwinism and in Nazi Germany. That experience has left many people extremely sensitive to and deeply cautious about any attempt to apply biology to social and political problems. Thus although evolutionary biology provides the best scientific understanding of where we came from and what we are biologically, the messages that evolutionary biologists try to communicate to society at large must be very carefully formulated, for even the best formulations are liable to be misinterpreted by many and misused by some. It will be a long time before Darwin's insights are fully assimilated.

Summary

This chapter discusses the status, scope, and prospects of evolutionary explanation.

- The mechanisms of evolution are well understood. It is strongly supported by many kinds of data, from fossils to molecules. It is as well confirmed as any major idea in science.
- Evolution has a complex causal structure that combines historical influences with selection and drift.
- Evolutionary change can be rapid. It is happening all around us.
- The scope of evolutionary explanation remains controversial. How far it can be applied to humans remains open, both because of resistance from established religions and because the scientific problem of understanding the interaction of biological evolution with cultural change has not been resolved.
- The major preoccupation of evolutionary biology in the first half of the twentieth century was demonstrating its consistency with Mendelian genetics. In the second half of the twentieth century, the

major preoccupation was the integration of evolution with molecular biology. In the twenty-first century, it is likely to be the connection of genotype with phenotype.

- There are two outstanding puzzles concerning the connection of genotype and phenotype: How did traits fixed within lineages become invariant? What are the effects of the fixed traits on the further evolution of the variable ones?
- Other unsolved problems are the huge variation in evolutionary rates, from very fast to almost zero; and the nature of species—reasons for the formation of clumps in phenotype space.
- Evolutionary biology has made successful predictions, but the limits to what can be predicted, in principle, are not clear. They will probably be pushed back.
- There are good reasons why it will take a long time for evolutionary thinking to be fully assimilated, both in academic circles and by the lay public.

Recommended reading

Barkow, L., Cosmides, L., and Tooby, J. (ed.) (1992). *The adapted mind.* Oxford University Press, Oxford.

Bettinger, R. L. (1991). *Hunter–gatherers. Archaeological and evolutionary theory.* Plenum Press, New York.

Betzig, L. (1997). *Human nature: A critical reader.* Oxford University Press, Oxford.

Cavalli-Sforza, L. L., Menozzi, P., and Piazza, A. (1994). *The history and geography of human genes.* Princeton University Press, Princeton.

Dunbar, R. I. M. (ed.) (1995). *Human reproductive decisions.* MacMillan, London.

Durham, W. (1991). *Coevolution: Genes, culture and human diversity.* Stanford University Press, Stanford.

Ewald, P. W. (1994). *Evolution of infectious diseases.* Oxford University Press, Oxford.

Hardy, S. B. (ed.) (1998). *Mother nature.* Pantheon, New York.

McGuire, M. T. and Troisi, A. (1998). *Darwinian psychiatry.* Harvard University Press, Cambridge.

Morbeck, M. E., Galloway, A., and Zihlman, A. L. (ed.) (1997). *The evolving female.* Princeton University Press, Princeton

Nesse, R. M. and Williams, G. (1995). *Why we get sick: The new science of Darwinian medicine.* Times Books, New York.

Rose, M. R. (1991). *The evolutionary biology of aging.* Oxford University Press, Oxford.

Rose, M. R. (1998). *Darwin's spectre: Evolutionary biology in the modern world.* Princeton University Press, Princeton.

Stearns, S. C. (ed.) (1999). *Evolution in health and disease.* Oxford University Press, Oxford.

Trevathan, W. R., McKenna, J. J., and Smith E. O. (ed.) (1999). *Evolutionary medicine.* Oxford University Press, Oxford.

Questions

17.1 Is the evolution of canalization necessary to account for the fixation of traits within lineages? Or is such fixation sufficiently explained by shared developmental control genes?

17.2 Was Darwin's prediction of a long-tongued moth needed to pollinate a deep-nectaried orchid just as impressive as Wallace's prediction of the change in frequency of a recessive lethal gene in a *Drosophila* population cage? Do qualitative and quantitative predictions have different value? How much can evolution predict without genetics?

17.3 Discuss the nature and limits of materialistic explanations of human behavior. How does biological evolution interact with cultural change?

Glossary

Adaptation. A state that evolved because it improved reproductive performance, to which survival contributes. Also the process that produces that state.

Adaptive evolution. The process of change in a population driven by variation in reproductive success that is correlated with heritable variation in a trait.

Additive genetic variance. The part of total genetic variance that can be modeled by allelic effects whose influence on the phenotype in heterozygotes is additive—halfway between the effects caused in the phenotype by the two homozygotes. This part of genetic variance determines the response to selection by quantitative traits.

Additive tree. A phylogenetic tree in which the branch lengths between two species add up to the distance between those species.

Allele. One of the different homologous forms of a single gene; at the molecular level, a different DNA sequence at the same place in the chromosome.

Allopatric. Occurring in two or more geographically separate places.

Allopolyploidy. An increase in the number of chromosomes involving hybridization followed by chromosome doubling. The offspring then contain chromosomes from both parental species.

Altruism. Behavior that benefits others at a cost to oneself.

Anagenetic. Evolutionary change occurring within a species between speciation events. Usually applied to fossils.

Ancestral. Originating prior in evolution to the derived state.

Anisogamy. Having gametes of different sizes, large eggs and small sperm.

Antagonistic pleiotropy. One gene has positive effects on fitness through its impact on one trait but negative effects on fitness through its impact on another trait.

Apomixis. A form of clonal propagation in which progeny genetically identical to the mother are formed without genetic recombination.

Apomorphic. Derived, relative to an ancestral, or plesiomorphic, state.

Autopolyploidy. Doubling of one's own entire chromosome set.

Autosome. A chromosome not involved in sex determination.

Bootstrapping. A statistical method to estimate our confidence in a pattern. One takes random samples of the original data (with replacement to get a data set of the same size) and repeats the calculations with the new, artificial data sets many times. The patterns that do not change, or only rarely change, receive our confidence.

Bootstrap value. The proportion of times a pattern is repeated in a bootstrapping procedure.

Breeding true. Displaying the same character state in the offspring as in the parents, an indication that the trait is genetically determined and that the parents are genetically similar.

Broad-sense heritability. The proportion of total phenotypic variation that can be ascribed to all the genetic differences among individuals in the population. For example, the amount of variation caused by differences among full-sib families.

Canalization. The limitation of phenotypic variation by developmental mechanisms. It can be demonstrated by disturbing developmental control to reveal the underlying genetic variability that had been canalized.

Centrosome. The cell organelle that becomes the spindle apparatus during mitosis and meiosis in eucaryotes.

Clade. A branch of the evolutionary tree containing all the species descended from a single common ancestor.

Cladogenetic. Evolutionary change occurring during speciation events. Usually applied to fossils.

Cline. A spatial gradient in trait values or gene frequencies.

Clocklike tree. A phylogenetic tree in which the length of the branches is proportional to the time elapsed between branching events.

Coalescence. A process occurring during the calculation of how long ago two alleles shared a common ancestor: the branches of a geneology coalesce as one works back in time.

Codon bias. Synonymous codons do not occur with equal frequency.

Coevolution. Evolutionary changes in one thing—genes, sexes, species—induce evolutionary changes in another, which in turn induce further evolutionary changes in the first, and so forth.

Colinearity. A property of homeobox genes: they have the same sequence along the chromosome as the segments of the body whose development they control.

Comparative trend analysis. The relations of two or more traits among higher taxa are analyzed with proper control for phylogeny and covariates such as body weight.

Conflict. Evolutionary conflict arises when two genes that interact with each other have different transmission patterns and therefore different evolutionary interests.

Conjugation. A form of mating in which two bacteria build a tube between them through which both plasmids and chromosomal DNA can be exchanged. Bacteria can be induced to conjugate by transposons and plasmids, which thereby achieve horizontal transmission.

Conserved function. A property of paralogous genes with high DNA sequence homology that code for proteins with similar function in distantly related organisms.

Convergence. Two species resemble each other not because they shared common ancestors but because evolution has adapted them to similar ecological conditions.

Cyclical parthenogenesis. A life cycle typical of aphids, rotifers, cladocerans, and some beetles, in which a series of asexual generations is interrupted by a sexual generation. The offspring of the sexual generation are often adapted to resist extreme conditions and to disperse.

Density-dependent selection. Selection that favors different things at different population densities.

Density-independent selection. Selection that favors the same things at all population densities, or that has always occurred at the same population density so that density effects did not occur.

Derived. Defined relative to ancestral: originating later in evolution than the ancestral state.

Diecy. Having separate sexes; individuals are either males or females; used for plants. See gonochorism.

Diplontic life cycle. A life cycle in which diploid somatic adults produce haploid gametes by meiosis that fuse to form diploid zygotes that develop into somatic adults.

Directional selection. Selection that always acts in a given direction; for example, always to increase the value of a trait, or always to decrease it.

Disruptive selection. Selection that favors the extremes and eliminates the middle of a frequency distribution of trait values; for example, increasing the frequency of small and large individuals and reducing the frequency of medium-sized individuals.

Divergence. Related species no longer resemble each other because evolution has adapted them to different ecological conditions.

Dominant allele. In diploids an allele is dominant if it, but not the recessive homolog, is expressed in the phenotype of heterozygotes.

Downstream gene. A gene under the control of a regulatory gene; the genes downstream from a regulatory gene constitute a regulatory pathway.

Drift. The random walk of gene frequencies that occurs in both large and small populations when variation in genes is not correlated with variation in reproductive success.

Duplication. Copying of a DNA sequence without loss of the original, increasing the size of the genome by the size of the sequence copied.

Effective population size. The size N_e of an ideal population consisting of individuals with equal reproductive output that would experience the same amount of genetic drift as a real population of size N. This defines effective population size for drift; there are other definitions for unequal family sizes, subdivided populations, differences in male versus female reproductive success, and variable population sizes.

Epigenetic inheritance. Somatic inheritance of the differentiated state of the cell through cycles of cell division.

Eucaryote. An organism with a cell nucleus surrounded by a nuclear membrane, usually with organelles, such as mitochondria and chloroplasts, that have their own circular DNA genome. Eucaryotes include the protists, fungi, plants, and animals.

Eusociality. A social system with nonreproductive workers.

Exon. The part of a eucaryotic gene whose DNA sequence is preserved in post-transcriptional splicing and is represented in the spliced mRNA and in the resulting amino acid sequence of the protein product. Exons occur in eucaryotes but not in procaryotes.

Fitness. For a start, relative lifetime reproductive success, which includes the probability of surviving to reproduce. In certain situations, other measures are more appropriate. The most important modifications to this definition include the inclusion in the definition of the effects of age-specific reproduction, and of fluctuations of density dependence.

Fixation probability. The probability that a new mutation will reach fixation.

Fixation time. The time that it takes new mutations that get fixed to reach fixation, in generations.

Founder effect. Major changes in gene frequencies that occur in a population founded with a small sample of a larger population.

Frequency-dependent selection. A mode of natural selection in which either rare types (negative frequency-dependent selection) or common types (positive frequency-dependent selection) are favored.

Genealogy. A tree describing the history of a single gene, as opposed to a phylogeny, which uses information from many genes or traits to reconstruct the history of a set of species.

Gene flow. Genes flow from one place to another when organisms born in one place have offspring in another place that survive to reproduce there.

Gene frequency. The frequency of an allele in a population. If there are 100 individuals in a population of diploid individuals, and we consider one locus (one gene) that is present in two forms (two alleles), *A* and *a*, then if 20 of the individuals carry two copies of *A* (they are *AA* homozygotes), 60 of the individuals are *Aa* heterozygotes, and the remaining 20 individuals are *aa* homozygotes, then the gene frequencies are calculated as the number of each allele divided by the total number, in this case, $(40 + 60)/200 = 0.50$ for both alleles.

Gene substitution. The process by which a new mutation becomes fixed in a population.

Genetic bottleneck. A reduction in population size to a low-enough level for long enough that many alleles are lost and others are fixed.

Genetic diversity. The probability that two homologous alleles chosen at random from a population differ.

Genetic drift. Random change in allele frequencies due to chance factors.

Genetic imprinting. Genes marked by DNA methylation in the germ line of parents; some are marked in mothers, others in fathers. Methylated genes are not expressed in the early development of the offspring.

Genetic separation. The separation of gene pools during speciation.

Genomic conflict. Occurs when genes affecting the same trait experience different selection pressures because they obey different transmission rules.

Genotype. In evolutionary biology, the information stored in the genes of one individual; in population genetics, the diploid combination of alleles at one locus present in an adult prior to meiosis.

Gonochorism. Having separate sexes; individuals are either males or females, not both; used for animals. See diecy.

Group selection. Selection generated by variation in the reproductive success of groups.

Gynodiecy. A condition in plant populations that consist of a mixture of male-sterile, or effectively female, individuals (gyno-), and hermaphroditic individuals bearing functional male and female flowers (-diecy).

Haplontic life cycle. A life cycle in which haploid adults produce haploid gametes by mitosis; the diploid zygotes immediately undergo meiosis to produce haploid individuals.

Heterogametic. The sex having two different sex chromosomes; for organisms with chromosomal sex determination; males are XY in humans.

Heterosis. The heterozygote is more fit than either homozygote.

Heterozygosity. The proportion of a population that is heterozygous at a locus; also the average proportion of loci heterozygous per individual.

Hitch-hiking. Changes in the frequencies of neutral traits that are pleiotropically linked to other traits that are under selection; or changes in the frequencies of neutral genes that are linked on chromosomes to changes in other genes that are under selection.

Homeobox. A 180 base-pair sequence in important regulatory genes that codes for a protein segment that binds to DNA as a key part of a transcription factor.

Homogametic. The sex having two similar sex chromosomes; for organisms with chromosomal sex determination; females are XX in humans.

Homology. Identity of one trait in two or more species by descent from a common ancestor.

Homoplasy. Similarity for any reason other than common ancestry. The commonest cause of homoplasy in morphological traits is probably convergence; in DNA sequences, simple mutation.

Inbreeding. The mating of related organisms.

Inbreeding depression. The reduction in the survival or reproduction of offspring of related parents caused by the expression as homozygotes of deleterious recessive genes that were present in the parents as heterozygotes.

Individual selection. Selection generated by variation in the reproductive success of individual organisms, affecting all their genes and traits.

Induced responses. A change in a phenotype that occurs in response to a specific environmental signal and that improves growth, survival, or reproduction. Without the signal, the change does not take place.

Infinite-allele model. A model in the neutral theory of molecular evolution that assumes that every mutation is unique in the sense that it does not already exist in the population; plausible for long DNA sequences.

Instar. A larval growth stage, usually used for insects and crustaceans.

Interactor. The organism in its ecological role, in which it develops, grows, acquires food and mates, survives, and reproduces.

Intron. A sequence within a gene that is removed after transcription and before translation by gene splicing; its DNA sequence is not represented in the RNA sequence of the spliced mRNA or the amino acid sequence of the resulting protein; introns occur in eucaryotes but not procaryotes.

Isogamy. Mating partners have gametes of the same size.

Iteroparous. Having several discrete reproductive events per lifetime.

Kin selection. Adaptive evolution of genes caused by relatedness; an allele causing an individual to act to benefit its relatives will increase in frequency if that allele is also found in the relatives and if the benefit to the relatives more than compensates the cost to the individual.

Lek. A traditional display site where males gather to defend mating territories and females come to mate; Swedish for sports field or display.

Lineage-specific developmental mechanisms. Developmental mechanisms found within all organisms of one lineage but not in other lineages, responsible for the morphology that characterizes the lineage. They limit thc genetic variation that can be expressed in the lineage.

Macroevolution. The pattern of evolution at and above the species level, including most of fossil history and much of systematics.

Mating types. Sets of potential mating partners. Matings can occur between partners of different type but not with partners of the same type.

Meiosis. Reduction division of diploid germ cells to yield haploid gametes.

Meiotic drive. Distortion of the fairness of meiosis by nuclear genes to increase their representation in the gametes at the expense of other alleles.

Mendelian lottery. A particular allele will or will not be represented in the offspring because of the segregation of alleles at meiosis and the random chance that any particular gamete will form a zygote. Most easily seen with small family sizes. Think about single children.

Method of independent contrasts. A comparative method that controls for the fact that character states in related organisms are not statistically independent because of shared ancestors. The basic idea is that *differences* between one pair of species are independent of *differences* between another pair of species even if both pairs are related.

Microevolution. The process of evolution within populations, including adaptive and neutral evolution.

Mitochondria. Intracellular organelles with their own genomes, derived from bacterial ancestors. The energy factories of the cell where ATP and the intermediate products of the Krebs cycle, used in the cytoplasm for energy release and biosynthesis, are made.

Molecular clock. The approximately constant rate of nucleotide substitution for particular genes and classes of genes within particular lineages. The constancy of the rate depends on the randomness with which particular nucleotides mutate and then drift to fixation.

Monecy. Individuals reproduce both as males and as females; hermaphrodites; used for plants.

Monophyletic. All species in a monophyletic

group are descended from a common ancestor, and all species descended from that ancestor are in that group.

Multigene family. Sets of multiple copies of genes derived by duplication from a common ancestor gene and retaining the same function.

Multilevel evolution. Adaptive evolution occurring simultaneously at several levels of a biological hierarchy, e.g. nuclear and cytoplasmic genes.

Mutation. Any change in the nucleic acid sequence of an organism, either a point mutation, a deletion, an insertion, or a chromosomal rearrangement.

Narrow-sense heritability. The fraction of total phenotypic variance in a trait that is accounted for by additive genetic variance; measures the potential response to selection.

Natural selection. Variation in reproductive success.

Neutral. Variation in state is not correlated with variation in reproductive success: states are equally fit.

Neutral evolution. The change and occasional fixation of alleles caused by the drift of alleles not correlated with reproductive success.

Nucleotide diversity. The average number of nucleotide differences per site between randomly chosen pairs of sequences.

Operational sex ratio. The local ratio of receptive females to sexually active males.

Orthology. DNA sequence homology.

Paralogy. DNA sequence homology plus conserved functions

Paraphyly. A group does not contain all species descended from the most recent common ancestor of its members.

Parsimony. A criterion used in cladistic tree-building: the best tree has the fewest changes in character states and the least homoplasy.

Parthenogenesis. Asexual reproduction from an egg cell that may or may not involve recombination, depending on the mechanism. In most cases the daughters are exact genetic copies of the mothers.

Phage. A virus that infects bacteria.

Phenotype. The material organism, or some part of it, as opposed to the information in the genotype that provides the blueprint.

Phenotypic differentiation. The differentiation of phenotypes in separated gene pools during and after speciation.

Phenotypic plasticity. Sensitivity of the phenotype to differences in the environment. Less precise than reaction norm.

Phylogenetic trait analysis. A comparative method in which one constructs a phylogenetic tree, plots character states (traits) on the tree, and infers transitions in character states from their position on the tree. Geographical locations of taxa can be plotted onto the tree to infer the location of ancestors.

Phylogeny. The history of a group of taxa described as an evolutionary tree with a common ancestor as the base and descendent taxa as branch tips.

Plastid. A cell organelle with its own bacteria-like genome: mitochondria and chloroplasts.

Pleiotropy. One gene has effects on two or more traits.

Plesiomorphic. Ancestral, relative to a derived, or apomorphic, state.

Point mutation. A change in a single DNA nucleotide, e.g. adenine mutates to thymine, or an insertion or deletion of a single nucleotide.

Polyphenism. A form of induced response in which the phenotypes are discrete.

Polyphyly. A group is polyphyletic if its species are descended from several ancestors that are also the ancestors of species classified into other groups.

Polyploidization. A doubling of the complete chromosome set.

Population genetics. The discipline that studies changes in frequencies of alleles in populations; issues include mutation, selection, inbreeding, assortative mating, gene flow, and drift; suitable when genetic differences at one locus can be detected as phenotypic differences.

Procaryotes. Organisms that lack a nucleus and organelles such as mitochondria or chloroplasts. The Eubacteria and the Archaea.

Protandry. Individuals are born as males, reproduce as males, then change sex and reproduce as females. In plants, individuals express male function prior to female function, producing pollen before being pollinated.

Protogyny. Individuals are born as female, reproduce as females, then change sex and reproduce as males. In plants, individuals express female function prior to male function, being pollinated before producing pollen.

Proximate causation. The mechanical determination of traits during the lifetime of an organism, including biochemistry, development, and physiology.

Pseudogene. A nonfunctional copy of a gene; it is not expressed.

Punctuated equilibrium. A pattern seen in many but not all lineages in the fossil record, in which a long period of stasis is broken by a short period of rapid change. In some cases the rapid change is associated with speciation.

Punctuation. A short period of rapid change breaking a long period of stasis in the fossil record.

Quantitative genetics. Studies changes in traits in populations when genetic differences at one locus are too small to detect in phenotypes and when many genes affect one trait; common themes are heritability, genetic covariance, response to selection.

Random. A word with many meanings. In evolution mutations are random with respect to the needs of the organism in which they occur. They are not distributed at random along the DNA sequence (some parts of genomes are more mutable than others) and they are not always random with respect to environmental conditions (in bacteria and fungi environmental stimuli can increase the mutation rate).

Reaction norm. A property of a genotype: how development maps the genotype into the phenotype as a function of the environment. Usually given for one genotype and one environmental factor. Populations and families can be described as having mean reaction norms.

Recessive allele. An allele is recessive if it is not expressed in the phenotype in the heterozygous diploid state.

Regulatory gene. A gene that turns another gene, or group of genes, on or off. Small changes in regulatory genes cause large changes in phenotypes.

Replicator. The organism in its role as information copier, the mechanism that copies the DNA sequence of the parent and passes it to the offspring.

Reproductive success. A measure of fitness defined as the number of offspring produced per lifetime. It can be extended through several generations; for example, one could define it as the number of grandchildren that survive to reproduce.

Reproductive value. The expected contribution of organisms in that stage of life to lifetime reproductive success.

Residual reproductive value. The remaining contribution to lifetime reproductive success after the current activity has made its contribution.

Secondary reinforcement. The reinforcement of prezygotic barriers to hybridization after secondary contact between isolated populations.

Segregation distortion. Deviation from the Mendelian ratios that give equal chances to homologous alleles in meiosis; unfair ratios can be caused by nuclear genes that interfere with meiosis or with the products of meiosis to improve their own chances at the expense of their homologs.

Selection differential. The difference between the population mean and the mean phenotype of the parents that produce the next generation.

Selection response. The difference between the mean of the parental population and the offspring mean.

Semelparous. Reproducing once, then dying.

Sex allocation. The allocation of reproductive effort to male versus female function in hermaphrodites and to male versus female offspring in species with separate sexes.

Sexual dimorphism. Males and females have different phenotypes.

Sexual selection. The component of natural selection that is associated with success in mating.

Sibling species. Species that are reproductively isolated but cannot be distinguished, or can be distinguished only with difficulty and by experts, using morphological criteria.

Species. Either a set of organisms that could share grandchildren (the biological species concept), or the smallest diagnosable cluster of individual organisms within which there is a parental pattern of ancestry and descent (the phylogenetic species concept).

Spindle apparatus. Pulls the chromosomes apart during meiosis and mitosis. It replicates, divides, and each copy anchors itself. Microtubules grow from the spindle to the centromeres of the chromosomes, attach, contract, and pull a set of chromosomes to one end of the dividing cell.

Stabilizing selection. Selection that eliminates the extremes of a distribution and favors the center.

Stasis. A long period without evolutionary change.

Strict consensus tree. A phylogenetic tree derived from a set of equally parsimonious trees and constructed by only including the groups that are supported in all the equally parsimonious trees.

Sympatry. Occurring in the same geographic area.

Synapomorphy. A shared, derived character state indicating that two species belong to the same group.

Synergism. A nonadditive interaction between two or more factors.

Synonymous mutation. A point mutation (change in a single nucleotide) that does not change the amino acid for which the DNA triplet codes.

Terrane. A piece of continental crust that did not originally belong to the continent on which it is found but moved there from elsewhere.

Trade-off. A change in one trait that increases fitness causes a change in the other trait that decreases fitness.

Transcription factor. A gene product that binds to DNA at a specific site and regulates the expression of genes downstream from that site.

Transduction. A virus that infects bacteria picks up some bacterial DNA from one host and transfers it to the next host, which may incorporate the DNA if it survives the infection.

Transformation. Bacteria take up DNA from the medium and incorporate it into their circular chromosome.

Truncation selection. Artificial selection in which only individuals with a value of a trait above (or below) some threshold are allowed to breed.

Ultimate causation. The evolutionary determination of the state of a trait at the level of a population or lineage through adaptive evolution or drift.

Wild type. A term used in classical genetics to designate the standard genotype in the population from which mutations formed rare deviations. Modern molecular data have destroyed the concept by revealing so much variation that the concept has become meaningless.

Literature cited

Abele, L. G., Kim, W., and Felgenhauer, B. E. (1989). Molecular evidence for inclusion of the Phylum Pentastomida in the Crustacea. *Molecular Biology and Evolution*, **6**, 685–91.

Akam, M., Holland, P., and Wray, G. (1994*a*). Preface. In *The evolution of developmental mechanisms*, (ed. M. Akam, P. Holland, P. Ingham, and G. Wray). Development 1994 Supplement, The Company of Biologists Limited, Cambridge.

Akam, M., Holland, P., Ingram, P., and Wray, G. (ed.) (1994*b*). *The evolution of developmental mechanisms.* Development Supplement. The Company of Biologists, Cambridge.

Alberch, P. (1989). The logic of monsters, evidence for internal constraint in development and evolution. *Geobios, mémoire spécial*, **12**, 21–57.

Albert, V. A., Williams, S. E., and Chase, M. W. (1992). Carnivorous plants, phylogeny and structural evolution. *Science*, **257**, 1491–5.

Alexander, R. D., Hoogland, J. L., Howard, R. D., Noonan, K. M., and Sherman, P.W. (1979). Sexual dimorphisms and breeding systems in pinnipeds, ungulates, primates, and humans. In *Evolutionary biology and human social behavior, an anthropological perspective*, (ed. N. A. Chagnon and W. Irons), pp. 402–35. Duxbury Press, North Scituate, Massachusetts.

Andersson, M. (1982). Female choice selects for extreme tail length in a widow bird. *Nature*, **299**, 818–20.

Andersson, M. (1994). *Sexual selection.* Princeton University Press, Princeton.

Antonovics, J. and Bradshaw, A. D. (1970). Evolution in closely adjacent populations, VIII. Clinal patterns at a mine boundary. *Heredity*, **25**, 349–62.

Antonovics, J. and Ellstrand, N. C. (1984). Experimental studies of the evolutionary significance of sexual reproduction. I. A test of the frequency-dependent selection hypothesis. *Evolution,* **38**, 103–15.

Avise, J. C. (1994). *Molecular markers, natural history and evolution.* Chapman & Hall, London.

Bakker, T. C.M. (1993). Positive genetic correlation between female preference and preferred male ornament in sticklebacks. *Nature,* **363**, 255–7.

Barbujani, G. and Excoffier, L. (1998). The history and geography of human genetic diversity. In *Evolution in health and disease,* (ed. S. C. Stearns), pp. 27–40. Oxford University Press, Oxford.

Barkow, L., Cosmides, L., and Tooby, J. (ed.) (1992). *The adapted mind.* Oxford University Press, Oxford.

Bateman, A. J. (1948). Intra-sexual selection in *Drosophila. Heredity,* **2**, 349–68.

Bateman, R. M., Crane, C. R., DiMichele, W. A., Kenrick, P. R., Rowe, N. P., Speck, T., and Stein, W. E. (1998). Early evolution of land plants, phylogeny, physiology, and ecology of the primary terrestrial radiation. *Annual Review of Ecology and Systematics,* **29**, 263–92.

Bell, G. (1982). *The masterpiece of nature. The evolution and genetics of sexuality.* University of California Press, Berkeley.

Bell, G. (1985). The origin and early evolution of germ cells as illustrated by the Volvocales. In *The origin and evolution of sex,* (ed. H. Halvarson and A. Mornoy), pp. 221–56. Alan Liss, New York.

Bell, G. (1996). *The basics of selection.* Chapman & Hall, New York.

Benton, M .J. (ed.) (1993). *The fossil record 2.* Chapman & Hall, London.

Berven, K. A., Gill, D. E., and Smith-Gill, S. J. (1979). Countergradient selection in the green frog, *Rana clamitans. Evolution,* **33**, 609–23.

Bettinger, R. L. (1991). *Hunter–gatherers. Archaeological and evolutionary theory.* Plenum Press, New York.

Betzig, L. (1997). *Human nature: A critical reader.* Oxford University Press, Oxford.

Bierzychudek, P. (1987). Resolving the paradox of sexual reproduction: A review of experimental tests. In *The evolution of sex and its consequences,* (ed. S. C. Stearns), pp. 163–74. Birkhäuser, Basel.

Blondel, J., Perret, P., Maistre, M., and Dias, P. (1992). Do harlequin Mediterranean environments function as source sink for Blue Tits (*Parus caeruleus* L.)? *Landscape Ecology,* **7**, 213–19.

Boraas, M. E., Seale, D. B., and Boxhorn, J. E. (1998). Phagotrophy by a flagellate selects for colonial prey, a possible origin of multicellularity. *Evolutionary Ecology*, **12**, 153–64.

Brakefield, P. M. and Larsen, T. B. (1984). The evolutionary significance of dry and wet season forms in some tropical butterflies. *Biological Journal of the Linnean Society*, **22**, 1–12.

Briggs, D. E. G. and Crowther, P. R. (ed.) (1990). *Paleobiology, a synthesis*. Blackwell Scientific, Oxford.

Brooks, D. R. and McLennan, D. A. (1991). *Phylogeny, ecology, and behavior. A research program in comparative biology*. University of Chicago Press, Chicago.

Brooks, D. R. and McLennan, D. A. (1993). *Parascript*. Smithsonian Institution Press, Washington.

Brooks, D. R., Thorson, T. B., and Mayes, M. A. (1981). Freshwater stingrays (Potamotrygonidae) and their helminth parasites, Testing hypotheses of evolution and coevolution. In *Advances in cladistics, Proceedings of the first meeting of the Willi Hennig Society*, (ed. V. A. Funk and D. R. Brooks), pp. 147–75. New York Botanical Garden, New York.

Bull, J. J. (1983). *Evolution of sex determining mechanisms*. Benjamin/Cummings, Menlo Park.

Bulmer, M. G. (1989). Structural instability of models of sexual selection. *Theoretical Population Biology*, **35**, 195–206.

Bush, G. L. (1994). Sympatric speciation in animals, new wine in old bottles. *Trends in Ecology and Evolution*, **9**, 285–8.

Cann, R. L., Stoneking, M., and Wilson, A. C. (1987). Mitochondrial DNA and human evolution. *Nature*, **325**, 31–6.

Carroll, R. L. (1997). *Patterns and processes of vertebrate evolution*. Cambridge University Press, Cambridge.

Carroll, S. B. (1994). Developmental regulatory mechanisms in the evolution of insect diversity. In *The evolution of developmental mechanisms*, (ed. M. Akam, P. Holland, P. Ingham, and G. Wray), pp. 217–23. Development 1994 Supplement, The Company of Biologists Limited, Cambridge.

Carroll, S. B., Gates, J., Keys, D. N., Paddock, S. W., Panganiban, G. E.F., Selegue, J. E., and Williams, J. A. (1994). Pattern formation and eyespot determination in butterfly wings. *Science*, **265**, 109–14.

Carroll, S. B., Weatherbee, S. D., and Langeland, J. A. (1995). Homeotic genes and the regulation and evolution of insect wing number. *Nature*, **375**, 58–61.

Cavalli-Sforza, L. L. and Bodmer, W. F. (1971). *The genetics of human populations*. Freeman, San Francisco.

Cavalli-Sforza, L. L., Menozzi, P., and Piazza, A. (1994). *The history and geography of human genes*. Princeton University Press, Princeton.

Charlesworth, B. (1980). *Evolution in age-structured populations.* Cambridge University Press, Cambridge.

Charlesworth, B. and Williamson, J. A. (1975). The probability of the survival of a mutant gene in an age-structured population and implications for the evolution of life-histories. *Genetical Research,* **26**, 1–10.

Charnov, E. L. (1982). *The theory of sex allocation.* Princeton University Press, Princeton.

Clarke, B. (1975). The contribution of ecological genetics to evolutionary theory, detecting the direct effects of natural selection on particular polymorphic loci. *Genetics,* **79**, 101–13.

Clutton-Brock, T. H. (1991). *The evolution of parental care.* Princeton University Press, Princeton.

Clutton-Brock, T. H. and Iason, G. R. (1986). Sex ratio variation in mammals. *Quarterly Review of Biology,* **61**, 339–74.

Clutton-Brock, T. H. and Vincent, A. C. J. (1991). Sexual selection and the potential reproductive rates of males and females. *Nature,* **351**, 58–60.

Cohen, M. L. (1992). Epidemiology of drug resistance, implications for a post-antimicrobial era. *Science,* **257**, 1050–5.

Conway Morris, S. (1998*a*). *The crucible of creation: The Burgess Shale and the rise of animals.* Oxford University Press, Oxford.

Conway Morris, S. (1998*b*). The evolution of diversity in ancient ecosystems, a review. *Philosophical Transactions of the Royal Society of London B,* **353**, 327–45.

Coyne, J. A. and Orr, H. A. (1989). Patterns of speciation in *Drosophila*. *Evolution,* **43**, 362–81.

Cracraft, J. (1983). Species concepts and speciation analysis. In *Current ornithology* (ed. R. F. Johnston), pp. 159–87. Plenum Press, New York.

Crow, J. F. (1986). *Basic concepts in population, quantitative, and evolutionary genetics*. Freeman, New York.

Crow, J. F. and Kimura, M. (1965). Evolution in sexual and asexual populations. *American Naturalist,* **99**, 439–50.

Crow, J. F. and Kimura, M. (1979). Efficiency of truncation selection. *Proceedings of the National Academy of Sciences, USA,* **76**, 396–9.

Curio, E. (1973). Towards a methodology of teleonomy. *Experientia,* **29**, 1045–59.

Daan, S., Dijkstra, C., and Tinbergen, J. M. (1990). Family planning in the kestrel (*Falco tinnunculus*), the ultimate control of covariation of laying date and clutch size. *Behavior,* **114**, 83–116.

Darnell, J., Lodish, H., and Baltimore, D. (1990). *Molecular cell biology,* (2nd edn). W.H. Freeman, New York.

Darwin, C. (1859). *On the origin of species by means of natural selection or the preservation of favoured races in the struggle for life.* John Murray, London.

Darwin, C. (1871). *The descent of man, and selection in relation to sex.* John Murray, London.

Dawkins, R. (1976). *The selfish gene.* Oxford University Press, Oxford.

Dawkins, R. (1982). *The extended phenotype. The gene as the unit of selection.* W.H. Freeman, New York.

Dawkins, R. (1986). *The blind watchmaker.* Longman, London.

De Robertis, E. M. and Sasai, Y. (1996). A common plan for dordsoventral patterning in Bilateria. *Nature,* **380**, 37–40.

Dobzhansky, T. (1937). *Genetics and the origin of species.* Columbia University Press, New York.

Dodson, S. I. (1989). Predator-induced reaction norms. *BioScience,* **39**, 447–52.

Doolittle, R. F., Feng, D.,Tsang, S., Cho, G., and Little, E. (1996). Determining divergence times of the major kingdoms of living organisms with a protein clock. *Science,* **271**, 470–7.

Doolittle, W. F. (1998.) A paradigm gets shifty. *Nature,* **392**, 15–16.

Dorit, R. L., Akashi, H., and Gilbert, W. (1995). Absence of polymorphism at the ZFY locus on the human Y chromosome. *Science,* **268**, 1183–5.

Dowton, M., Austin, A. D., Dillon, N., and Bartowsky, E. (1997). Molecular phylogeny of the apocritan wasps, the Proctotrupomorpha and Evaniomorpha. *Systematic Entomology,* **22**, 245–55.

Dowton, M., Austin, A. D., and Antolin, M. F. (1998). Evolutionary relationships among the Braconidae (Hymenoptera, Icheumonoidea) inferred from partial 16S rDNA gene sequences. *Insect Molecular Biology,* **7**,129–50.

Doyle, J. A. (1998). Phylogeny of vascular plants. *Annual Review of Ecology and Systematics,* **29**, 567–99.

Dudley, J. W. (1977). 76 generations of selection for oil and protein percentage in maize. In *Proceedings of the international conference on quantitative genetics,* (ed. E. Pollack, O. Kempthorne, and T. B. Bailey), pp. 459–73. Iowa State University Press, Ames, Iowa.

Dunbar, R. I. M. (ed.) (1995). *Human reproductive decisions.* MacMillan, London.

Durham, W. (1991). *Coevolution: Genes, culture and human diversity.* Stanford University Press, Stanford.

Eberhard, W. G. (1996). *Female control, sexual selection by cryptic female choice.* Princeton University Press, Princeton.

Ebert, D. (1994). A maturation size threshold and phenotypic plasticity of age and size at maturity in *Daphnia magna*. *Oikos*, **69**, 309–17.

Eigen, M. (1971). Self-organization of matter and the evolution of biological macromolecules. *Naturwissenschaften*, **58**, 465–523.

Elgar, M. A. (1992). Sexual cannibalism in spiders and other invertebrates. In *Cannibalism, ecology and evolution among diverse taxa*, (ed. M. A. Elgar and B. J. Crespi), pp. 128–55. Oxford University Press, Oxford.

Elinson, R. P. (1989). Egg evolution. In *Complex organismal functions, integration and evolution in vertebrates*, (ed. D. B. Wake and G. Roth), pp. 251–62. Dahlem Conference Report, John Wiley and Sons, New York.

Endler, J. A. (1977). *Geographic variation, speciation, and clines*. Princeton University Press, Princeton.

Endler, J. A. (1986). *Natural selection in the wild*. Princeton University Press, Princeton.

Ewald, P. W. (1994. *Evolution of infectious diseases*. Oxford University Press, Oxford.

Ewens, W. J. (1993). Beanbag genetics and after. In *Human population genetics. A centennial tribute to J. B. S. Haldane*, (ed. P. P. Majumder), pp. 7–29. Plenum Press, New York.

Eyre-Walker, A. and Keightly, P. D. (1999). High genomic deleterious mutation rates in hominids. *Nature*, **397**, 344–7.

Felsenstein, J. (1985). Phylogenies and the comparative method. *American Naturalist*, **125**, 1–15.

Felsenstein, J. (1988). Phylogenies from molecular sequences, inference and reliability. *Annual Review of Genetics*, **22**, 521–65.

Fisher, R. A. (1918). The correlation between relatives on the supposition of Mendelian inheritance. *Transactions of the Royal Society of Edinburgh*, **52**, 399–433.

Fisher, R. A. (1930). *The genetical theory of natural selection*. Oxford University Press, Oxford.

Ford, E. B. (1975). *Ecological genetics*, (4th edn). Chapman & Hall, London.

Forey, P. L., Humphries, C. J., Kitching, I. L., Scotland, R. W., Siebert, D. J., and Williams, D. M. (1992). *Cladistics*. The Systematics Association. Clarendon Press, Oxford.

Forster, L. M. (1992). The stereotyped behaviour of sexual cannibalism in *Latrodectus hasselti* Thorell (Araneae, Theridiidae), the Australian redback spider. *Australian Journal of Zoology*, **40**, 1–11.

Franco, M. G., Rubini, P. G., and Vecchi, M. (1982). Sex determinants and their distribution in various populations of *Musca domestica* L. of Western Europe. *Genetical Research*, **40**, 279–93.

Frohnhöfer, H. G., and Nüsslein-Volhard, C. (1986). Organisation of anterior pattern in the *Drosophila* embryo by the maternal gene *bicoid*. *Nature*, **324**, 120–5.

Futuyma, D. (1998). *Evolutionary biology*, (3rd edn). Sinauer, Sunderland, Massachusetts.

Gerhart, J. and Kirschner, M. (1997). *Cells, embryos and evolution*. Blackwell Science, Oxford.

Ghiselin, M. T. (1969). The evolution of hermaphroditism among animals. *Quarterly Review of Biology*, **44**, 189–208.

Gibson, G. and Hodgness, D. S. (1996). Effect of polymorphism in the *Drosophila* regulatory gene Ultrabithorax on homeotic stability. *Science*, **271**, 200–3.

Gillespie, J. H. (1991). *The causes of molecular evolution*. Oxford University Press, Oxford.

Gingerich, P. D. (1983). Rates of evolution: effects of time and temporal scaling. *Science*, **222**, 159–61.

Givnish, T. J., Sytsma, K. J., Smith, J. F., and Hahn, W. J. (1995). Molecular evolution, adaptive radiation, and geographic speciation in *Cyanea* (Campanulaceae, Lobelioideae). In *Hawaiian biogeography*, (ed. W. T. Wagner and V. A. Funk), pp. 288–337. Smithsonian Institution Press, Washington,

Glesener, R. R. and Tilman, D. (1978). Sexuality and the components of environmental uncertainty: Clues from geographic parthenogenesis in terrestrial animals. *American Naturalist*, **112**, 659–73.

Godfray, H. C. J. (1994). *Parasitoids. Behavioral and evolutionary ecology*. Princeton University Press, Princeton.

Goldsmith, T. H. (1990). Optimization, constraint, and history in the evolution of eyes. *Quarterly Review of Biology*, **65**, 281–322.

Gould, J. L. and Keeton, W. T. (1996). *Biological science*, (6th edn). Norton & Co. New York

Gould, S. J. (1990). *Wonderful life: The Burgess Shale and the nature of history*. W. W. Norton & Co., New York.

Grant, B. R. and Grant, P. R. (1989). *Evolutionary dynamics of a natural population. The Large Cactus Finch of the Galápagos*. University of Chicago Press, Chicago.

Grant, P. R. (1986). *Ecology and evolution of Darwin's finches*. Princeton University Press, Princeton.

Greenwood, P. H. (1974). *Cichlid fishes of Lake Victoria, East Africa*. Natural History Museum, London.

Griffin, D. R. (1958). *Listening in the dark*. Yale University Press, New Haven.

Griffiths, A. J. F., Miller, J. H., Suzuki, D. T., Lewontin, R. C., and

Gelbart, W. M. (1996). *An introduction to genetic analysis*, (6th edn). W. H. Freeman, New York.

Grubb, P. J. and Metcalfe, D. J. (1996). Adaptation and inertia in the Australian tropical lowland rain-forest flora: contradictory trrends in intergeneric and intrageneric comparisons of seed size in relation to light demand. *Functional Ecology*, **10**, 512–20.

Gwynne, D. T. and Simmons, L. W. (1990). Experimental reversal of courtship roles in an insect. *Nature*, **346**, 171–4.

Haig, D. (1992). Genomic imprinting and the theory of parent–offspring conflict. *Seminars in Developmental Biology*, **3**, 153–60.

Haldane, J. B. S. (1990). *The causes of evolution*, (reprint edn). Princeton University Press, Princeton.

Halder, G., Callaerts, P., and Gehring, W. J. (1995). Induction of ectopic eyes by targeted expression of the eyeless gene in *Drosophila*. *Science*, **267**, 1788–92.

Hall, B. K. (1998). *Evolutionary developmental biology.* Kluwer Academic, Leiden.

Hamilton, W. D. (1964). The genetical theory of social behaviour. I, II. *Journal of Theoretical Biology*, **7**, 1–52.

Hamilton, W. D. and Zuk, M. (1982). Heritable true fitness and bright birds. A role for parasites? *Science*, **218**, 384–7.

Harborne, J. B. (1973). *Phytochemical methods, a guide to modern techniques of plant analysis.* Chapman & Hall, London.

Hartl, D. L. (1988). *A primer of population genetics*, (2nd edn). Sinauer, Sunderland, Massachusetts.

Hartl, D. L. and Clark, A. G. (1989). *Principles of population genetics*, (2nd edn). Sinauer Associates, Sunderland, Massachusetts.

Harvey, P. H. and Pagel, M. D. (1991). *The comparative method in evolutionary biology.* Oxford University Press, Oxford.

Hass, C. A., Hoffman, M. A., Densmore, L. D. III, and Maxson, L. R. (1992). Crocodilian evolution, insights from immunological data. *Molecular Phylogenetics and Evolution*, **1**, 193–201.

Haukioja, E. and Neuvonen, S. (1985). Induced long-term resistance of birch foliage against defoliators, defensive or incidental? *Ecology*, **66**, 1303–8.

Hennig, W. (1966). *Phylogenetic systematics.* University of Illinois Press.

Hillis, D. M., Moritz, C., and Mable, B. K. (ed.) (1996). *Molecular systematics*, (2nd edn). Sinauer Associates, Sunderland, Massachusetts.

Holloway, G. J., Brakefield, P. M., and Kofman, S. (1993). The genetics of wing pattern elements in the polyphenic butterfly, *Bicyclus anynana*. *Heredity*, **70**, 179–86.

Hori, M. (1993). Frequency-dependent natural selection in the handedness of scale-eating cichlid fish. *Science,* **260**, 216–19.

Houle, D., Hoffmaster, D. K., Assimacopoulos, S., and Charlesworth, B. (1992). The genomic mutation rate for fitness in *Drosophila. Nature,* **359**, 58–60.

Howard, D. J. and Berlocher, S. H. (ed.) (1998). *Endless forms: Species and speciation.* Oxford University Press, Oxford.

Hrdy, S. B. (ed.) (1998). *Mother nature.* Pantheon, New York.

Hudson, R. R. (1990). Gene genealogies and the coalescent process. *Oxford Surveys in Evolutionary Biology,* **7**, 1–44.

Hughes, A. L. and Nei, M. (1989). Nucleotide substitution at major histocompatibility complex class II loci, Evidence for overdominant selection. *Proceedings of the National Academy of Sciences, USA,* **86**, 958–62.

Hurst, L. D. and Peck, J. R. (1996). Recent advances in understanding of the evolution of sex and maintenance of sex. *Trends in ecology and evolution,* **11**, 46–52.

Hurst, L. D., Atlan, A., and Bengtsson, B. O. (1996). Genetic conflicts. *Quarterly Review of Biology,* **71**, 317–64.

IUGS (International Union of Geological Sciences) (1989). Global stratigraphic chart. *Episodes,* June 1989.

Jablonka, E. and Szathmáry, E. (1995). The evolution of information storage and heredity. *Trends in Ecology and Evolution,* **10**, 206–11.

Jackson, J. B. C. and Cheetham, A. H. (1999). Tempo and mode of speciation in the sea. *Trends in Ecology and Evolution,* **14**, 72–7.

Jacob, F. (1977). Evolution and tinkering. *Science,* **196**, 1161–6.

Jones, D. A. (1972). Cyanogenic glycosides and their function. In *Phytochemical ecology,* (ed. J. B. Harborne), pp. 103–24. Academic Press, London.

Jones, J. S., Ebert, D., and Stearns, S. C. (1992). Life history and mechanical constraints on reproduction in genes, cells and waterfleas. In *Genes in ecology,* (ed. R. J. Berry, T. J. Crawford, and G. M. Hewitt), pp. 393–404. Blackwell Scientific, Oxford.

Juchault, P., Rigaud, T., and Mocquard, J. P. (1993). Evolution of sex determination and sex ratio variability in wild populations of Armadillidium vulgare (Latr.) (Crustacea, Isopoda), a case study in conflict resolution. *Acta Œcologia,* **14**, 547–62.

Kaneshiro, K. Y. and Boake, C. R. B. (1987). Sexual selection and speciation, issues raised by Hawaiian drosophilids. *Trends in Ecology and Evolution,* **2**, 207–12.

Karn, M. N. and Penrose, L. S. (1951). Birth weight and gestation time in relation to maternal age, parity and infant survival. *Annals of Eugenics,* **16**, 147–64.

Kempenaers, B., Verheyen, G. R., Van den Broeck, M., Burke, T., Van Broeckhoven, C. V., and Dhondt, A. (1992). Extra-pair paternity results from female preference for high-quality males in the blue tit. *Nature,* **357**, 494–6.

Kenrick, P. and Crane, P. R. (1997). The origin and early evolution of plants on land. *Nature,* **389**, 33–9.

Kimura, M. (1968). Evolutionary rate at the molecular level. *Nature,* **217**, 624–6.

Kimura, M. (1983). *The neutral theory of molecular evolution.* Cambridge University Press, Cambridge.

Kirkwood, T. B. L. (1987). Immortality of the germ line versus disposability of the soma. In *Evolution of longevity in animals,* (ed. A. D. H. Woodhead and K. H. Thompson), pp. 209–18. Plenum, New York.

Klein, J. and Klein, D. (1992). *Molecular evolution of the major histocompatibility complex.* Springer Verlag, Berlin.

Klein, J., Gutknecht, J., and Fischer, N. (1990). The major histocompatibility complex and human evolution. *Trends in Genetics,* **6**, 7–11.

Klein, J., Takahata, N., and Ayala, F. (1993). MHC polymorphism and human origins. *Scientific American,* December, 46–51.

Klein, J., Sato, A., Nagl, S., and O'hUigin, C. (1998). Molecular trans-species polymorphism. *Annual Review of Ecology and Systematics,* **29**, 1–21.

Knoll, A. H., Bambach, R. K., Canfield, D. E., and Grotzinger, J. P. (1996). Comparative earth history and Late Permian mass extinction. *Nature,* **273**, 452–7.

Koehn, R. K., Zera, A. J., and Hall, J. G. (1983). Enzyme polymorphism and natural selection. In *Evolution of genes and proteins,* (ed. M. Nei and R. K. Koehn), pp. 115–36. Sinauer, Sunderland, Massachusetts.

Koelewijn, H. P. (1993). On the genetics and ecology of sexual reproduction in *Plantago coronopus.* Ph.D. thesis, University of Utrecht.

Kondrashov, A. S. (1982). Selection against harmful mutations in large sexual and asexual populations. *Genetical Research,* **40**, 325–32.

Kondrashov, A. S. (1993). Classification of hypotheses on the advantage of amphimixis. *Journal of Heredity,* **84**, 372–80.

Kozlowski, J. (1992). Optimal allocation of resources to growth and reproduction, implications for age and size at maturity. *Trends in Ecology and Evolution,* **7**, 15–19.

Kumar, S. and Hedges, S. B. (1998). A molecular time scale for vertebrate evolution. *Nature,* **392**, 917–20.

Labandeira, C. C. (1998*a*). How old is the flower and the fly? *Science*, **280**, 57–9.

Labandeira, C. C. (1998*b*). Early history of arthropod and vascular plant associations. *Annual Review of Earth and Plantary Science*, **26**, 329–77.

Lack, D. (1947). *Darwin's finches.* Cambridge University Press, Cambridge.

Laroche, J., Li, P., and Bousquet, J. (1995). Mitochondrial DNA and monocot–dicot divergence time. *Molecular Biology and Evolution*, **12**, 1151–6.

Lewontin, R. C. (1974). *The genetic basis of evolutionary change.* Columbia University Press, New York.

Li, W.-H. and Graur, D. (1991). *Fundamentals of molecular evolution.* Sinauer, Sunderland, Massachusetts.

Lively, C. M. (1986). Predator-induced shell dimorphism in the acorn barnacle *Chthamalus anisopoma. Evolution*, **40**, 232–42.

Lively, C. M. (1992). Parthenogenesis in a freshwater snail: reproductive assurance versus parasite release. *Evolution*, **46**, 907–13.

Lowe, C. J. and Wray, G. A. (1997). Radical alterations in the roles of homeobox genes during echinoderm evolution. *Nature*, **389**, 718–20.

Lyttle, T. W. (1991). Segregation distorters. *Annual Review of Genetics*, **25**, 511–57.

McCune, A. R. (1996). Biogeographic and stratigraphic evidence for rapid speciation in semionotid fishes. *Paleobiology*, **22**, 34.

McGuire, M. T. and Troisi, A. (1998). *Darwinian psychiatry.* Harvard University Press, Cambridge.

McPhee, J. (1998). *Annals of the former world.* Farrar Straus and Giroux, New York.

Mallet, J. (1995). A species definition for the Modern Synthesis. *Trends in Ecology and Evolution*, **10**, 294–9.

Margulis, L. (1970). *Origin of eukaryotic cells.* Yale University Press, New Haven.

Margulis, L. (1981). *Symbiosis in cell evolution.* W.H. Freeman, San Francisco.

Margulis, L. and Schwartz, K. V. (1982). *Five kingdoms, An illustrated guide to the phyla of life on earth.* W.H. Freeman, San Francisco.

Martin, W. and Müller, M. (1998). The hydrogen hypothesis for the first eukaryote. *Nature*, **392**, 37–41.

Martin, W., Lydiate, D., Brinkmann, H., Forkmann, G., Saedler, H., and Cerff, R. (1993). Molecular phylogenies in angiosperm evolution. *Molecular Biology and Evolution*, **10**, 140–62.

Mason, S. F. (1992). *Chemical evolution.* Oxford University Press, Oxford.

Maynard Smith, J. (1964). Group selection and kin selection, a rejoinder. *Nature,* **201**, 1145–7.

Maynard Smith, J. (1978). *The evolution of sex.* Cambridge University Press, Cambridge.

Maynard Smith, J. (1998). *Evolutionary genetics,* (2nd edn). Oxford Univesity Press, Oxford.

Maynard Smith, J. and Szathmáry, E. (1995). *The major transitions in evolution.* W.H. Freeman, New York.

Maynard Smith, J. and Szathmáry, E. (1999). *The origins of life: From the birth of life to the origin of language.* Oxford University Press, Oxford.

Maynard Smith, J., Dowson, G. C., and Spratt, B. G. (1991). Localized sex in bacteria. *Nature,* **349**, 29–31.

Mayr, E. (1942). *Systematics and the origin of species.* Columbia University Press, New York.

Mayr, E. (1963). *Animal species and evolution.* Harvard University Press, Cambridge, Massachusetts.

Michod, R. E. (1979). Evolution of life histories in response to age-specific mortality factors. *American Naturalist,* **113**, 531–50.

Michod, R. E. and Levin, B. R. (ed.) (1988). *The evolution of sex.* Sinauer, Sunderland, Massachusetts.

Miles, D. B. and Dunham, A. E. (1993). Historical perspectives in ecology and evolutionary biology: The use of phylogenetic comparative analyses. *Annual Review of Ecology and Systematics,* **24**, 587–619.

Milinski, M. and Bakker, T. C. M. (1990). Female sticklebacks use male coloration in mate choice and hence avoid parasitized males. *Nature,* **344**, 330–3.

Miller, S. L. (1953). A production of amino acids under possible primitive Earth conditions. *Science,* **117**, 528–9.

Mock, D. W. and Parker, G. A. (1997). *The evolution of sibling rivalry.* Oxford University Press, Oxford.

Mohr, P. (1983). Ethiopian flood basalt province. *Nature,* **303**, 577–84.

Møller, A. P. (1994). *Sexual selection and the barn swallow.* Oxford University Press, Oxford.

Monteiro, A. F., Brakefield, P. M., and French, V. (1994). The evolutionary genetics and developmental basis of wing pattern variation in the butterfly *Bicyclus anynana. Evolution,* **48**, 1147–57.

Morbeck, M. E., Galloway, A., and Zihlman, A. L. (ed.) (1997). *The evolving female.* Princeton University Press, Princeton

Mousseau, T. A. and Roff, D. A. (1987). Natural selection and the heritability of fitness components. *Heredity,* **59**, 181–97.

Moxon, E. R., Rainey, P. B., Nowak, M. A., and Lenski, R. E. (1994). Adaptive evolution of highly mutable loci in pathogenic bacteria. *Current Biology*, **4**, 24–33.

Mukai, T. (1964). The genetic structure of natural populations of *Drosophila melanogaster.* I. Spontaneous mutation rate of polygenes controlling viability. *Genetics*, **50**, 1–19.

Mukai, T., Chigusa, S. I., Mettler, L. E., and Crow, J. F. (1972). Mutation rate and dominance of genes affecting viability in *Drosophila melanogaster. Genetics*, **72**, 335–55.

Muller, H. J. (1964). The relation of recombination to mutational advance. *Mutation Research*, **1**, 2–9.

Nanney, D. L. (1982). Genes and phenes in *Tetrahymena. BioScience*, **32**, 783–8.

Nei, M. (1987). *Molecular evolutionary genetics.* Columbia University Press, New York.

Nesse, R. M. and Williams, G. (1995). *Why we get sick: The new science of Darwinian medicine.* Times Books, New York.

Neu, H. C. (1992). The crisis in antibiotic resistance. *Science*, **257**, 1064–73.

Newman, R. A. (1988). Genetic variation for larval anuran (*Scaphiophus couchii*) development time in an uncertain environment. *Evolution*, **42**, 763–73.

Newton, I. (1988). Age and reproduction in the Sparrowhawk. In *Reproductive success*, (ed. T. H. Clutton-Brock), pp. 201–19. University of Chicago Press, Chicago.

Nijhout, H. F. (1991). *The development and evolution of butterfly wing patterns.* Smithsonian Institution Press, Washington.

Niklas, K. J., Tiffney, B. H., and Knoll, A. H. (1983). Patterns in vascular land plant diversification. *Nature*, **303**, 614–16.

Nilsson, L. A. (1988). The evolution of flowers with deep corolla tubes. *Nature*, **334**, 147–9.

Nisbet, E. G. and Piper, D. J. W. (1998). Giant submarine landslides. *Nature*, **392**, 329–30.

Nordskog, A. W. (1977). Success and failure of quantitative genetic theory in poultry. In *Proceedings of the international conference on quantitative genetics*, (ed. E. Pollack, O. Kempthorne, and T. B. Bailey), pp. 569–85. Iowa State University Press, Ames, Iowa.

O'Neil, P. (1997). Selection on genetically correlated phenological characters in *Lythrum salicaria* L. (Lythraceae). *Evolution*, **51**, 267–74.

Otte, D. and Endler, J. (ed.) (1989). *Speciation and its consequences.* Sinauer, Sunderland, Massachusetts.

Raff, R. A. (1996). *The shape of life. Genes, development, and the evolution of animal form.* University of Chicago Press, Chicago.

Raup, D. M. and Jablonski, D. (ed.) (1986). *Patterns and processes in the history of life.* Dahlem Konferenzen. Springer-Verlag, Berlin.

Raup, D. M. and Stanley, S. M. (1978). *Principles of paleontology,* (2nd edn). W.H. Freeman, New York.

Read, A. F. and Harvey, P. H. (1989). Life history differences among the eutherian radiations. *Journal of Zoology, London,* **219**, 329–53.

Reznick, D. A., Bryga, H., and Endler, J. A. (1990). Experimentally induced life-history evolution in a natural population. *Nature,* **346**, 357–9.

Rice, W. R. and Hostert, E. E. (1993). Laboratory experiments on speciation, what have we learned in 40 years? *Evolution,* **47**, 1637–53.

Ridley, M. (1996). *Evolution,* (2nd edn). Blackwell, Oxford.

Roff, D. A. (1981). On being the right size. *American Naturalist,* **118**, 405–22.

Roff, D. A. (1992). *The evolution of life histories.* Chapman & Hall, London.

Roff, D. A. (1997). *Evolutionary quantitative genetics.* Chapman & Hall, London.

Romer, A. S. (1962). *The vertebrate body.* W.B. Saunders, Philadelphia.

Rose, M. R. (1991). *Evolutionary biology of aging.* Oxford University Press, Oxford.

Rutherford, S. L. and Lindquist, S. (1998). Hsp90 as a capacitor for morphological evolution. *Nature,* **396**, 336–42.

Ryan, M. J. (1985). *The tungara frog, a study in sexual selection and communication.* University of Chicago Press, Chicago.

Savard, L., Li, P., Strauss, S. H., Chase, M. W., Michaud, M., and Bousquet, J. (1994). Chloroplast and nuclear gene sequences indicate late Pennsylvanian time for the last common ancestor of extant seed plants. *Proceedings of the National Academy of Sciences, USA,* **91**, 5163–7.

Schaal, S. and Ziegler, W. (ed.) (1992). *Messel. An insight into the history of life and of the earth.* Oxford University Press, Oxford.

Schlichting, C. D. and Pigliucci, M. (1998). *Phenotypic evolution. A reaction norm perspective.* Sinauer Associates, Sunderland, Massachusetts.

Schmitt, J. and Wulff, R. D. (1993). Light spectral quality, phytochrome and plant competition. *Trends in Ecology and Evolution,* **8**, 47–51.

Schopf, J. W. (1993). Microfossils of the Early Archaean Apex Chert, new evidence of the antiquity of life. *Science,* **260**, 640–6.

Schuster, P. (1993). RNA based evolutionary optimization. *Origin of life and Evolution of the Biosphere,* **23**, 373–91.

Schuster, P., Fontana, W., Stadler, P. F., and Hofacker, I. L. (1994). From sequences to shapes and back, a case study in RNA secondary structures. *Proceedings of the Royal Society of London B*, **255**, 279–84.

Seehausen, O., Van Alphen, J. J. M., and Witte, F. (1997). Cichlid fish diversity threatened by eutrophication that curbs sexual selection. *Science*, **277**, 1808–11.

Selker, E. (1990). Premeiotic instability of repeated sequences in *Neurospora crassa*. *Annual Review of Genetics*, **24**, 579–613.

Sepkoski, J. J. Jr (1984). A kinetic model of Phanerozoic taxonomic diversity. III. Post-Paleozoic families and mass extinctions. *Paleobiology*, **10**, 246–67.

Shaw, R. F. and Mohler, J. D. (1953). The selective advantage of the sex ratio. *American Naturalistt*, **87**, 337–42.

Shine, R. (1985). The evolution of viviparity in reptiles, an ecological analysis. In *Biology of the reptilia*, (ed. C. Gans and F. Billet), pp. 605–94. Academic Press, New York.

Simmons, J.A. (1973). The resolution of target range by echolocating bats. *Journal of the Acoustical Society of America*, **54**, 157–73.

Simoons, F. J. (1978). The geographic hypothesis and lactose malabsorption. A weighing of the evidence. *Digestive Diseases*, **23**, 963–80.

Singh, N., Agrawal, S., and Rastogi, A. K. (1997). Infectious diseases and immunity, special reference to Major Histocompatibility Complex. *Emerging Infectious Disease*, **3**, http//www.cdc.gov/ncidod/EID/vol3no1/singh.htm

Skelton, P. W. (ed.) (1993). *Evolution, a biological and palaeontological approach*. Addison-Wesley, New York.

Smith, H. (1995). Physiological and ecological function within the phytochrome family. *Annual Review of Plant Physiology*, **46**, 289–315.

Stearns, S. C. (1983). The influence of size and phylogeny on patterns of covariation among life-history traits in the mammals. *Oikos*, **41**, 173–87.

Stearns, S. C. (ed.) (1987). *The evolution of sex and its consequences*. Birkhäuser, Basel.

Stearns, S. C. (1992). *The evolution of life histories*. Oxford University Press, Oxford.

Stearns, S. C. (1994). The evolutionary links between fixed and variable traits. *Acta Paleontologica Polonica*, **38**, 215–32.

Stearns, S. C. (ed.). (1999). *Evolution in health and disease*. Oxford University Press, Oxford.

Stearns, S. C., Kaiser, M., and Blarer, A. (1996). A case study in experimental evolution: Reproductive effort and induced responses in *Drosophila melanogaster*. *Plant Species Biology*, **11**, 97–105.

Stearns, S. C. and Koella, J. (1986). The evolution of phenotypic plasticity in life-history traits: Predictions for norms of reaction for age- and size-at-maturity. *Evolution,* **40**, 893–913.

Stearns, S. C., de Jong, G., and Newman, R. (1991). The effects of phenotypic plasticity on genetic correlations. *Trends in Ecology and Evolution,* **6**, 122–6.

Stouthamer, R., Luck, R. F., and Hamilton, W. D. (1990). Antibiotics cause parthenogenetic *Trichogramma* (Hymenoptera/Trichogrammatidae) to revert to sex. *Proceedings of the National Academy of Sciences, USA,* **87**, 2424–7.

Stryer, L. (1988). *Biochemistry,* (3rd edn). W.H. Freeman, New York.

Travisano, M., Vasi, F., and Lenski, R. E. (1995). Long-term experimental evolution in *Escherichia coli.* III. Variation among replicate populations in correlated responses to novel environments. *Evolution,* **49**, 189–200.

Trevathan, W. R., McKenna, J. J., and Smith E. O. (ed.) (1999). *Evolutionary medicine.* Oxford University Press, Oxford.

Trivers., R. L. and Willard, D. E. (1973). Natural selection of parental ability to vary the sex ratio of offspring. *Science,* **179**, 90–2.

van Hinsberg, A. (1997). Morphological variation in *Plantago lanceolata* L., effects of light quality and growth regulators on sun and shade populations. *Journal of Evolutionary Biology,* **10**, 687–701.

Vavilov, N. I. (1922). The law of homologous series in variation. *Journal of Genetics,* **12**, 47–89.

Vigilant, L., Stoneking, M., Harpending, H., Hawkes, K., and Wilson, A. C. (1991). African populations and the evolution of human mitochondrial DNA. *Science,* **253**, 1503–7.

de Visser, J. A. G. M., Hoekstra, R. F., and van den Ende, H. (1997). An experimental test for synergistic epistasis and its application in *Chlamydomonas. Genetics,* **145**, 815–19.

Vogel, F. and Motulsky, A. G. (1979). *Human genetics.* Springer Verlag, Berlin.

Vulic, M., Dionisio, F., Taddei, F., and Radman, M. (1997). Molecular keys to speciation, DNA polymorphism and the control of genetic exchange in enterobacteria. *Proceedings of the National Academy of Science USA,* **94**, 9763–7.

Waddington, C. H. (1942). Canalization of development and the inheritance of acquired characters. *Nature,* **150**, 563–5.

Wagner, G. P. (1989). The biological homology concept. *Annual Review of Ecology and Systematics,* **20**, 51–69.

Wake, D. B. and Larson, A. (1987). Multidimensional analysis of an evolving lineage. *Science,* **238**, 42–8.

Wallace, B. (1963). The elimination of an autosomal recessive lethal from an experimental population of *Drosophila melanogaster. American Naturalist,* **97**, 65–6.

Weigensberg, I. and Roff, D. (1996). Natural heritabilities: Can they be reliably estimated in the laboratory? *Evolution,* **50**, 2149–57.

Weiss, K. M. (1993). *Genetic variation and human disease.* Cambridge University Press, Cambridge.

Whitelam, G. C., Patel, S., and Devlin, P. F. (1998). Phytochromes and photomorphogenesis in *Arabidopsis. Philosophical Transactions of the Royal Society of London B,* **353**, 1445–53.

Wilkins, A. S. (1993). *Genetic analysis of animal development,* (2nd edn). Wiley-Liss, New York.

Williams, G. C. (1966). *Adaptation and natural selection.* Princeton University Press, Princeton.

Williams, G. C. (1992). *Natural selection. Domains, levels, and challenges.* Oxford University Press, Oxford.

Windig, J. J. (1993). The genetic background of plasticity in wing pattern of *Bicyclus* butterflies. Ph.D. thesis, Leiden.

Woese, C. R., Kandler, O., and Wheelis, M. L. (1990). Towards a natural system of organisms, Proposal for the domains Archaea, Bacteria and Eucarya. *Proceedings of the National Academy of Sciences, USA,* **87**, 4576–9.

Woltereck, R. (1909). Weitere experimentelle Untersuchungen über Artveränderung, speziell über das Wesen quantitativer Artunterschiede bei Daphniden. *Verhandlungen der Deutschen Zoologischen Gesellschaft,* 1909, 110–72.

Wynn-Edwards, V. C. (1962). *Animal dispersion in relation to social behaviour.* Oliver and Boyd, Edinburgh.

Yoo, B. H. (1980). Long-term selection for a quantitative character in large replicate populations of *Drosophila melanogaster.* I. Response to selection. *Genetical Research,* **35**, 1–17.

Index

Note page numbers in *italics* refer to figures and tables